IET MATERIALS, CIRCUITS AND DEVICES SERIES 43

Negative Group Delay Devices

Other volumes in this series:

Volume 2 **Analogue IC Design: The current-mode approach** C. Toumazou, F.J. Lidgey and D.G. Haigh (Editors)
Volume 3 **Analogue-Digital ASICs: Circuit techniques, design tools and applications** R.S. Soin, F. Maloberti and J. France (Editors)
Volume 4 **Algorithmic and Knowledge-based CAD for VLSI** G.E. Taylor and G. Russell (Editors)
Volume 5 **Switched Currents: An analogue technique for digital technology** C. Toumazou, J.B.C. Hughes and N.C. Battersby (Editors)
Volume 6 **High-frequency Circuit Engineering** F. Nibler *et al.*
Volume 8 **Low-power High-frequency Microelectronics: A unified approach** G. Machado (Editor)
Volume 9 **VLSI Testing: Digital and mixed analogue/digital techniques** S.L. Hurst
Volume 10 **Distributed Feedback Semiconductor Lasers** J.E. Carroll, J.E.A. Whiteaway and R.G.S. Plumb
Volume 11 **Selected Topics in Advanced Solid State and Fibre Optic Sensors** S.M. Vaezi-Nejad (Editor)
Volume 12 **Strained Silicon Heterostructures: Materials and devices** C.K. Maiti, N.B. Chakrabarti and S.K. Ray
Volume 13 **RFIC and MMIC Design and Technology** I.D. Robertson and S. Lucyzyn (Editors)
Volume 14 **Design of High Frequency Integrated Analogue Filters** Y. Sun (Editor)
Volume 15 **Foundations of Digital Signal Processing: Theory, algorithms and hardware design** P. Gaydecki
Volume 16 **Wireless Communications Circuits and Systems** Y. Sun (Editor)
Volume 17 **The Switching Function: Analysis of power electronic circuits** C. Marouchos
Volume 18 **System on Chip: Next generation electronics** B. Al-Hashimi (Editor)
Volume 19 **Test and Diagnosis of Analogue, Mixed-signal and RF Integrated Circuits: The system on chip approach** Y. Sun (Editor)
Volume 20 **Low Power and Low Voltage Circuit Design with the FGMOS Transistor** E. Rodriguez-Villegas
Volume 21 **Technology Computer Aided Design for Si, SiGe and GaAs Integrated Circuits** C.K. Maiti and G.A. Armstrong
Volume 22 **Nanotechnologies** M. Wautelet *et al.*
Volume 23 **Understandable Electric Circuits** M. Wang
Volume 24 **Fundamentals of Electromagnetic Levitation: Engineering sustainability through efficiency** A.J. Sangster
Volume 25 **Optical MEMS for Chemical Analysis and Biomedicine** H. Jiang (Editor)
Volume 26 **High Speed Data Converters** Ahmed M.A. Ali
Volume 27 **Nano-Scaled Semiconductor Devices** E.A. Gutiérrez-D (Editor)
Volume 29 **Nano-CMOS and Post-CMOS Electronics: Devices and modelling** Saraju P. Mohanty and Ashok Srivastava
Volume 30 **Nano-CMOS and Post-CMOS Electronics: Circuits and design** Saraju P. Mohanty and Ashok Srivastava
Volume 32 **Oscillator Circuits: Frontiers in design, analysis and applications** Y. Nishio (Editor)
Volume 33 **High Frequency MOSFET Gate Drivers** Z. Zhang and Y. Liu

Negative Group Delay Devices

From concepts to applications

Edited by
Blaise Ravelo

The Institution of Engineering and Technology

Published by The Institution of Engineering and Technology, London, United Kingdom

The Institution of Engineering and Technology is registered as a Charity in England & Wales (no. 211014) and Scotland (no. SC038698).

First published 2018

The Institution of Engineering and Technology
Michael Faraday House
Six Hills Way, Stevenage
Herts, SG1 2AY, United Kingdom

www.theiet.org

British Library Cataloguing in Publication Data
A catalogue record for this product is available from the British Library

ISBN 978-1-78561-640-2 (hardback)
ISBN 978-1-78561-641-9 (PDF)

Typeset in India by MPS Limited
Printed in the UK by CPI Group (UK) Ltd, Croydon

Contents

Foreword

The destiny of science, engineering, and technology is determined by the regular contradiction between conventional and unconventional knowledge. One of the conventional notions in contemporary engineering is that group delays of input/output systems are positive and typically can only be positive. This perception leads to the illusion that negative group delay (NGD) is an impossible phenomenon, and thus unintentionally limits possible engineering concepts. However, when looking back at the present and past century in science and engineering, it is evident that technological progress has been evolving with many phenomena initially perceived as impossible. In this sense, one might wonder if the NGD concept could follow the same destiny. Despite the few hundreds of NGD-related papers published, NGD remains one of the least familiar concepts of contemporary science and engineering. To be adopted as conventional knowledge, some more educational effort is necessary. Toward this aim, the present handbook offers basic and applied knowledge on the NGD concept from research and engineering groups worldwide.

The chapter authors of this book report their various viewpoints of NGD, matching their different skills in conventional science and engineering. The different ways developed by the authors constitute this unique and original handbook dedicated to the spreading the knowledge of the NGD concept. Certainly, because of the familiarity of the group delay parameter, most of the chapters are describing the NGD concept applied to electronics, RF, and microwave engineering. However, as the NGD effect is a universal parameter, it can be exploited in various other fields, for example, in neural engineering. The simple feasibility of some of the counterintuitive effects, such as prediction and forecasting, will be illustrated with the vision of its application to human-machine interfaces.

The different chapters illustrate the existence, design methods, demonstrations, and applications of the NGD function.

The present book chapters present the basic principle of the counterintuitive NGD principle. The analytical definition of the NGD function will be introduced in different context, for example, with different electronics and electromagnetic structures. The meaning and principle of the NGD function seen by the worldwide active research team about the topic are described in this book. The various technical concepts of the NGD phenomenon will be explored with different physics based on the transfer function approach. The different NGD synthesizers in different areas from electronics, RF/microwave, and optics engineering will be described. The existence of this unfamiliar and unconventional NGD concept will be illustrated from theoretical to experimental demonstrations. The meaning of the

NGD concept for the future academic teaching and industrial applications is briefly introduced.

The readers of the book can start and can focus on any chapter depending on their own interests or their feeling to understand the counter-intuitive NGD. The present book could be a pioneering reference which could guide the under and graduate students, amateurs, scientific, engineers, academic, and industrial staffs to discover and practice the NGD concept.

The authors believe that in the forthcoming decades outstanding NGD concept applications will be developed and will help to solve multiple engineering system problems for the benefit of mankind.

Introduction on the context and NGD theoretical bases

Blaise Ravelo[1]

Abstract: This introductory part synthesizes the content of the next 11 chapters of the book. It illustrates the diversity of the negative group delay (NGD) analysis, design method, demonstration, and application. The investigations are performed by different worldwide groups of the coauthors sharing their expertise about the nonfamiliar NGD topic. This introduction summarizes the main chapter content as well. The overview about the NGD concept explored is presented.
Keywords: NGD principle, NGD concept, NGD theory, NGD function, NGD application, theoretical bases, experimental demonstrators.

Context of the NGD concept

The book combines different NGD analyses applied to various analog and digital circuits. It is organized in 11 chapters. Each chapter describes the outstanding expertise of the coauthors about the NGD concept. Various ways to theorize, analyze, synthesize, and characterize the NGD function are elaborated. The NGD studies cover large area from the most basic theoretical to the practical contexts of application. The following sections state the abstract of the book chapters.

Summary of Chapter 1

Author(s): Yongchae Jeong[1] and Girdhari Chaudhary[1]

Authors' Affiliation:

[1]Division of Electronics and Information Engineering, Chonbuk National University, Republic of Korea

Keywords: Coupling matrix synthesis, coupled line, cross-cancellation power amplifier, defected ground structure, defected microstrip structure, feed-forward amplifier, filter synthesis, finite unloaded-quality factor resonators, negative group delay circuit, transmission line resonators.
Abstract: In this chapter, the synthesis of the several types of microstrip line passive negative group delay circuits (NGDCs) will be presented. The different

[1]ESIGELEC, France

transmission line resonators with combination of resistor will be considered for the designing various topologies of transmission- and reflection-type NGDCs. Similarly, the defected transmission line structures such as defected ground structure (DGS) and defected microstrip structure (DMS) resonators will also be used to synthesize the transmission-type NGDCs. In addition, filter synthesis approach will be presented to design NGDC with the predefined group delay including the coupling matrix synthesis with finite unloaded-quality resonators.

The major drawback of passive NGDCs is their inherent signal attenuation. Therefore, the novel approach and topologies for NGDC with very small signal attenuation will be also presented in this chapter. Finally, the applications of NGDC in the field of communication systems will be explored. The applications include the efficiency enhancement of feed-forward amplifier and cross-cancellation power amplifier, bandwidth enhancement of analog feedback amplifier, etc.

Summary of Chapter 2

Author(s): Girdhari Chaudhary[1] and Yongchae Jeong[1]

Authors' Affiliation:

[1]Division of Electronics and Information Engineering, Chonbuk National University, Republic of Korea

Keywords: Arbitrary power division ratio, negative group delay, positive group delay, power divider, rat-race coupler.
Abstract: Power dividers have been widely adopted as basic building components in microwave circuits and systems and extensively studied over the past decades including the unequal and tunable power division ratio, circuit miniaturization, and multiband operation. Because the group delay (GD) influences the performances of RF/electronic circuits and systems, understanding effect of GD has become critical for communication systems. However, GD analysis of power divider has rarely been considered in previous works and the positive GDs have only been provided through different transmission paths. In this chapter, several topologies of power divider and coupler with the predefined negative group delay characteristics will be presented including equal and unequal power division ratios. Similarly, coupling matrix synthesis with finite unloaded-quality resonators will be also presented in this chapter to design the power divider with the predefined negative group delay.

Summary of Chapter 3

Author(s): Henning U. Voss[1]

Author's Affiliation:

[1]Department of Radiology, Citigroup Biomedical Imaging Center, USA

Keywords: Prediction, forecasting, negative group delay, human motor control, man-machine interfacing.

Abstract: Signal prediction is one of the most basic tasks of signal analysis. In the majority of cases, signal prediction relies on a prior analysis of the signal to be predicted, and a subsequent model fit. In the present contribution, an alternative approach to signal prediction and time series forecasting, based on NGD, is presented.

It has been shown recently that NGD can occur in linear input/output systems with time-delayed feedback. This type of negative group delay is called *delay-induced negative group delay*, or *DINGD*. It exists for both continuous and discrete time. Both are described in this chapter. The continuous-time DINGD predictor is modeled as a non-autonomous linear delay-differential equation in the variable y with input x, i.e., $\dot{y}(t) = -ay(t) + bx(t) - cy(t-\tau)$. The DINGD predictor output is supposed to predict the signal in the following sense: $y(t) \approx x(t+\delta)$, where $\delta > 0$ is the prediction horizon. In other words, the output of the filter is a real-time estimate of future signal values. In contrast to the well-studied case of conventional NGD filters without time delay, the group delay here is caused by the delayed-feedback term; it is a *delay-induced group delay*. It can be positive or negative, depending on the model parameters. If the group delay is negative, this model constitutes a predictor. DINGD predictors do not need to model the signal generating system. Rather, their coefficients only depend on the desired prediction horizon and basic spectral properties of the input signal. Generally, DINGD prediction works for band-limited signals. The DINGD predictor differs from most other predictors or forecasting models by using past predicted rather than past input values for prediction.

There are immediate applications of this model. For example, it is evident that this model constitutes nothing but a leaky integrator with time-delayed feedback, which can be implemented easily by electronic circuits, and also has direct relevance in the neurosciences: Time delays are a typical component of biological neuronal networks, and it is reasonable to hypothesize a possible relevance of DINGD prediction in neuronal computations.

The theory of DINGD predictors is outlined and demonstrated on various types of broadband input data simulations. The DINGD theory will be executed by an algorithm that predicts, in real time, the trajectory of smooth voluntary manual movements of a computer input device such as a mouse or stylus on a graphics pad.

Since the DINGD predictor does not require a specific model of the signal to be predicted, it is one of the most parsimonious descriptions of a real-time predictor conceivable. The simplicity of the DINGD predictor and the fact that it accomplishes real-time prediction without the need for model fitting or adaptation invites its use in engineering applications, machine learning, and artificial intelligence, despite its generally limited prediction horizon and accuracy if compared with model-based or adaptive prediction schemes.

Summary of Chapter 4

Author(s): Chung-Tse Michael Wu[1]

Author's Affiliation:

[1]Department of Electrical and Computer Engineering, Rutgers University, USA

Keywords: Anomalous dispersion, dispersion engineering, negative group delay, non-Foster elements, recursive filters, transversal filters.
Abstract: This chapter presents some latest developments in microwave transversal and recursive-filter-based NGD circuits as well as their potential applications in realizing non-Foster elements. By using transversal and recursive-filter approaches, the desired amount of NGD can be synthesized with an increased bandwidth. The fully distributed topology of the proposed NGD circuits makes it feasible to be further scaled up to higher frequency ranges such as millimeter waves. In addition, it is shown that the proposed transversal-filter-based NGD circuit is able to create non-Foster elements exhibiting unconditional stability, enabling a new perspective of realizing non-Foster elements.

Summary of Chapter 5

Author(s): Blaise Ravelo[1]

Author's Affiliation:

[1]ESIGELEC, France

Keywords: Negative group delay (NGD), hybrid coupler, circuit theory, characterization method, S-parameter modeling.
Abstract: A theory on hybrid coupler-based NGD topology is investigated. The passive NGD cell under study is constituted by a hybrid coupler with one of output and isolated access ways connected in feedback with a lossy transmission line (TL). Two configurations of cell with interconnected isolated-coupled and isolated-direct accesses are investigated. The equivalent *S*-parameter model of the proposed structure is established. It is derived mathematically that the fully distributed hybrid coupler based topology is susceptible to exhibit NGD phenomenon. The fundamental properties of the innovative NGD topology as the NGD level-bandwidth product are formulated. The synthesis relations enabling to determine the NGD circuit parameters in function of the desired NGD value are established. The relation between the NGD level and the insertion loss is elaborated. Parametric numerical analyses in function of the TL attenuation and time-delay were performed. The NGD concept feasibility is highlighted by a very good correlation between the theoretical calculation and the numerical simulations. In addition, the NGD phenomena also occurred with the consideration of microstrip lines in feedback with 90°- and 180°-hybrid couplers. The NGD topology under study is potentially useful for the signal delay correction in the RF system.

Summary of Chapter 6

Author(s): Blaise Ravelo[1]

Author's Affiliation:

[1]ESIGELEC, France

Keywords: Negative group delay (NGD), Hilbert filter, phase shifter, microwave circuit theory, synthesis method.
Abstract: An original circuit theory on analog microwave Hilbert filter is established in this chapter. The Hilbert filter is a particular circuit presenting transfer function defined by $H(j\omega) = j$. The generality on the proposed Hilbert filter as phase shifter topology is described. The Hilbert filter synthesizer topology consists of a passive LC cell in cascade with NGD active cell. The circuit synthesis methodology is elaborated by assuming that the transmission coefficient is ideally equal to Hilbert filter transfer function $H(s) = \exp(j \cdot 90°)$. It is demonstrated that the analytical equations enabling to determine the LC- and NGD-cell parameters are established based on the *S*-parameter modeling. The microwave Hilbert filter concept feasibility is validated by three proofs of concept (POC) circuit operating around 2.45 GHz. After *S*-parameter simulations, transmission phase 90±1° flatness's are generated within the relative frequency better than 50% and ±1 dB gain flatness's within 23% bandwidth. The developed analog Hilbert filter topology is potentially useful for the RF/microwave transceiver system architecture designs, notably for image intermediate frequency rejection.

Summary of Chapter 7

Author(s): Fayu Wan[1] and Junxiang Ge[1]

Authors' Affiliation:

[1]Nanjing University of Information Science and Technology, China

Keywords: Negative group delay (NGD) synthesizer, NGD topologies, NGD applications.
Abstract: This chapter addresses a microwave circuit theory on the NGD principle. The characterization methodology of this unfamiliar NGD function is described. The basic NGD topologies are typically active cells composed of RF low-noise amplifier (LNA). Typically, low-pass NGD topologies are identified and characterized. The low-pass NGD circuits consisted of simple first-order RL-series and RC-parallel networks. Similar to the filter theory, band-pass NGD circuits are introduced from low-pass to band-pass transform. The proposed NGD circuit theory is established on the *S*-parameter approach. The proposed NGD topology main characteristics and properties as the NGD level, cut-off frequencies or bandwidth, central frequency and figure-of-merit are established. Given the LNA *S*-parameter model, the NGD synthesis and design method of the low- and band-pass NGD cells in function of the specified NGD values are presented. The NGD cell parameter calculations in function of the expected NGD level, insertion loss and reflection coefficient are introduced. NGD proof of concept (PoC) is synthesized, designed and simulated in order to understand and validate the proposed NGD low-pass and band-pass functions. As expected, low-pass and band-pass NGD aspects are obtained. The low-pass NGD circuits present an NGD level of about −2 ns over the bandwidth of about 70 MHz. Then, band-pass NGD circuits were synthesized to

operate at about 0.5 GHz over the bandwidth of about 140 MHz with NGD level of about −4 ns. It is worth mentioning that the NGD circuits fulfill the requirements of RF and microwave specifications.

Summary of Chapter 8

Author(s): Miodrag Kandic[1] and Greg E. Bridges[2]

Authors' Affiliation:

[1]AC/DC System Planning Group, Manitoba, Canada
[2]Department of Electrical and Computer Engineering, University of Manitoba, Canada

Keywords: Passive and active circuits, distributed circuit, NGD limitation.
Abstract: The chapter discusses about active and passive distributed element transmission line circuits exhibiting negative group delay (NGD). The first section introduces resonator-based circuit negative delay circuit topologies and shows how active elements are employed for gain compensation. The asymptotic limits of these circuits are then derived while considering issues of gain, bandwidth, in-band distortion, and out-of-band gain. The relationship between the time-domain transients and out-of-band gain will be analyzed and discussed in the context of modulated signals. Figures-of-merit for characterizing active NGD circuits will be introduced. The next section will extend the theory to negative group delay phenomena in general linear causal media. The final chapter will describe bidirectional active NGD transmission line circuits and describe the limitations and stability of these circuits. Experimental results of fabricated designs demonstrating the concepts will be put forward throughout the chapter.

Summary of Chapter 9

Author(s): Ahmed H. Dorrah[1] and Mo Mojahedi[1]

Authors' Affiliation:

[1]Edward S. Rogers Sr. Department of Electrical and Computer Engineering, University of Toronto, Canada

Keywords: Negative group delay (NGD), negative group velocity (NGV), superluminal information.
Abstract: Energy exchange between an electromagnetic pulse and dispersive media may result in complicated, yet an interesting phenomenon in which group velocity becomes abnormal (i.e., superluminal or negative). For such cases, detection of signal velocity (velocity of detectable information) remains debatable, as it is not a straightforward task to determine by how much the information content of the pulse, in a superluminal channel, experiences "fast" propagation and whether or not such a fast propagation can be used to reduce the channel latency. As such, the following

fundamental question remains unresolved: Can superluminal propagation in a noisy dispersive channel reduce signal latency? A careful literature survey shows the following: a group of researchers have considered "information" to be encoded in the discontinuities (sudden transitions) of a pulse which always propagate at the speed of light in vacuum; and in such cases information has been regarded as a strictly luminal quantity. Other researchers relate the arrival of information at the receiver with the time instant at which the signal-to-noise ratio (SNR) exceeds the 0-dB threshold (when the signal and noise levels are comparable). Using this definition for signal velocity, some studies have reported the possibility of achieving fast signaling in dispersive channel, whereas other researchers have indicated that fast signaling is not possible. In this chapter, we aim to present a unified approach within which the diverse points-of-views of different researchers can be consolidated under one umbrella. We present a systematic study that can be applied to pulse propagation in any dispersive medium in order to quantify the detectable information content and calculate its speed at different propagation distances while accounting for pulse reshaping effects, noise generated in the medium, and the detector's noise. Accordingly, we present an operational context within which the constraints of superluminal signaling and its potential applications are shown. Furthermore, we explain the fundamental limitations imposed by the medium length, beyond which superluminal propagation cannot be achieved. Such analysis explores the fundamental limitations and capabilities of a broad range of fast (and slow) light applications.

Summary of Chapter 10

Author(s): Joaquim J. Barroso[1], Joaquim P. Leite Neto[2], Pedro J. Castro[2], Ugur C. Hasar[3], and José Edimar B. Oliveira[1]

Authors' Affiliation:

[1]Instituto Tecnológico de Aeronáutica, Brazil
[2]Instituto Nacional de Pesquisas Nacionais, Brazil
[3]University of Gaziantep, Turkey

Keywords: Metamaterials, split-ring resonator, negative group velocity, retrieval of constitutive parameters.
Abstract: Analytical discussions and experimental results are presented for an X-band (8.2–12.4 GHz) waveguide loaded with an array of split-ring resonators (SRR) designed to exhibit NGD and negative refractive index in the 2.8–3.6 GHz frequency range, thus below the cutoff frequency (6.557 GHz) of the empty waveguide. Negative-group-velocity effects arise from resonant absorption losses of the split-rings, whereby such effects are demonstrated by time-domain simulations of propagating modulated Gaussian pulses. It is shown that an injected pulse of 80-ns time width and central frequency of 3.26 GHz experiences a negative delay of −14 ns after traversing the waveguide while retaining its Gaussian shape although with strong attenuation. A retrieval procedure is used to determine the complex propagation factor for which the role of the attenuation constant in reshaping the output spectrum amplitude is

discussed. The SRR-loaded waveguide examined here has attractive features, as the system is easily scalable to higher frequencies and proves to show potential applications for the delay compensations of signal in microwave waveguide systems by keeping a constant group delay to avoid signal distortion.

Summary of Chapter 11

Author(s): Richard Xian-Ke Gao[1], Si-Ping Gao[2], Hui Min Lee[1], and En-Xiao Liu[1]

Authors' Affiliation:

[1]Institute of High Performance Computing, Agency for Science, Technology and Research (A*STAR), Singapore
[2]National University of Singapore, Singapore

Keywords: Metamaterials, split-ring resonator, negative group velocity, retrieval of constitutive parameters.
Abstract: The ever increasing data transmission rate that runs to tens of Gb/s or beyond in the high-speed electronic circuits nowadays makes common-mode noise become a critical issue that could introduce electromagnetic interference (EMI) and adversely affect signal integrity (SI) and power integrity (PI). Together with miniaturization requirement of circuit design, it becomes requisite to explore newly effective common-mode filtering technology to tackle the common-mode noise associated EMI problems. In this chapter, the metamaterial-based common-mode filter design is elaborated. Through the study of a set of compact planar common-mode filters using defect ground structure (DGS), the fundamental filtering methodology and equivalent circuit models are elucidated with the presence of reasonable insertion loss performance. One of the intrinsic characteristics of the filters, termed NGD, has yet been paid much attention before, is also evaluated and discussed in this chapter for addressing SI/PI and EMI issues in electronic circuits.

Ending remark

The coauthors are inviting the readers to explore the NGD concept in different ways and depending on the areas of interests, the readers of this book can start with any chapter independently.

Chapter 1

Microstrip line negative group delay circuits and applications to communication systems

Yongchae Jeong[1] *and Girdhari Chaudhary*[1]

Abstract

In this chapter, a synthesis of microstrip line negative group delay (NGD) circuits is presented. Microstrip line resonators are considered for designing various topologies of transmission- and reflection-type NGD circuits. Similarly, the defected microstrip line structures such as defected ground structure (DGS) and defected microstrip structure (DMS) are also used to synthesize the transmission-type NGD circuits. In addition, a filter synthesis based on finite unloaded-quality factor resonators is also presented to design NGD circuit with the arbitrarily prescribed group delay. Finally, the applications of NGD circuit in communication systems are explored. The applications of NGD circuit include the efficiency enhancement of feed-forward amplifier and bandwidth enhancement of analog feedback amplifier.

Keywords

Coupling matrix synthesis, coupled line, cross-cancellation power amplifier, defected ground structure, defected microstrip structure, feed-forward amplifier, filter synthesis, finite unloaded-quality factor resonators, negative group delay circuit, transmission line resonators.

1.1 What is negative group velocity and negative group delay?

The group delay (GD) is defined as the transit time of the reflected or transmitted pulse envelope when a pulse envelope is influenced by the circuit. Mathematically, the GD can be expressed using the phase (φ) variation of transmission S-parameter as (1.1):

$$\tau = -\frac{1}{2\pi}\frac{d\varphi}{df}. \tag{1.1}$$

[1]Division of Electronics Engineering, Chonbuk National University, Republic of Korea

The group velocity (v_g) is another relevant parameter to illustrate the propagation of electromagnetic waves. Propagation of electromagnetic waves in dispersive media can be defined by the phenomena incorporating abnormal phase, GDs, and velocities [1]. Abnormal group velocities incorporate superluminal and even negative group velocities (NGVs). The NGD and NGV influence each other and they convey that peak of a pulse envelope appears from the medium at an instant before the peak of the pulse penetrates the medium. Such an ostensibly anticausal phenomenon does not breach the principle of causality because turn-on and turn-off points of the wave packet propagate with a positive delay in agreement with the causality requirements [2,3].

To recognize the reality of the NGD, the group velocity (v_g) can be represented as (1.2) for homogeneous media or effective homogeneous periodic structures described by a propagation constant (β) and refractive index n, where c and ω are the speed of light in vacuum and angular frequency, respectively [1].

$$v_g = \frac{\partial \omega}{\partial \beta} = \frac{c}{n + \omega \frac{\partial n}{\partial \omega}} \tag{1.2}$$

From (1.2), it is evident that if n decreases rapidly with respect to ω, the group velocity and the GD become negative. In short, NGV or NGD exists near an absorption line or signal attenuation (SA) condition, where "anomalous" wave propagation effects can occur [1–3]. Typically, n of the RF circuits is difficult to handle, therefore, it is only possible to achieve the NGD through the SA condition. The SA of NGD circuits can be easily recompensed with the small signal gain amplifier without reducing a significant amount of NGD.

1.2 Reflection-type NGD circuits using microstrip line resonators

In electronic circuits, the NGD phenomenon has been signified using simple low-pass and bandpass operational amplifiers at low frequency [4,5]. In microwave regime, lossy bandstop structures are used to achieve NGD circuits because the NGD exists at a range of frequencies where the absorption or the signal attenuation (SA) is maximum. Therefore, various researchers have utilized RLC lumped resonators to implement active/passive NGD circuits which could be difficult to scale to higher frequencies [6–16]. To overcome the finite accessible problem of lumped elements, the NGD circuits using the microstrip line resonators have been investigated. These NGD circuits can be classified into reflection- and transmission-type circuits.

Since reflection-type NGD circuits can be accomplished by terminating coupled and through ports of the 3-dB hybrid coupler with one-port NGD circuits, the GD response of resulting two-port network is computed as the addition of GD responses of the hybrid and one-port terminating NGD circuit [6–10]. These NGD circuits have excellent input/output return loss characteristics at the expense of an extra component (such as hybrid coupler). Similarly, the NGD phenomena in the transmission-type NGD circuits can be determined by transmitting a signal through lossy bandstop resonators [11–17]. However, these circuits can be designed with excellent input/output return loss characteristics at the expense of extra matching networks.

1.2.1 Realization of two-port reflection-type NGD circuits using hybrid coupler

The one-port termination units of the reflection-type NGD circuits can be combined to form a two-port circuit by using hybrid coupler as shown in Figure 1.1. Presuming the coupled and through responses of the 3-dB hybrid coupler as S_{21}^H and S_{31}^H, respectively, the *S*-parameters of two-port reflection-type NGD circuit can be computed as (1.3) and (1.4) [7]:

$$S_{11} = \left[\left(S_{21}^H\right)^2 + \left(S_{31}^H\right)^2\right]\Gamma_{\text{NGD}} = R^H\Gamma_{\text{NGD}} \tag{1.3}$$

$$S_{21} = 2S_{21}^H S_{31}^H \Gamma_{\text{NGD}} = T^H\Gamma_{\text{NGD}} \tag{1.4}$$

The combined two-port NGD circuit would reveal perfect input/output matching ($S_{11} = S_{22} = 0$) and transmission ($S_{21} = 1$), respectively, only if the hybrid coupler bespeaks the *S*-parameters such that $|R^H| = |(S_{21}^H)^2 + (S_{31}^H)^2| = 0$ and $|T^H| = |2S_{21}^H S_{31}^H| = 1$. Therefore, the effect of hybrid can be quantified by adding GD of hybrid to GD of reflection termination unit.

1.2.2 Reflection-type NGD circuits using λ/4 transmission line resonator

Figure 1.2(a) delineates the structure of reflection-series NGD circuit which comprises series resistor R_{RS} and an open-circuited λ/4 transmission line resonator with

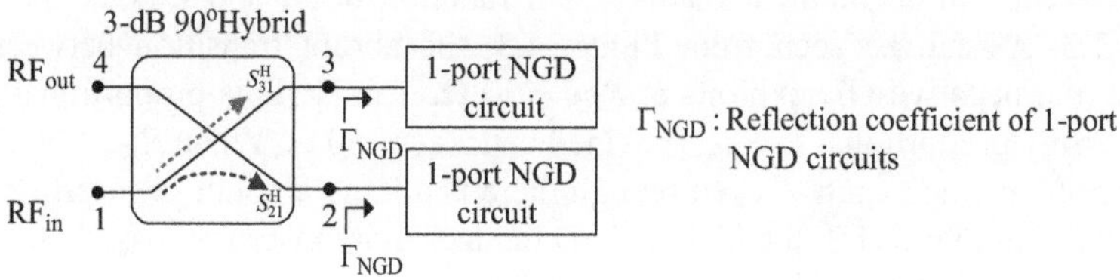

Figure 1.1 Schematic of two-port reflection-type negative GD circuit using a hybrid coupler and one-port reflection termination circuit

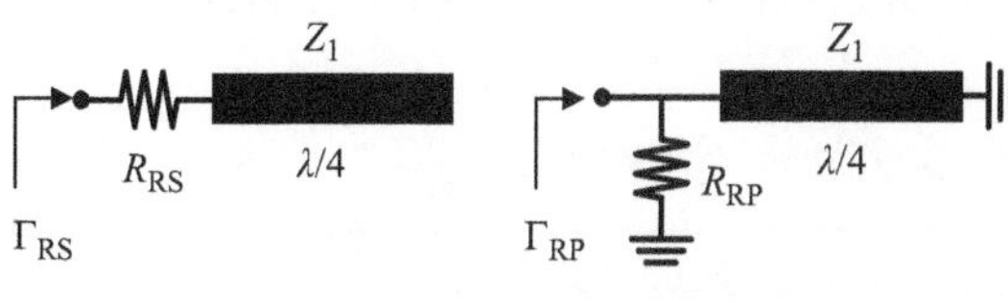

Figure 1.2 Schematic of λ/4 microstrip line-based one-port reflection-type negative GD circuits: (a) reflection-series and (b) reflection-parallel

characteristic impedance of Z_1. The expression of the input reflection coefficient (Γ_{RS}) of the one-port circuit shown in Figure 1.2(a) is expressed as (1.5):

$$\Gamma_{RS} = \sqrt{\frac{(R_{RS}-Z_0)^2 + Z_1^2 \cot^2 \frac{\pi f}{2f_0}}{(R_{RS}+Z_0)^2 + Z_1^2 \cot^2 \frac{\pi f}{2f_0}}} \angle \tan^{-1}\left(\frac{Z_1 \cot \frac{\pi f}{2f_0}}{R_{RS}+Z_0}\right) - \tan^{-1}\left(\frac{Z_1 \cot \frac{\pi f}{2f_0}}{R_{RS}-Z_0}\right) \tag{1.5}$$

where Z_0, f, and f_0 are port impedance, operating, and design center frequencies, respectively. Using the phase of Γ_{RS}, the GD of this circuit can be expressed as (1.6):

$$\tau_{RS} = -\frac{Z_1}{4f_0}\left\{\frac{(R_{RS}-Z_0)\csc^2 \frac{\pi f}{2f_0}}{(R_{RS}-Z_0)^2 + Z_1^2 \cot^2 \frac{\pi f}{2f_0}} - \frac{(R_{RS}+Z_0)\csc^2 \frac{\pi f}{2f_0}}{(R_{RS}+Z_0)^2 + Z_1^2 \cot^2 \frac{\pi f}{2f_0}}\right\}. \tag{1.6}$$

Subsequently, the magnitude and GD at f_0 can be further simplified as (1.7) using the relation $f = f_0$ in (1.5) and (1.6):

$$|\Gamma_{RS}|_{f=f_0} = \left|\frac{R_{RS}-Z_0}{R_{RS}+Z_0}\right| \tag{1.7a}$$

$$\tau_{RS}\Big|_{f=f_0} = -\frac{Z_0 Z_1}{2f_0\left(R_{RS}^2 - Z_0^2\right)} \tag{1.7b}$$

For the intuitive demonstration of analytical analysis, (1.7a) and (1.7b) are represented in a graphical manner as a function of Z_1 and R_{RS} as shown in Figure 1.3. As can be seen from Figure 1.3, the abrupt transition between the positive and negative GD exhibits at $R_{RS} = 50\ \Omega$. The NGD is proportional to Z_1 and inversely proportional to R_{RS}, provided that $R_{RS} > 50\ \Omega$. When $R_{RS} < 50\ \Omega$, the GD cannot become negative even though there is a large amount of insertion loss. As observed in Figure 1.3, the higher NGD induces more insertion loss, therefore, a trade-off occurs between NGD and insertion loss.

Figure 1.2(b) outlines the schematic of reflection-parallel NGD circuit consisting of parallel-connected shunt resistor R_{RP} and short-circuited $\lambda/4$ microstrip line resonator with a characteristic impedance of Z_1. Using Figure 1.2(b), the reflection coefficient of the circuit can be derived as (1.8):

$$\Gamma_{RP} = \sqrt{\frac{(R_{RP}-Z_0)^2 + Z_0^2 R_{RP}^2/Z_1 \cot \frac{\pi f}{2f_0}}{(R_{RP}+Z_0)^2 + Z_0^2 R_{RP}^2/Z_1 \cot \frac{\pi f}{2f_0}}} \angle \varphi_{RP}, \tag{1.8}$$

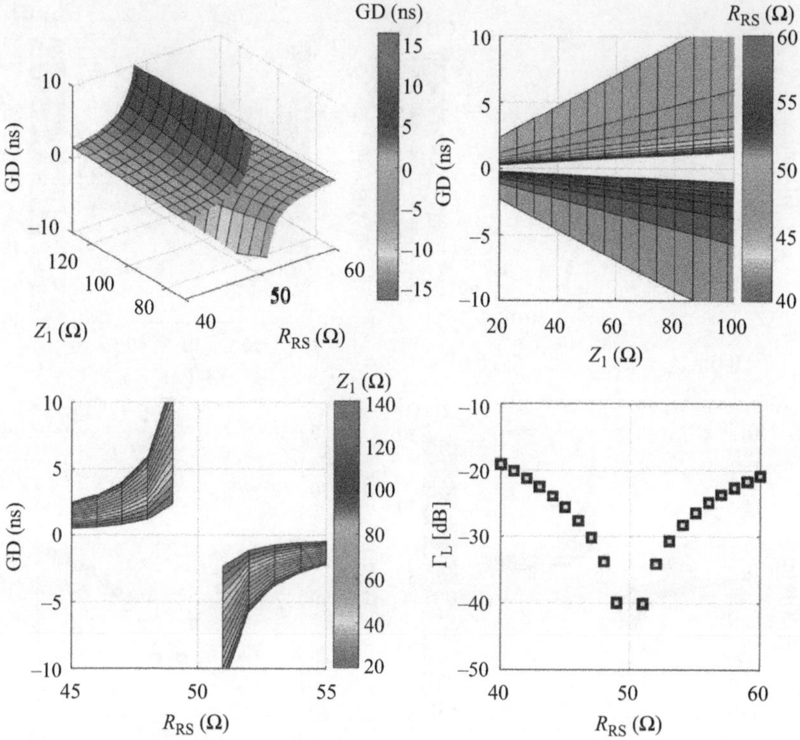

Figure 1.3 Computed GD and insertion loss of one-port reflection-series NGD circuit as a function of R_{RS} and Z_1

where

$$\varphi_{RP} = \tan^{-1}\left\{\frac{Z_0 R_{RP}}{Z_1(R_{RP}-Z_0)}\cot\frac{\pi f}{2f_0}\right\} + \tan^{-1}\left\{\frac{Z_0 R_{RP}}{Z_1(R_{RP}+Z_0)}\cot\frac{\pi f}{2f_0}\right\}. \tag{1.9}$$

Using the phase of Γ_{RP}, the GD of this circuit can be expressed as (1.10):

$$\tau_{RP} = \frac{Z_0 R_{RP}}{4Z_1 f_0}\left\{\frac{(R_{RP}-Z_0)\csc^2\frac{\pi f}{2f_0}}{(R_{RP}-Z_0)^2 + Z_0^2 R_{RP}^2/Z_1^2\cot^2\frac{\pi f}{2f_0}} + \frac{(R_{RP}+Z_0)\csc^2\frac{\pi f}{2f_0}}{(R_{RP}+Z_0)^2 + Z_0^2 R_{RP}^2/Z_1^2\cot^2\frac{\pi f}{2f_0}}\right\} \tag{1.10}$$

Using $f=f_0$ in (1.8) and (1.10), the magnitude and GD can be further simplified as (1.11)

$$|\Gamma_{RP}|_{f=f_0} = \left|\frac{R_{RP}-Z_0}{R_{RP}+Z_0}\right| \tag{1.11a}$$

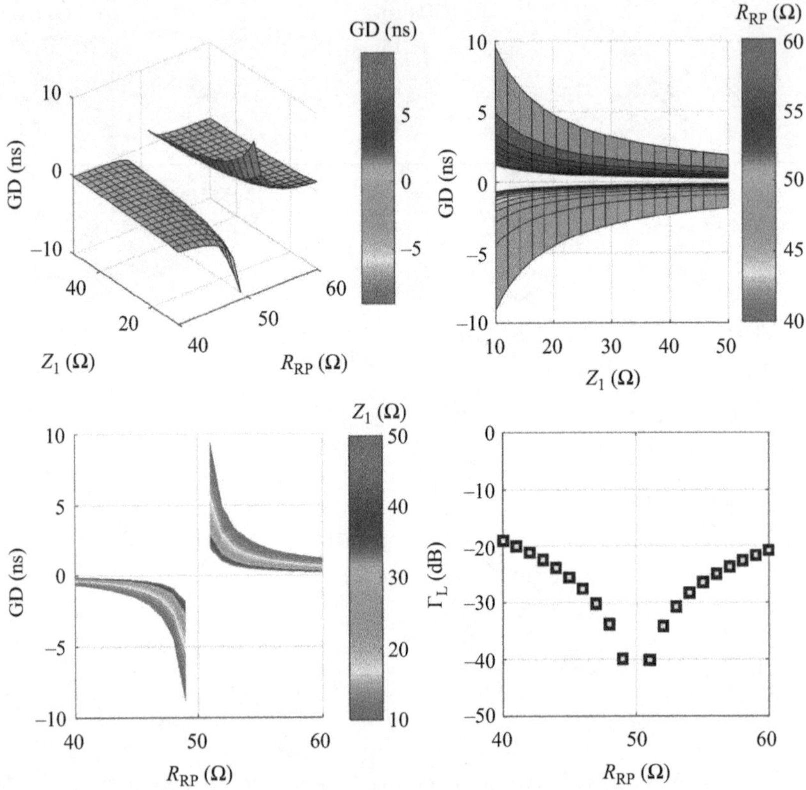

Figure 1.4 Computed GD and insertion loss of one-port reflection-parallel NGD circuit as a function of R_{RP} and Z_1

$$\tau_{RP}\Big|_{f=f_0} = \frac{Z_0 R_{RP}^2}{2Z_1 f_0 \left(R_{RP}^2 - Z_0^2\right)} \tag{1.11b}$$

Figure 1.4 shows the graphical representations of (1.11a) and (1.11b) for intuitive understanding. As can be seen from graphs, an abrupt transition between the positive and negative GDs occurs at $R_{RP} = 50\ \Omega$. Similarly, the NGD becomes higher when Z_1 is low, provided that $R_{RP} < 50\ \Omega$. In contrast to reflection-series structure, the NGD cannot be achieved when $R_{RS} > 50\ \Omega$, even though there is a large amount of insertion loss. In addition, a high NGD results in high insertion loss which challenges trade-off between NGD and insertion loss to the designer.

1.2.3 Low insertion loss reflection-type NGD circuits using coupled lines

The major drawback of $\lambda/4$ microstrip line resonators based on NGD circuits is their inherent insertion loss. Even though insertion loss can be resolved by a gain

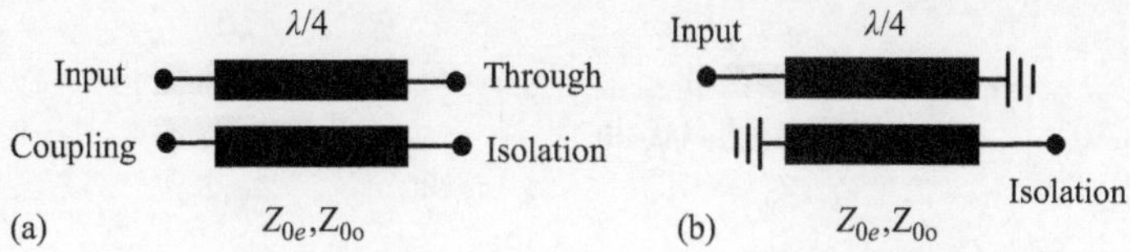

Figure 1.5 (a) Four-port coupled line and (b) two-port short-circuited coupled line

compensation amplifier, other problems such as increased noise due to out-of-band gain and other-tradeoffs issues (such as stability problems) occur.

In recent years, researchers have concentrated on the design of low insertion loss NGD circuits. In [18], a composite NGD circuit with low insertion loss is demonstrated, however, this structure necessitates parallel lumped elements (such as capacitors and inductors) between two transmission lines. Similarly, the microstrip line NGD circuit with reduced insertion loss is demonstrated in [19] and [20]; however, further improvement in the insertion loss is difficult in these structures because of the difficulty in implementing high characteristic impedance (such as 140 Ω) of the microstrip line in PCB technology. In addition, the NGD circuit requires two microstrip line resonators (low and high characteristic impedance microstrip line resonators) which increase the circuit size.

A general structure of four-port $\lambda/4$ coupled line is depicted in Figure 1.5(a) where odd- and even-mode impedances are represented as Z_{0_o}, Z_{0_e}, respectively. The four-port coupled line is converted into a two-port line as shown in Figure 1.5(b) by shorting coupled and through ports. Expressing the electrical length of a coupled line as $\theta = \pi f/2f_0$, the admittance parameters (Y-parameters) of the short-circuited two-port coupled line are expressed as (1.12) by presuming transverse electromagnetic wave propagation and symmetrical transmission lines [21].

$$[Y] = \begin{bmatrix} -j\dfrac{Z_{0_e}+Z_{0_o}}{2Z_{0_e}Z_{0_o}}\cot\dfrac{\pi f}{2f_0} & -j\dfrac{Z_{0_e}-Z_{0_o}}{2Z_{0_e}Z_{0_o}}\csc\dfrac{\pi f}{2f_0} \\ -j\dfrac{Z_{0_e}-Z_{0_o}}{2Z_{0_e}Z_{0_o}}\csc\dfrac{\pi f}{2f_0} & -j\dfrac{Z_{0_e}+Z_{0_o}}{2Z_{0_e}Z_{0_o}}\cot\dfrac{\pi f}{2f_0} \end{bmatrix} \tag{1.12}$$

Furthermore, the input impedance of the coupled line is expressed as (1.13) when isolation port is terminated with a load $Z_L = 1/Y_L$.

$$Z_{in} = \frac{1}{Y_{in}} = \left[Y_{11} - \frac{Y_{12}Y_{21}}{Y_{22}+Y_L}\right]^{-1} \tag{1.13}$$

1.2.4 Type-I: short-circuited coupled line with resistor-terminated isolation port

The schematic of one-port reflection type-I low loss NGD circuit is depicted in Figure 1.6(a), which constitutes a coupled line with an isolation port terminated with load $Z_L = R_L$ [21].

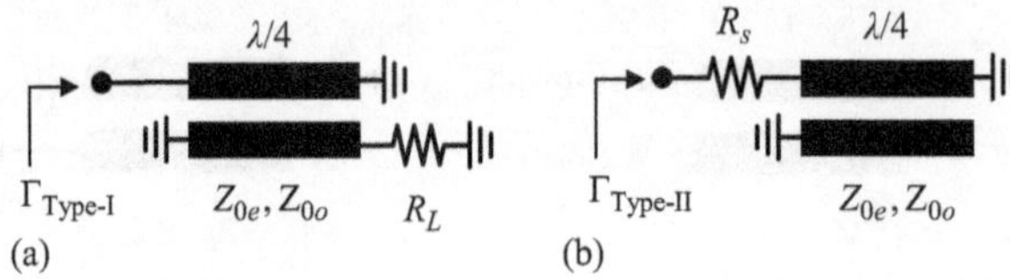

Figure 1.6 Structure of low insertion loss negative GD circuits using short-circuited coupled line: (a) type-I and (b) type-II

Assuming the reference port impedance $Z_0 = 1/Y_0$, the input reflection coefficient is expressed as (1.14):

$$\left|\Gamma_{\text{Type-I}}\right| \angle \varphi_{\text{Type-I}} = \left|\frac{X_1 + X_2}{X_3 + X_4}\right| \angle \left(\tan^{-1}\frac{X_2}{X_1} + \tan^{-1}\frac{X_4}{X_3}\right), \tag{1.14}$$

where

$$X_1 = Z_c^2 C - C^2 R_L Z_0 \csc^2\left(\frac{\pi f}{2f_0}\right) + R_L Z_0 \cot^2\left(\frac{\pi f}{2f_0}\right) \tag{1.15a}$$

$$X_2 = Z_c C(Z_0 - R_L)\cot\left(\frac{\pi f}{2f_0}\right) \tag{1.15b}$$

$$X_3 = Z_c^2 C^2 + C^2 R_L Z_0 \csc^2\left(\frac{\pi f}{2f_0}\right) - R_L Z_0 \cot^2\left(\frac{\pi f}{2f_0}\right) \tag{1.15c}$$

$$X_4 = Z_c C(Z_0 + R_L)\cot\left(\frac{\pi f}{2f_0}\right) \tag{1.15d}$$

Likewise, Z_c and C are characteristic impedance and coupling coefficient of the coupled line, respectively, which are expressed as (1.16):

$$Z_c = \frac{2Z_{0_e}Z_{0_o}}{Z_{0_e} - Z_{0_o}} = Z_{0_e}\frac{1-C}{C} = Z_{0_o}\frac{1+C}{C}, \tag{1.16a}$$

$$C = \frac{Z_{0_e} - Z_{0_o}}{Z_{0_e} + Z_{0_o}}, \tag{1.16b}$$

The GD expression of the one-port reflection type-I NGD circuit can be determined as (1.17) from (1.14) and (1.15) as:

$$\tau_{\text{Type-I}} = -\frac{d\varphi_{\text{Type-I}}}{d\omega} = \frac{X_2X_1' - X_1X_2'}{X_1^2 + X_2^2} + \frac{X_4X_3' - X_3X_4'}{X_3^2 + X_4^2} \tag{1.17}$$

where

$$X_1' = \frac{R_L Z_0}{2f_0}(C^2 - 1)\csc^2\left(\frac{\pi f}{f_0}\right)\cot\left(\frac{\pi f}{f_0}\right) \tag{1.18a}$$

$$X_2' = -\frac{Z_c C}{2f_0}(Z_0 - R_L)\csc^2\left(\frac{\pi f}{f_0}\right) \tag{1.18b}$$

$$X_3' = -\frac{R_L Z_0}{2f_0}(C^2 - 1)\csc^2\left(\frac{\pi f}{f_0}\right)\cot\left(\frac{\pi f}{f_0}\right) \tag{1.18c}$$

$$X_4' = -\frac{Z_c C}{2f_0}(Z_0 + R_L)\csc^2\left(\frac{\pi f}{f_0}\right) \tag{1.18d}$$

The GD and insertion loss can be further simplified at $f = f_0$ as (1.19):

$$\left|\Gamma_{\text{Type-I}}\right|_{f=f_0} = \left|\frac{Z_c^2 - Z_0 R_L}{Z_c^2 + Z_0 R_L}\right| \tag{1.19a}$$

$$\tau_{\text{Type-I}}\Big|_{f=f_0} = -\frac{Z_c Z_0 \left(R_L^2 - Z_c^2\right)}{2f_0 C\left(Z_c^4 - R_L^2 Z_0^2\right)} \tag{1.19b}$$

As noted from (1.19a), the insertion loss at $f = f_0$ depends on Z_c and R_L. Therefore, the insertion loss can be decreased by increasing Z_c. Similarly, the GD at $f = f_0$ is a function of Z_c, R_L, and C, which yields three degrees of freedom to manage the GD. For graphical illustration, the calculated GD and insertion loss at $f_0 = 2.14$ GHz are depicted in Figure 1.7 for distinct Z_c, C, and R_L. As observed from the graphs, the insertion loss can be improved by controlling C of the coupled line for same GD. Therefore, small C is preferable for the low insertion loss.

1.2.5 Type-II: short-circuited coupled line with open-circuited isolation port and series-connected resistor at input port

The structure of one-port reflection type-II NGD circuit is depicted in Figure 1.16(b). The circuit consists of a series-connected resistor (R_s) at the input port of the short-circuited coupled lines with an open-circuited isolation port. As described previously, the input reflection coefficient of this circuit is determined as (1.20):

$$\left|\Gamma_{\text{Type-II}}\right| \angle \varphi_{\text{Type-II}} = \frac{|X_5 + jX_6|}{|X_7 + jX_8|} \angle \left(\tan^{-1}\frac{X_5}{X_6} - \tan^{-1}\frac{X_7}{X_8}\right), \tag{1.20}$$

where

$$X_5 = 2(R_s - Z_0)\left(\cos^2\left(\frac{\pi f}{2f_0}\right) - C^2\right) \tag{1.21a}$$

$$X_6 = X_8 = Z_c C \sin\left(\frac{\pi f}{f_0}\right) \tag{1.21b}$$

$$X_7 = 2(R_s + Z_0)\left(\cos^2\left(\frac{\pi f}{2f_0}\right) - C^2\right) \tag{1.21c}$$

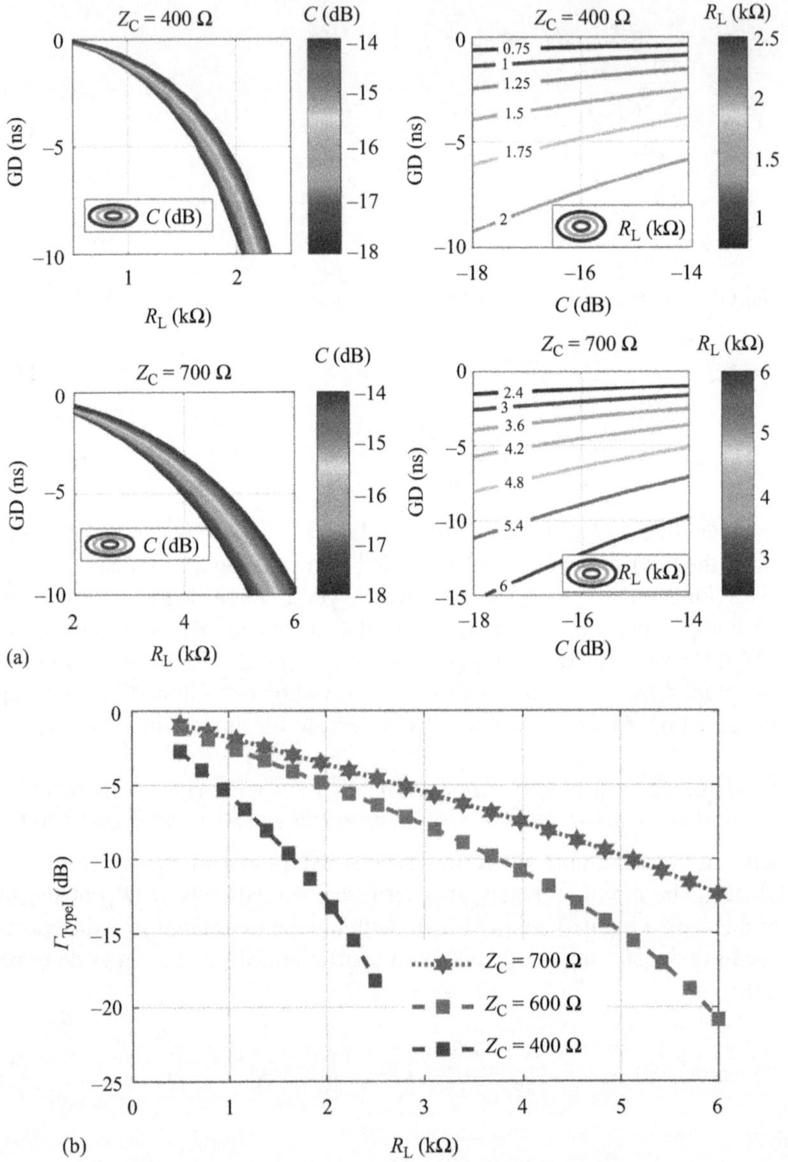

Figure 1.7 Calculated GD and magnitude of the one-port type-I low-loss NGD circuit at $f_0 = 2.14$ GHz: (a) GD and (b) magnitude

Likewise, the GD expression of one port type-II NGD circuit is given as in (1.22):

$$\tau_{\text{Type-II}} = -\frac{d\varphi_{\text{Type-II}}}{d\omega} = \frac{X_5 X_6' - X_6 X_5'}{X_5^2 + X_6^2} + \frac{X_7 X_8' - X_8 X_7'}{X_7^2 + X_8^2}, \tag{1.22}$$

where

$$X_5' = -\frac{(R_s - Z_0)}{2f_0}\sin\left(\frac{\pi f}{f_0}\right) \tag{1.23a}$$

$$X_6' = X_8' = \frac{Z_c C}{2f_0}\cos\left(\frac{\pi f}{f_0}\right) \tag{1.23b}$$

$$X_7' = -\frac{(R_s + Z_0)}{2f_0}\sin\left(\frac{\pi f}{f_0}\right) \tag{1.23c}$$

Furthermore, the magnitude and GD at $f = f_0$ are simplified as (1.24):

$$\left|\Gamma_{\text{Type-II}}\right|_{f=f_0} = \frac{|R_s - Z_0|}{|R_s + Z_0|} \tag{1.24a}$$

$$\tau_{\text{Type-II}}\Big|_{f=f_0} = -\frac{Z_c Z_0}{2f_0 C\left(R_s^2 - Z_0^2\right)}. \tag{1.24b}$$

Similar to type-I NGD circuit, the insertion loss is a function of R_s whereas GD depends on Z_c, C, and R_s. The insertion loss at $f = f_0$ can be decreased by increasing R_s. To get NGD, the R_s should be $> Z_0$ as shown in (1.24b). For an intuitive understanding of (1.24), the GD and insertion loss at $f_0 = 2.14$ GHz is plotted in Figure 1.8 as a function of Z_c, C, and R_s. As observed from these graphs, the GD increases toward negative as R_s decreases for fixed values of Z_c and C. Similarly, when R_s is fixed, the higher NGD can be obtained for high Z_c and low C. Therefore, the insertion loss can be improved by increasing Z_c and decreasing C for the same NGD.

1.2.6 Implementation of low insertion loss NGD circuits

For verification, the low-loss NGD circuits are designed and fabricated at $f_0 = 2.14$ GHz on a substrate with a dielectric constant (ε_r) of 2.2 and a thickness (h) of 31 mils. The one-port NGD circuit is converted into the two-port circuit as described in Section 1.2.1 by using a quadrature surface mount 3-dB hybrid coupler. The circuit parameters of NGD circuits (type-I and type-II) for GD of −7 ns are given as $Z_c = 630\ \Omega$, $C = -18.06$ dB, $Z_{0_e} = 90\ \Omega$, and $Z_{0_o} = 70\ \Omega$. Similarly, the values of R_L and R_s for type-I and type-II NGD circuits are given as 3,800 Ω and 105 Ω, respectively.

The simulated and measured results of type-I single-stage NGD circuit are shown in Figure 1.9(a). From the measurement, the GD and insertion loss at f_0 are determined as −7.27 ns and 7.43 dB, respectively. Similarly, the magnitude of input and output reflection coefficients (S_{11} and S_{22}) are greater than 20 dB. The phase slope of NGD circuit is positive at a range of frequencies where NGD occurs as shown in Figure 1.9(a).

The simulated and measured performances of the type-II single-stage NGD circuit are shown in Figure 1.9(b) where the measurement results are in agreement with the simulation. The measured GD and insertion loss at f_0 are given as −6.30 ns and 9.23 dB, respectively. From the measurement, the magnitude of input and output reflection coefficients (S_{11} and S_{22}) are > 19 dB. The measured slope of S_{21} is also positive over a certain frequency range which signifies the presence of NGD.

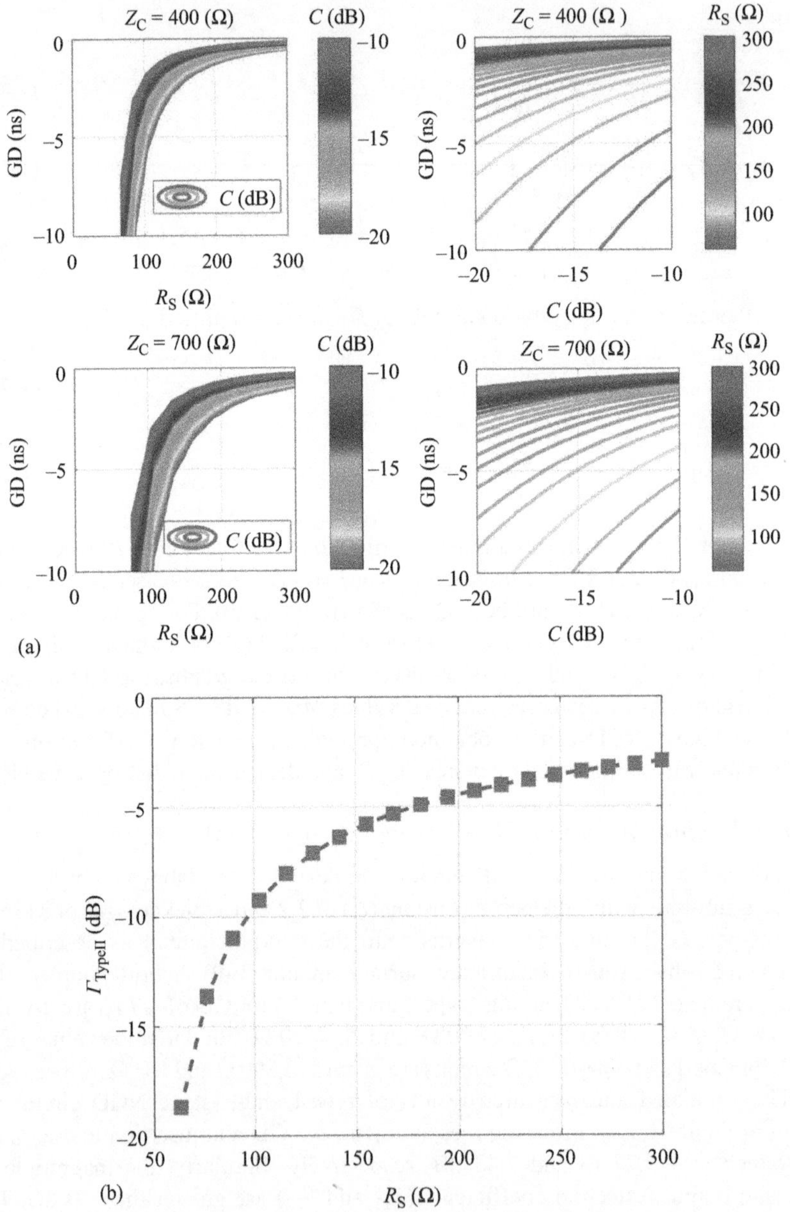

Figure 1.8 Calculated GD and magnitude of the one-port type-II low NGD circuit at $f_0 = 2.14$ GHz: (a) GD and (b) magnitude

As noted from the results of single-stage NGD circuits, the flat NGD and magnitude bandwidths are narrow which is not practically applicable for commercial RF systems. One way to increase the NGD bandwidth (which is defined as bandwidth when GD < 0) and magnitude flatness is to combine two units of NGD

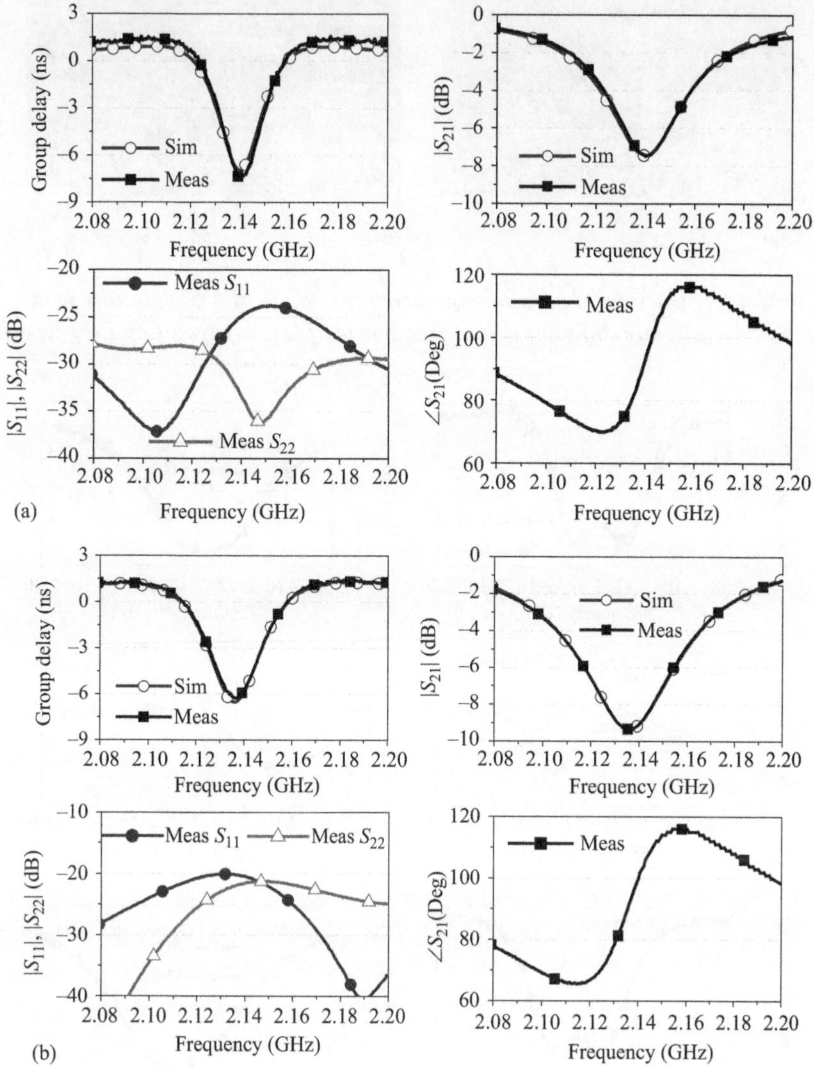

Figure 1.9 Simulation and measurement results of single-stage NGD circuits [21]: (a) type-I and (b) type-II

circuits operating at slightly different center frequencies in parallel, as shown in Figure 1.10. For demonstration, the NGD circuits are designed at $f_{01} = 2.125$ GHz and $f_{02} = 2.155$ GHz for GD of -7 ns. The circuit parameters are same as single-stage circuit except for the center frequencies.

Figure 1.11(a) shows the performances of parallel-connected two-stage type-I NGD circuit where the measured GD and insertion loss are determined as -6.02 ns and 14.70 dB at 2.135 GHz, respectively. Similarly, Figure 1.12(b) shows the simulated and measured results of parallel-connected two-stage type-II NGD

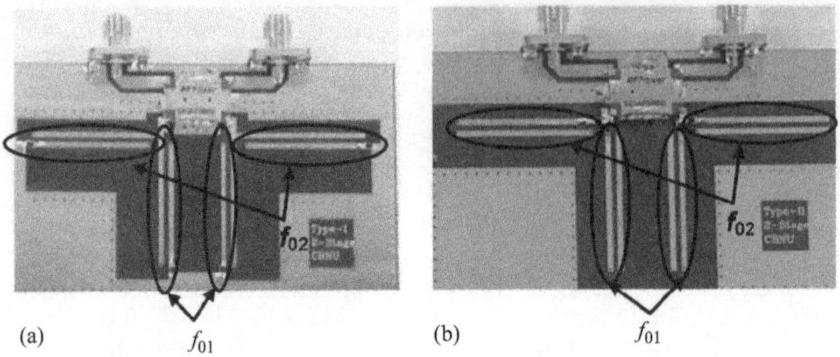

Figure 1.10 Structure of parallel-connected two-stage of NGD circuits with slightly different center frequencies [21]: (a) type-I and (b) type-II

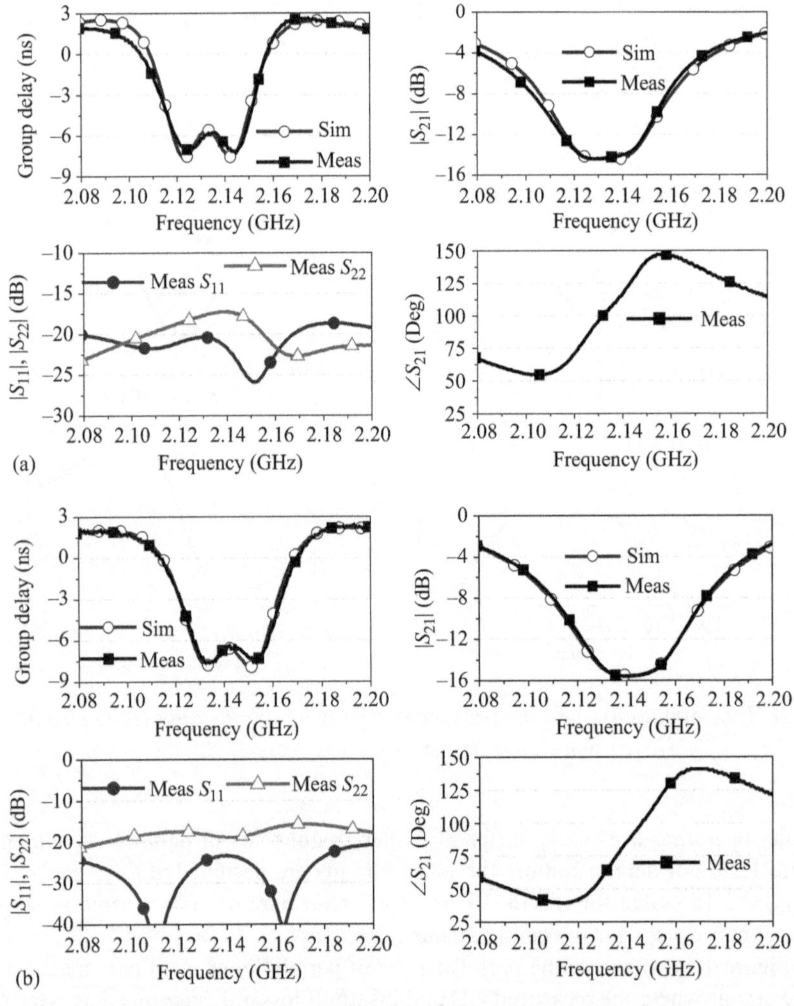

Figure 1.11 Simulation and measurement results of parallel-connected two-stage NGD circuits for bandwidth enhancement [21]: (a) type-I and (b) type-II

circuits. From the measurement, the GD and insertion loss at $f_0 = 2.14$ GHz are determined as -6.10 ns and 15.38 dB, respectively. As observed from these results, the acceptable flat NGD and magnitude bandwidths of parallel connected two-stage NGD circuits are wider than the single-stage NGD circuit demonstrating the practical applicability of the RF system.

1.3 Transmission-type NGD circuits

Various transmission-type NGD circuits have been theoretically and experimentally validated using lumped element RLC resonators [14–16]. In this section, periodic structures such as defected ground structure (DGS) and defected microstrip structure (DMS) are used to realize transmission-type NGD circuits.

1.3.1 Realization of transmission-type NGD circuits using DGS and DMS

Recently, it is of growing interest in the periodic structures such as microstrip photonic bandgap (PBG), DGS, and DMS. These structures provide the SA characteristics at a certain resonant frequency and have been appertained successfully in diverse applications. Since the equivalent circuits of DGS and DMS are expressed as parallel RLC resonator, these structures can be used to design transmission-type NGD circuits [22–26].

Figure 1.12 delineates the layout of transmission-type NGD circuits based on U-shaped DGS and DMS, respectively, where external lumped capacitor (C) and resistor (R) are connected across the DGS and DMS slot. The equivalent circuits of DGS- and DMS-based NGD circuits [22–24] are outlined in Figure 1.13. The lumped element values of the equivalent DGS- and DMS-based NGD circuits can be obtained by performing EM simulations as (1.25):

$$C_i = \frac{f_c}{4\pi Z_0\left(f_0^2 - f_c^2\right)}, \quad \text{where } i = \text{DGS or DMS}, \tag{1.25a}$$

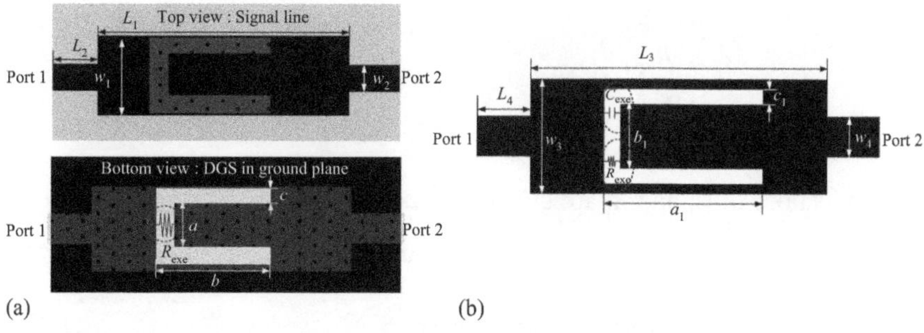

Figure 1.12. Structures of transmission-type NGD circuits: (a) DGS- and (b) DMS-based NGD

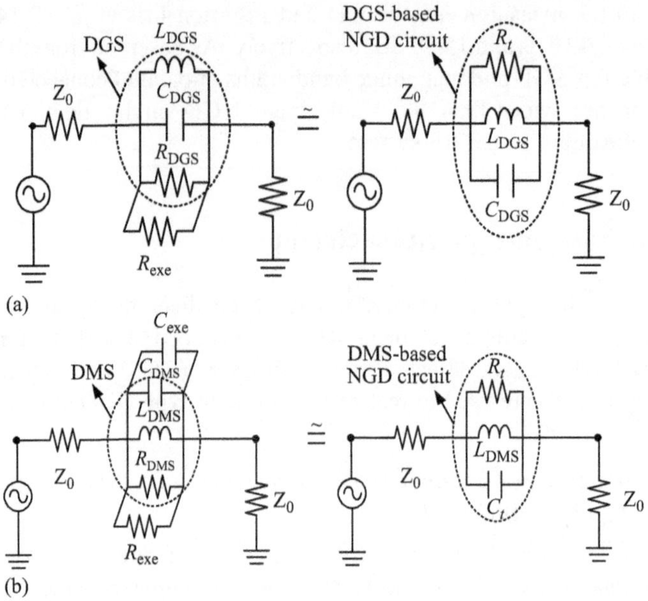

Figure 1.13 The equivalent circuits of DGS- and DMS-based NGD circuits:(a) DGS and (d) DMS

$$L_i = \frac{1}{4\pi^2 f_0^2 C_i} \tag{1.25b}$$

$$R_i = 2Z_0 \frac{1 - S_{21}(f_0)}{S_{21}(f_0)} \tag{1.25c}$$

where f_c, f_0, and S_{21} are the 3-dB cut-off frequency, resonant frequency, and the transmission coefficient magnitude at the resonant frequency, respectively.

From the equivalent circuits of DGS- and DMS-based NGD circuits, the S-parameter magnitudes and GD can be calculated as (1.26):

$$|S_{11}|_{f=f_0} = \frac{R_t}{2Z_0 + R_t} \tag{1.26a}$$

$$|S_{21}|_{f=f_0} = \frac{2Z_0}{2Z_0 + R_t} \tag{1.26b}$$

$$\tau\Big|_{f=f_0} = -\frac{1}{2\pi}\frac{d\angle S_{21}}{df} = -\frac{2R_t^2 C_t}{2Z_0 + R_t} \tag{1.26c}$$

where the values of C_t and R_t are given as (1.27).

$$C_t = C_i + C_{exe}, \quad \text{where } i = \text{DGS or DMS} \tag{1.27a}$$

$$R_t = \frac{R_i R_{exe}}{R_i + R_{exe}}, \quad \text{where } i = \text{DGS or DMS} \tag{1.27b}$$

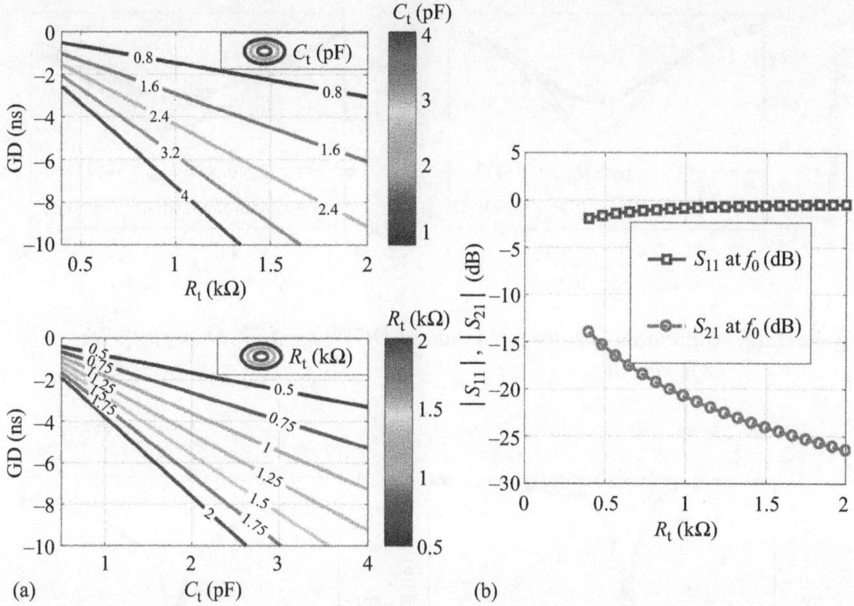

Figure 1.14 Calculated GD and magnitude of transmission-type NGD using DGS and DMS: (a) GD and (b) magnitude

As can be seen from (1.26) and (1.27), the GD can be controlled by R_t and C_t. For intuitive understanding, Figure 1.14 shows graphical representations of (1.26) and (1.27) using MATLAB®. It can be inferred from the above analysis that higher NGD induces higher insertion loss and poor return loss.

For the demonstration of DGS-based NGD circuit, firstly, the U-shaped DGS is simulated using 3-D full-wave solver with a dielectric constant (ε_r) of 2.2 and thickness (h) of 31 mils and following physical dimensions: $w_1 = 4$, $w_2 = 2.4$, $L_1 = 20$, $L_2 = 2$, a = 2.4, $b = 16$, and $c = 0.4$ (all units are in mm). From the simulation of the U-shaped DGS, a 3-dB cut-off and resonant frequencies are obtained as 3.065 and 3.505 GHz, respectively. Therefore, the extracted values of equivalent circuit of DGS are given as $C_{DGS} = 1.6874$ pF, $L_{DGS} = 1.2219$ nH, and $R_{DGS} = 3.9414$ kΩ. Figure 1.15 depicts the simulation results of DGS-based NGD circuit under different R_{exe} where GD and insertion loss are mainly controlled by external R_{exe}.

Similar to DGS, firstly, the U-shaped DMS is simulated with following dimensions: $w_3 = 2.4$, $w_4 = 4$, $L_3 = 2$, $L_4 = 20$, $a_1 = 6.8$, $b_1 = 2.8$, and $c_1 = 0.4$ (all units are in mm). From the simulation, the 3-dB cut-off and resonant frequencies are determined at 5.66 and 7.56 GHz [23]. Using (1.26), the equivalent circuit values of DMS are extracted as $C_{DMS} = 0.3586$ pF, $L_{DMS} = 1.2389$ nH, and $R_{DMS} = 3.5119$ kΩ. By connecting $C_{exe} = 4.15$ pF, the resonant frequency is moved toward lower frequency around 2.14 GHz and the insertion loss is controlled by connecting $R_{exe} = 1.1$ kΩ as shown in Figure 1.16.

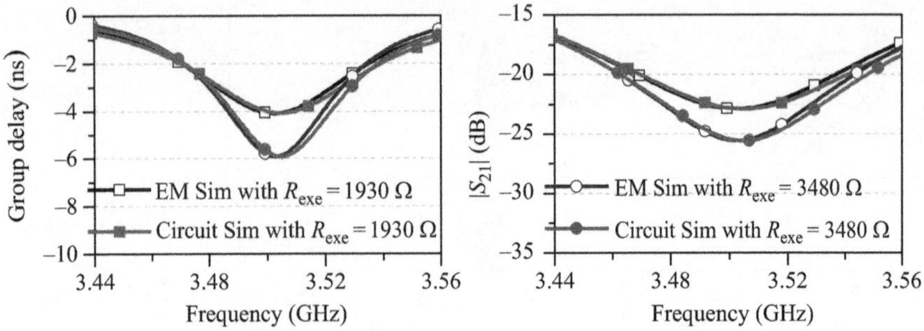

Figure 1.15 Simulated results of U-shaped DGS-based NGD circuit with different R_{exe}

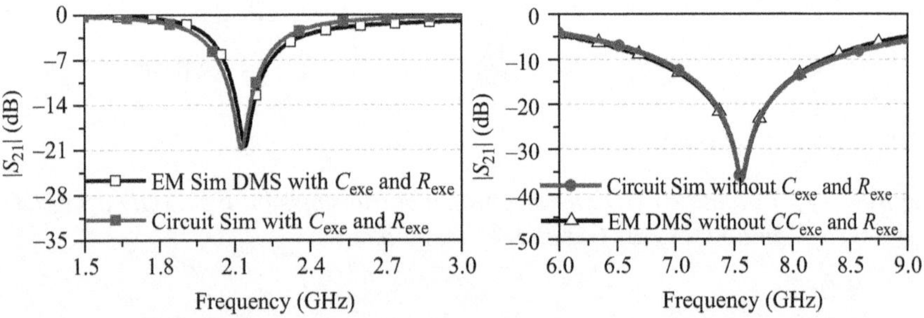

Figure 1.16 Simulated results of U-shaped DMS with and without external lumped elements

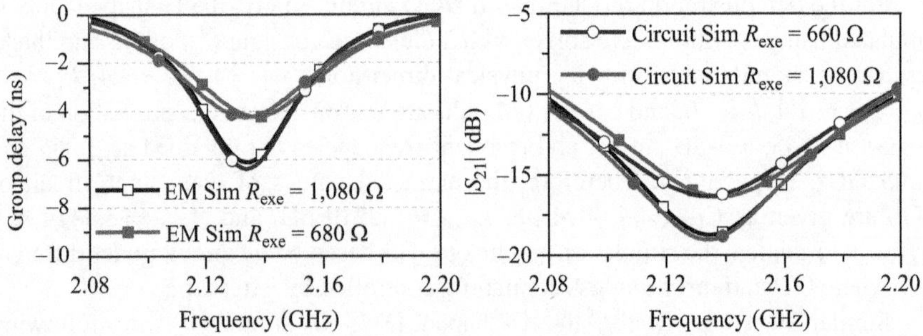

Figure 1.17 Simulated results of U-shaped DMS NGD circuit with different R_{exe}

Similar to DGS-based NGD circuit, the GD and insertion loss are controlled by external R_{exe} as shown in Figure 1.17. From these results, it can be noted that external C_{exe} is necessary to shift the design center frequency whereas R_{exe} is for obtaining specified GD. However, the external C_{exe} can be removed in case of higher design center frequency similar to the DGS-based NGD circuit.

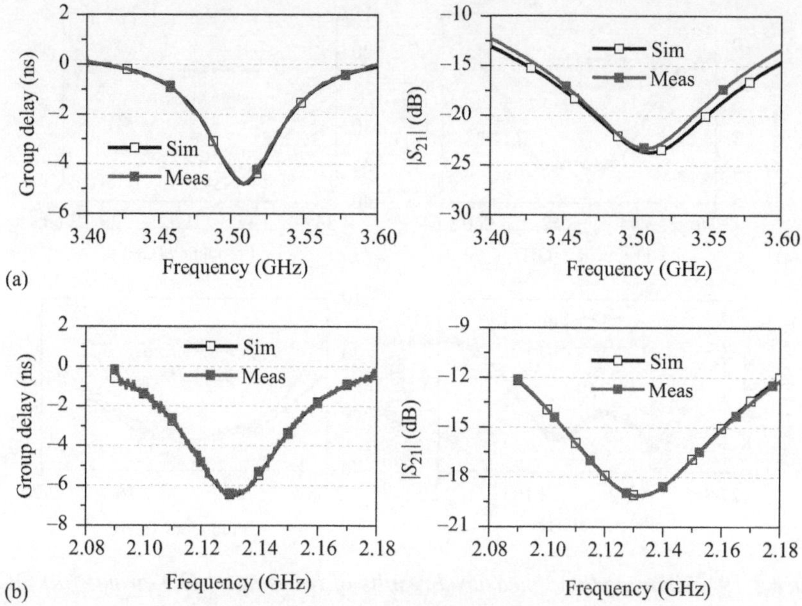

Figure 1.18 Simulated and measured results of single-stage NGD circuits: (a) DGS- and (b) DMS-based NGD circuit

For experimental demonstration, the transmission-type NGD circuits using DGS and DMS are designed and measured. The simulated and measured results of DGS-based NGD circuit are depicted in Figure 1.18(a) with $R_{exe} = 2.4$ kΩ. From this experiment, the GD and insertion loss at $f_0 = 3.506$ GHz are determined as −4.6 ns and 22.68 dB, respectively. Similarly, Figure 1.18(b) shows the simulation and measurement results of DMS-based NGD circuit with $C_{exe} = 4.14$ pF and $R_{exe} = 2.4$ kΩ. The measured GD and insertion loss are given as −6.4 ns and 19.8 dB, respectively.

Two-stage NGD circuits with slightly different center frequencies are designed for enhancing the acceptable flat NGD and magnitude bandwidths. However, insertion loss of two-stage NGD circuits is increased as compared to one-stage as shown in Figure 1.19. Therefore, a trade-off occurs between NGD bandwidth flatness and insertion loss.

1.4 NGD circuit using finite unloaded quality factor resonators

Generally, reflection- and transmission-type NGD circuits are realized by utilizing the lumped resistors along with different kinds of microstrip line resonators. These structures suffer from high insertion loss, narrow flat magnitude, and NGD bandwidth. To overcome these problems, the NGD circuits are designed using different methods such as cross-coupling between resonators, increasing number of resonators, and transversal-filter topologies [27–29]. Moreover, these NGD circuits also used a resistor to achieve NGD, which restrained fully distributed circuit implementation.

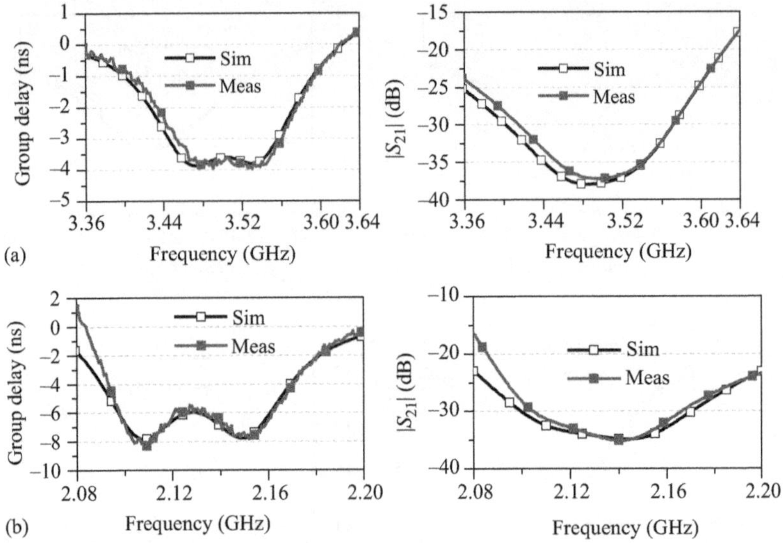

Figure 1.19 Simulated and measured results of two-stage NGD circuit: (a) DGS- and (b) DMS-based NGD circuit

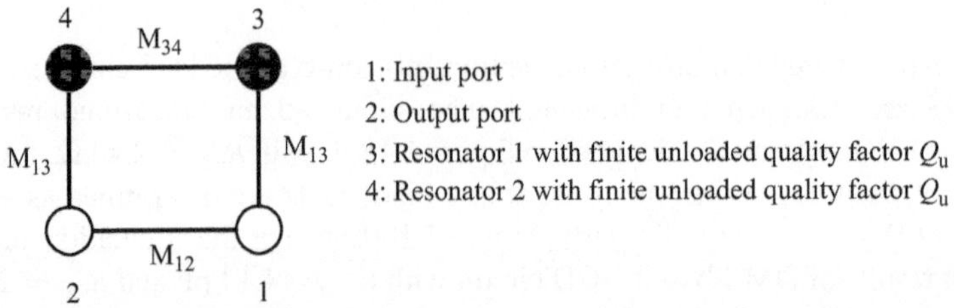

Figure 1.20 Coupling diagram of the NGD circuit using finite unloaded quality factor resonators

In this section, the NGD circuit using finite unloaded quality factor resonators will be presented [30–32]. This NGD topology does not necessitate any resistor to achieve NGD.

1.4.1 Coupling matrix analysis with finite unloaded quality factor resonators: S-parameters and GD

Figure 1.20 depicts the coupling diagram of the NGD circuit using finite unloaded quality-factor (Q_u) resonators where both external couplings are presumed as equal ($M_{13} = M_{31}$). Similarly, the self-coupling values ($M_{11} = M_{33}$), the source-to-load coupling, and load-to-source coupling ($M_{12} = M_{21}$) have the same magnitude. The $(N+2) \times (N+2)$ coupling matrix proportional to the coupling diagram is expressed as (1.28):

$$M = \begin{bmatrix} 0 & M_{13} & 0 & M_{12} \\ M_{13} & M_{11} & M_{34} & 0 \\ 0 & M_{34} & M_{44} & M_{13} \\ M_{SL} & 0 & M_{13} & 0 \end{bmatrix}, \tag{1.28}$$

where the subscripts 1, 2, 3, and 4 denote the source, load, the first resonator, and the second resonator, respectively. The reflection and transmission coefficients of this NGD circuit can be found by applying filter theory as (1.29):

$$S_{11} = 1 - 2A_{1,1}^{-1} \tag{1.29a}$$

$$S_{21} = 2A_{N+2,1}^{-1} \tag{1.29b}$$

where

$$A = R + j\Omega W + M \quad \text{with} \quad \Omega = \frac{\omega_0}{\Delta}\left(\frac{\omega_0}{\omega} - \frac{\omega}{\omega_0}\right) \tag{1.30}$$

and R is an $(N+2) \times (N+2)$ matrix where only nonzero entries are $R_{1,1} = R_{N+2,N+2} = 1$. W is similar to the $(N+2) \times (N+2)$ identity matrix except that $W_{1,1} = W_{N+2,N+2} = 0$. Similarly, M is the main coupling matrix given in (1.28). Δ is the 3-dB fractional bandwidth (FBW) of resonator. In addition, ω and ω_0 are angular operating and design center frequencies, respectively. Presuming self-couplings of resonators as $M_{11} = M_{33} = -j/Q_u\Delta$, reflection and transmission coefficients of NGD circuit can be derived as (1.31) using (1.28)–(1.30) in the following:

$$S_{11} = 1 - \frac{\dfrac{1}{\Delta^2}\left(\dfrac{\omega}{\omega_0} - \dfrac{\omega_0}{\omega}\right)^2 - \dfrac{M_{13}^2}{Q_u\Delta} - M_{34}^2 - \dfrac{1}{Q_u^2\Delta^2} - j\dfrac{2}{Q_u\Delta^2}\left(\dfrac{\omega}{\omega_0} - \dfrac{\omega_0}{\omega}\right) - j\dfrac{M_{13}^2}{\Delta}\left(\dfrac{\omega}{\omega_0} - \dfrac{\omega_0}{\omega}\right)}{\left\{\begin{array}{l} 0.5\left(M_{12}^2 + 1\right)\left\{\dfrac{1}{\Delta^2}\left(\dfrac{\omega}{\omega_0} - \dfrac{\omega_0}{\omega}\right)^2 - \dfrac{1}{Q_u^2\Delta^2} - M_{34}^2\right\} - \dfrac{M_{13}^2}{Q_u\Delta} \\ +M_{34}M_{13}^2M_{12} - 0.5M_{13}^4 - j\dfrac{1}{\Delta}\left(\dfrac{\omega}{\omega_0} - \dfrac{\omega_0}{\omega}\right)\left(M_{13}^2 + \dfrac{1}{Q_u\Delta} + \dfrac{M_{12}^2}{Q_u\Delta}\right) \end{array}\right\}} \tag{1.31a}$$

$$S_{21} = \frac{2\left\{\dfrac{M_{12}}{\Delta^2}\left(\dfrac{\omega}{\omega_0} - \dfrac{\omega_0}{\omega}\right)^2 - M_{12}M_{34}^2 + M_{34}M_{13}^2 - \dfrac{M_{12}}{Q_u^2\Delta^2} - j\dfrac{2M_{12}}{Q_u\Delta^2}\left(\dfrac{\omega}{\omega_0} - \dfrac{\omega_0}{\omega}\right)\right\}}{\left[\begin{array}{l} \left(2M_{13}^2 + \dfrac{2M_{12}^2}{Q_u\Delta} + \dfrac{2}{Q_u\Delta}\right)\dfrac{1}{\Delta}\left(\dfrac{\omega}{\omega_0} - \dfrac{\omega_0}{\omega}\right) + j\left(M_{12}^2 + 1\right)\left\{\begin{array}{l} \dfrac{1}{\Delta^2}\left(\dfrac{\omega}{\omega_0} - \dfrac{\omega_0}{\omega}\right)^2 \\ -\dfrac{1}{Q_u^2\Delta^2} - M_{34}^2 \end{array}\right\} \\ -j\left(\dfrac{2M_{13}^2}{Q_u\Delta} - 2M_{34}M_{13}^2M_{12} + M_{13}^4\right) \end{array}\right]} \tag{1.31b}$$

By equating S_{11} to zero at $\omega = \omega_0$ for matched input/output ports, the relationship among M_{12}, M_{13} and M$_{34}$ can be established. For arbitrary M_{13}, the relationship among M_{12}, M_{13}, and M_{34} can be found as (1.32), where the positive sign for $M_{13} > 1$ and the negative sign for $M_{13} < 1$.

$$M_{12} = \frac{M_{34}M_{13}^2 \mp \sqrt{M_{34}^2M_{13}^4 + \left(M_{34}^2 + \frac{1}{Q_u^2\Delta^2}\right)\left(M_{34}^2 - M_{13}^4 + \frac{1}{Q_u^2\Delta^2}\right)}}{M_{34}^2 + \frac{1}{Q_u^2\Delta^2}} \tag{1.32}$$

Furthermore, the GD of the NGD circuit can be obtained using (1.33) as:

$$\tau = -\frac{d\angle S_{21}}{d\omega} = -\frac{d}{d\omega}\left\{\tan^{-1}\frac{\mathrm{Im}(S_{21})}{\mathrm{Re}(S_{21})}\right\} = \frac{\mathrm{Re}(S_{21})\frac{d}{d\omega}\mathrm{Im}(S_{21}) - \mathrm{Im}(S_{21})\frac{d}{d\omega}\mathrm{Re}(S_{21})}{[\mathrm{Re}(S_{21})]^2 + [\mathrm{Im}(S_{21})]^2} \tag{1.33}$$

where Im(S_{21}) and Re(S_{21}) represent the imaginary and real parts of the transmission coefficient (S_{21}), respectively. Furthermore, the magnitude of transmission coefficient and GD at $\omega = \omega_0$ can be further derived as (1.34):

$$|S_{21}|_{\omega=\omega_0} = \frac{2\left\{M_{12}M_{34}^2 - M_{34}M_{13}^2 + \frac{M_{12}}{Q_u^2\Delta^2}\right\}}{(M_{12}^2+1)\left(\frac{1}{Q_u^2\Delta^2} + M_{34}^2\right) + M_{13}^2\left(\frac{2}{Q_u\Delta} - 2M_{34}M_{12} + M_{13}^2\right)} \tag{1.34a}$$

$$\tau\Big|_{\omega=\omega_0} = \left\{\begin{array}{l} \dfrac{\dfrac{2M_{12}}{Q_u\Delta}}{M_{13}^2M_{34} - M_{12}M_{34}^2 - \dfrac{M_{12}}{Q_u^2\Delta^2}} \\ + \dfrac{M_{13}^2 + \dfrac{M_{12}^2+1}{Q_u\Delta}}{\dfrac{M_{13}^4}{2} + \dfrac{(1+M_{12}^2)(1+M_{34}^2Q_u^2\Delta^2)}{2Q_u^2\Delta^2} + \dfrac{M_{13}^2}{Q_u\Delta} - M_{13}^2M_{34}M_{12}} \end{array}\right\} \tag{1.34b}$$

To the intuitive understanding of the derived analytical expressions, the computed responses of the NGD circuit are depicted in Figure 1.21 for distinct M_{34}. As noted from these results, the input/output ports are perfectly matched with reference impedance Z_0 at ω_0. Similarly, insertion loss and GD at ω_0 are reduced as M_{34} increases. However, the NGD is enlarged when M_{34} is altered from 0.2738 to 0.5476. In addition, strong coupling between resonator 1 and 2 is necessitated by a wider NGD bandwidth, however, in-band GD increases. The in-band GD ripple

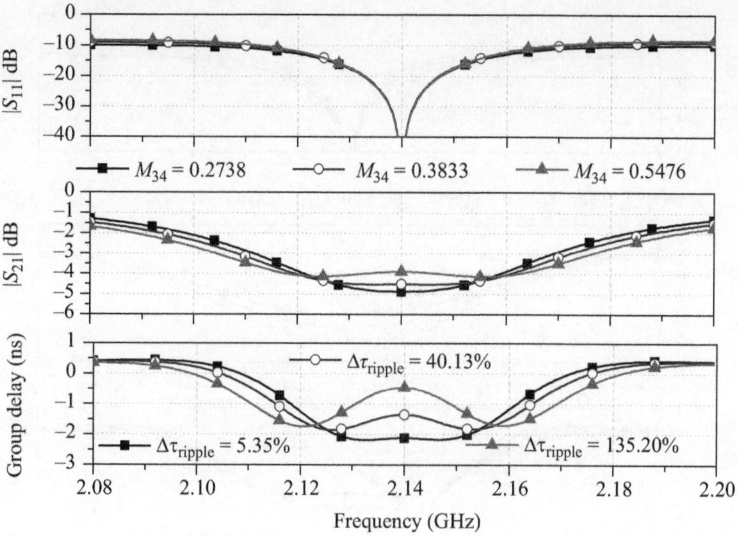

Figure 1.21 Calculated frequency responses of NGD circuit with $Q_u = 50$, $\Delta = 2\%$, $M_{13} = 0.74$, and different M_{34}

($\Delta\tau_{\text{ripple}}$) is mathematically defined as (1.35):

$$\Delta\tau_{\text{ripple}} = \left|\frac{\tau_{\omega_0} - \tau_{\min}}{\tau_{\omega_0} + \tau_{\min}}\right| \times 200\% \tag{1.35}$$

As observed in Figure 1.21, the $\Delta\tau_{\text{ripple}}$ is also controlled by M_{34}. If the higher $\Delta\tau_{\text{ripple}}$ is acceptable, the NGD bandwidth can be increased, however, GD at ω_0 is decreased. Higher NGD at ω_0 can be obtained by decreasing in-band GD ripple, however, insertion loss increases. Therefore, a trade-off exists among NGD, in-band GD ripple, insertion loss, and NGD bandwidth.

Similarly, to illustrate the effect of Q_{u}, Figure 1.22 depicts the computed magnitude and GD responses of NGD circuit for different Q_{u} and M_{34}. The NGD and insertion loss are increased, but the $\Delta\tau_{\text{ripple}}$ is slightly increased as Q_{u} increases and M_{34} decreases.

Figure 1.23 shows the calculated GD with different ripple and insertion loss. As seen from these figures, the higher NGD can be obtained by decreasing M_{13}, however, insertion loss is also increased. The NGD bandwidth can be increased with higher $\Delta\tau_{\text{ripple}}$, however, the NGD at f_0 is decreased.

1.4.2 Experimental performances

For experimental demonstration purposes, the NGD circuit is fabricated for GD of -1 ns at $f_0 = 2.14$ GHz on an FR-4 epoxy substrate with a dielectric constant (ε_r) of 4.4, thickness (h) of 0.787 mm, and loss tangent of 0.02. The coupling matrix of NGD circuit is determined as $M_{13} = 0.69$, $M_{12} = 0.3468$, and $M_{34} = -0.7677$ with $Q_{\text{u}} = 50$ and $\Delta = 2\%$.

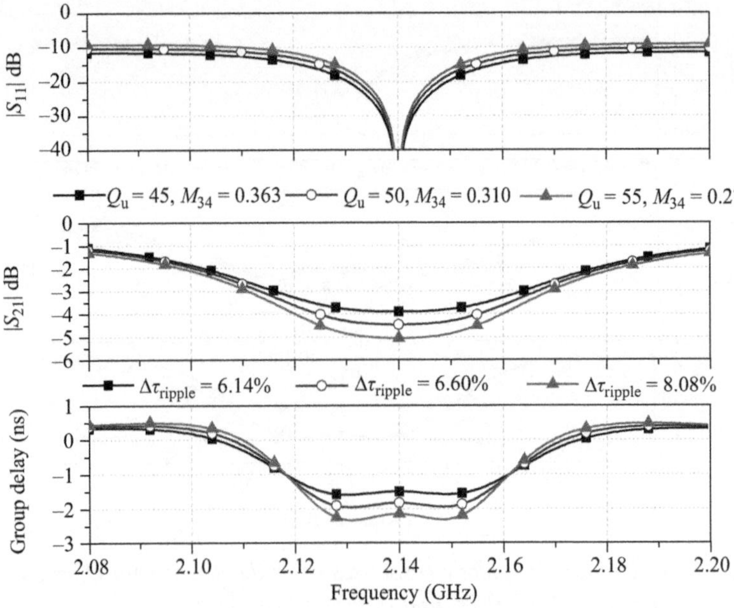

Figure 1.22 Calculated frequency responses of NGD circuit with $\Delta = 2\%$, $M_{13} = 0.72$, and different Q_u

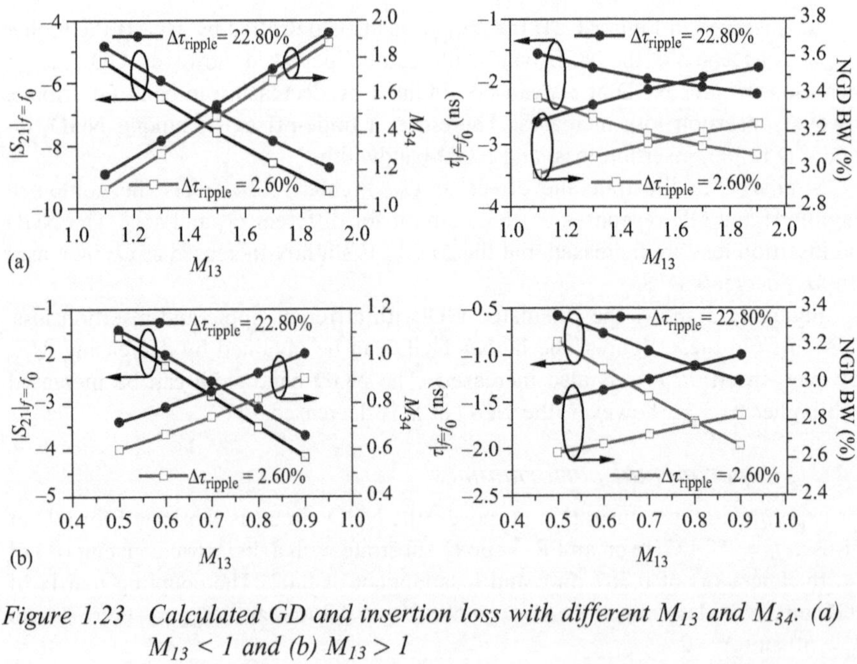

Figure 1.23 Calculated GD and insertion loss with different M_{13} and M_{34}: (a) $M_{13} < 1$ and (b) $M_{13} > 1$

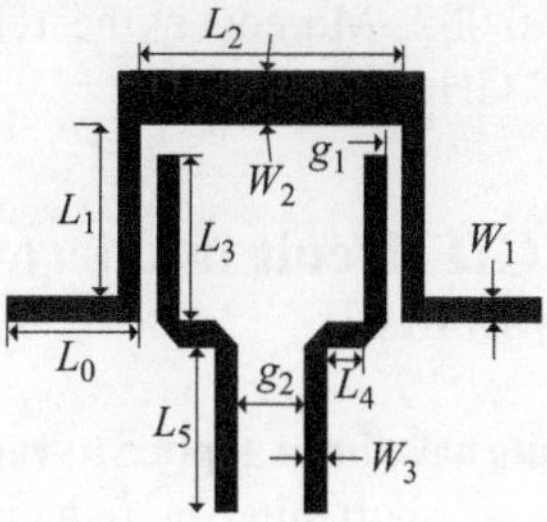

Figure 1.24 Implemented NGD circuit with physical dimensions: $L_0 = 6.8$, $L_1 = 12.3$, $L_2 = 17.4$, $L_3 = 14.1$, $L_5 = 19.9$, $W_1 = 1.8$, $W_2 = 3.8$, $W_3 = 1.8$, $g_1 = 0.6$, and $g_2 = 6$ (unit: millimeter)

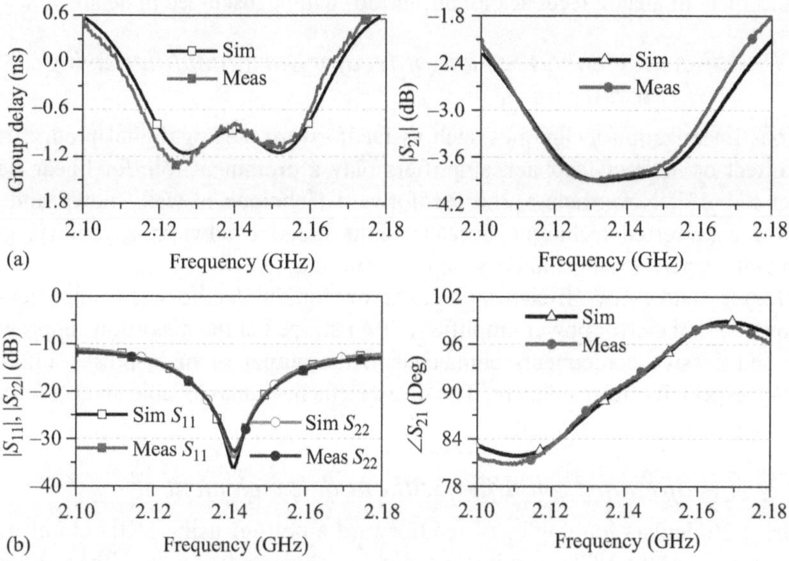

Figure 1.25 Simulation and measured results of NGD circuit: (a) magnitude/GD and (b) return loss/phase characteristics

The finite Q_u microstrip line resonators are implemented with an open-circuited $\lambda/2$ line, where Q_u of the $\lambda/2$ resonator in the FR-4 epoxy substrate is approximated to be around 50 using HFSS Eigen-mode simulation. Likewise, the coupling between the source and load is accomplished by a step-impedance $3\lambda/4$ line. The physical dimensions of implemented NGD circuit are shown in Figure 1.24 where coupling coefficients between input/output and resonators are administered by changing L_3 and g_1. Subsequently, the inter-resonator coupling coefficient is accomplished by varying L_5 and g_2.

The simulated and measured results are depicted in Figure 1.25. The measured insertion loss and GD at $f_0 = 2.14$ GHz are found as 3.82 dB and −0.98 ns, respectively. The NGD bandwidth is approximated as 60 MHz, providing the

NGD-bandwidth product of 0.062. Moreover, the return loss is >12 dB in the frequency range of 2.10–2.18 GHz.

1.5 Applications of NGD circuit in microwave communication systems

In recent years, the NGD circuits have been applied to various practical applications in communication systems such as shortening or reducing delay lines, beam-squint minimization in phased-array antenna systems, realization of non-Foster reactive elements such as negative capacitance and negative inductance and eliminating phase variation with frequency in phase shifters [33–40]. In this section, applications of NGD circuits in efficiency enhancement of feedforward amplifier [35] and bandwidth enhancement of analog feedback amplifier [36] will be discussed in detail.

1.5.1 Efficiency enhancement of feedforward amplifier using NGD circuit

Various linearization techniques such as feedforward, analog/digital predistortion and direct or indirect feedback amplifiers play a prominent role for linear transmitter [41–43]. Nevertheless, the feedforward technique is well known and has become a preferred technique because of its broad bandwidth capability, good linearization performance, and its stable operations.

Major sources of efficiency degradation for the feedforward amplifier are discovered in the error power amplifier (EPA) utilized at the distortion suppression loop and passive components connected to the output of main power amplifier (MPA), especially delay element. The delay element is unavoidable and essential to broadband signal suppression.

1.5.2 Feedforward amplifier without delay element

Figure 1.26 depicts topologies of feedforward amplifier using NGD circuit [35]. In this topology, the NGD circuit is inserted into the common path of the two loops. Therefore, the delay elements from the carrier suppression loop (DELAY 1) as well

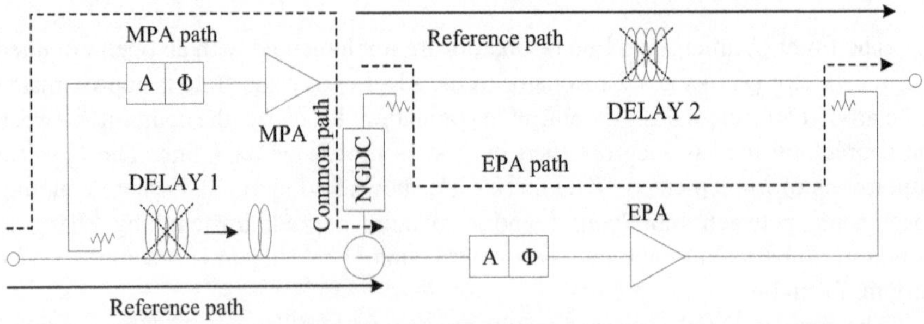

Figure 1.26 Feedforward amplifier topologies with NGD circuit employed at the common path

as the intermodulation (IMD) suppression loop (DELAY 2) can be reduced or eliminated. If the GDs of the MPA and the EPA path are same, the complete elimination of the two delay elements is possible. Even though the delay element is eliminated, additional power consuming circuits must be added to the NGD circuit. However, an important difference is that the newly adopted NGD circuit consumes a very small DC power as compared to the power lost by the delay elements at the output of the MPA. Since the delay elements are typically composed of bulky delay line filters or high power coaxial cables with the inevitable size, weight, and cost penalty, the NGD circuit can be integrated into the EPA module.

1.5.3 Signal suppression loop analysis for GD mismatch

Considering amplitude, out-of-phase, and GD mismatches, the simple signal suppression loop model of feedforward amplifier can be represented in Figure 1.27.

A portion of the input signal (V_{IN}) is coupled into the delay path as a reference signal (V_A) while the remaining portion of the V_{IN} is amplified by the power amplifier (PA) resulting in the amplified carrier with IMD components. A portion of the output of PA (V_B) is then coupled into the subtraction circuit, which is then destructively combined with the V_A, generating an error signal (V_{ERR}) that theoretically does not include any carrier components.

Amplitude and out-of-phase mismatches

Assuming the sinusoidal reference signal having an amplitude (V_1) and a phase (θ) including the amplitude mismatch (ΔV) and the out-of-phase mismatch ($\Delta\theta$) factors, the resultant suppressed signal can be denoted as a ratio of the average power of V_{ERR} to V_A as (1.36) [35]:

$$S(\text{dB}) = 10\log\left[1 + \left(\frac{V_1 + \Delta V}{V_1}\right)^2 - 2\left(\frac{V_1 + \Delta V}{V_1}\right)\cdot\cos(\Delta\theta)\right] \tag{1.36}$$

For an intuitive understanding of amplitude and phase mismatch effects in loop suppression, Figure 1.28 shows the calculated loop suppression assuming perfectly GD match. As observed from this figure, loop suppression of >-30 dB can be obtained for amplitude mismatch of ± 0.2 dB and an out-of-phase mismatch of $\pm 1°$.

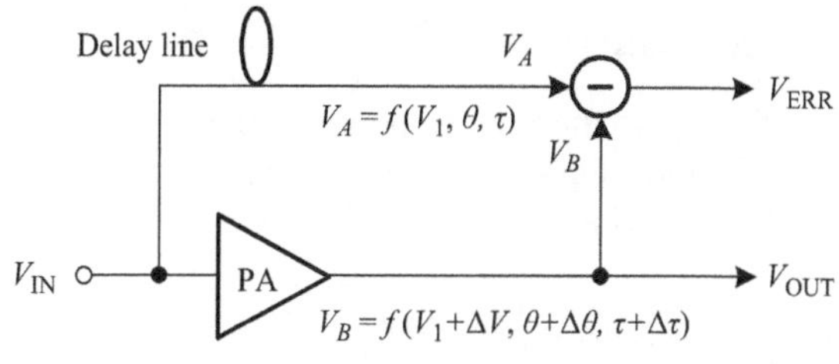

Figure 1.27 Simple signal suppression loop model

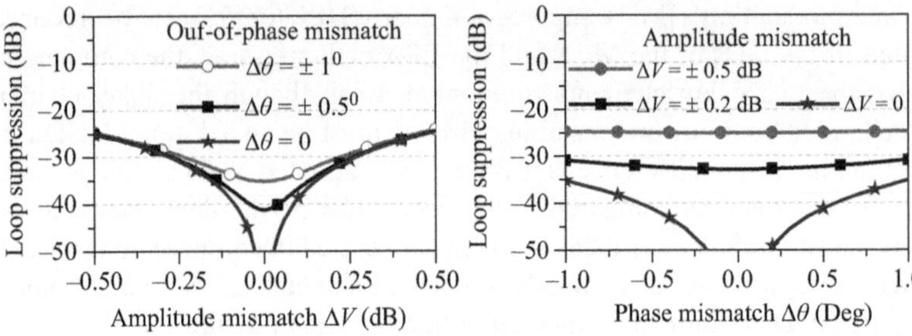

Figure 1.28 Loop suppression according to the amplitude and out-of-phase mismatches

GD mismatch

For broadband signal cancellation, it is crucial to control the GD mismatch ($\Delta\tau$) between the main path and reference path. For analysis, the condition of a unit amplitude ($V_1 = 1$), a perfectly matched amplitude and out-of-phase ($\Delta V = \Delta\theta = 0$) are assumed. The value of V_{ERR} can be expressed as (1.37) where ω_0 is an angular frequency.

$$\begin{aligned} V_{ERR} &= V_A + V_B \\ &= \cos(\omega_0\tau + \theta)(1 - \cos(\omega_0\Delta\tau)) + \sin(\omega_0\tau + \theta)\sin(\omega_0\Delta\tau) \end{aligned} \tag{1.37}$$

To compute the average power over one period, $|V_{ERR}|^2$ can be determined by using (1.38):

$$\begin{aligned} |V_{ERR}|^2 = \Bigg| & \frac{(1 - \cos\omega_0\Delta\tau)^2 + (1 - \cos\omega_0\Delta\tau)^2\cos 2(\omega_0\tau + \theta)}{2} \\ & + \frac{\sin^2(\omega_0\Delta\tau) + \sin^2(\omega_0\Delta\tau)\sin^2(\omega_0\tau + \theta)}{2} \\ & + \sin 2(\omega_0\tau + \theta)(1 - \cos\omega_0\Delta\tau)(\sin\omega_0\Delta\tau)\Bigg| \end{aligned} \tag{1.38}$$

By integrating (1.38) over an arbitrary period T_0, the average power of V_{ERR} is expressed by (1.39) as:

$$P_{ERR,avg} = 1 - \cos(\omega_0\Delta\tau) \tag{1.39}$$

Since the ratio of (1.39) to the average power of the V_A ($P_A = 1/2$) is defined as the loop suppression, the loop suppression in a dB scale including GD mismatching condition is depicted in (1.40) as:

$$S_{\Delta t} = 10\log\left[1 - \cos\left\{\omega_0\Delta\tau\left(1 - \frac{f}{f_0}\right)\right\}\right] + 3, \tag{1.40}$$

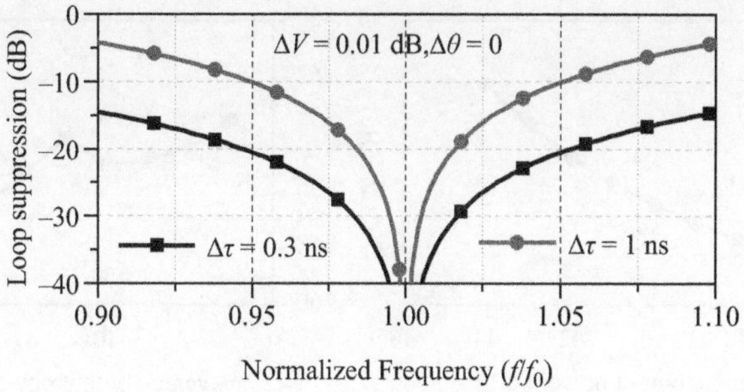

Figure 1.29 Calculated loop suppression for different GD mismatches

where the term $1-(f/f_0)$ is inserted to derive the equation with respect to the normalized frequency when f_0 represents the center frequency. Finally, from (1.36) and (1.40) the loop suppression expression considering the amplitude, out-of-phase, and delay mismatches can be expressed (1.41) as a function of ΔV and $\Delta\tau$:

$$S_{\text{total}} = 10\log\left[1+\left(\frac{V_1+\Delta V}{V_1}\right)^2 - 2\left(\frac{V_1+\Delta V}{V_1}\right)\cdot\cos\left\{\omega_0\Delta\tau\left(1-\frac{f}{f_0}\right)\right\}\right] \quad (1.41)$$

Figure 1.29 shows the calculated loop suppression performance for different GD mismatches. In this calculation, the amplitude and phase mismatches are assumed as 0.01 dB and 0°, respectively. As observed from this figure, the FBW of 20 dB cancellation will be 10% if the GD mismatching is less than 0.3 ns. Similarly, the bandwidth is considerably decreased to 3.2% when the GD mismatch is increased from 0.3 to 1 ns. This analysis confirms that the GD matching is very crucial for a high level of signal cancellation over wideband.

1.5.4 Experimental results

The feedforward amplifier is fabricated using MPA that operate up to an output power of 49.18 dBm with a gain of 54.8 dB at 28 V and EPA with an output power of 41 dBm. Since employed NGD circuit has a frequency-limited characteristic which can amplify an unwanted out-of-band noise, two-stage quarter-wavelength coupled line BPF is designed with a microstrip line and integrated in front of the NGD circuit to avoid possible instabilities.

Figure 1.30 shows the measured adjacent channel leakage ratio (ACLR) and power-added efficiency (PAE) performances according to the average output power. The measured ACLR and PAE (or drain efficiency) are −53.2 dBc and 19.4% (or 19.5%), respectively, at the average output power of 44 dBm. For the same ACLR level with and without NGD circuit feedforward topology, the PAE of the system is increased from 5.1% to 19.4%.

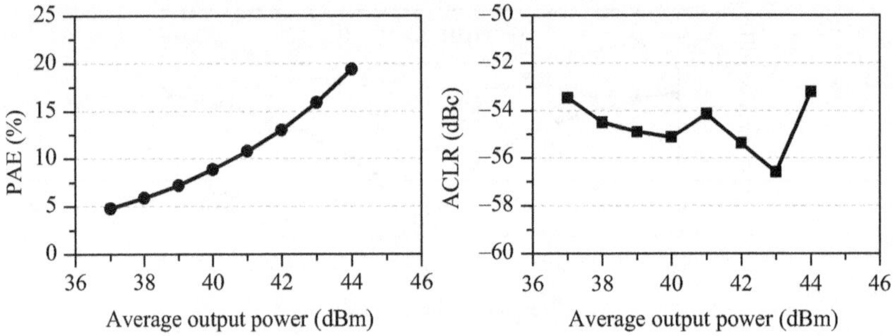

Figure 1.30 Measured PAE and ACLR performance of feedforward amplifier using negative GD circuit

1.5.5 Bandwidth enhancement of analog feedback amplifier using NGD circuit

In addition to the feedforward method, Seidel also introduced a feedback amplifier [44,45], which utilized the main and error amplifiers with three signal coupling devices to degenerate the error signal generated by the main amplifier by introducing a feedback loop. McRory *et al.* [46] mathematically analyzed the same structure based on the Volterra series. Kim *et al.* [47] proposed a modified version of the feedback topology and named "feedback predistortion." Qiang *et al.* [48] also analyzed the structure introduced by Seidel based on power series expansion. There are notable advantages in the feedback amplifier when compared to the feedforward amplifier: (1) EPA requires lower output power since the error signal is injected to the input of MPA, and (2) the RF output loss is smaller because there is no lossy GD element at the output of the MPA. Regardless of all these advantages, an extremely narrow operating bandwidth, which is only capable of covering a few megahertz, was the key limiting factor that discouraged the utilization of the feedback topology. This is especially true in the recent wireless communication environment, which utilizes several digital modulation schemes involving modulated signals with a broad bandwidth, such as wideband code division multiple access (WCDMA) and worldwide interoperability for microwave access, and orthogonal frequency division multiplexing. This limited bandwidth of feedback amplifier originates from the GD mismatching generated in the feedback loop. In the conventional researches, the minimum feedback GD is assumed, but in a practical situation the assumption is not valid because of the GDs of the MPA, EPA, and other signal adjusting devices including the band-pass filter [49,50].

1.5.6 Analog feedback amplifier using NGD: operating principle

The block diagram of the analog feedback amplifier employing the NGD circuit is shown in Figure 1.31. Firstly, a portion of the input signal (RF_{IN}) is inserted to the MPA and experiences the full gain of the amplifier. Secondly, the remaining input signal is utilized as a reference (A) against a portion of the distorted MPA output signal (B). Any difference between the reference signal and the output signal due to noise or distortion is identified as an error signal (C). The error signal is amplified

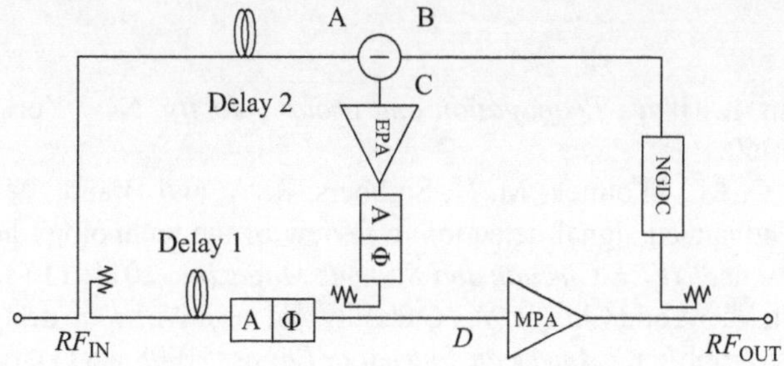

Figure 1.31 Block diagram of analog feedback amplifiers using NGD circuit

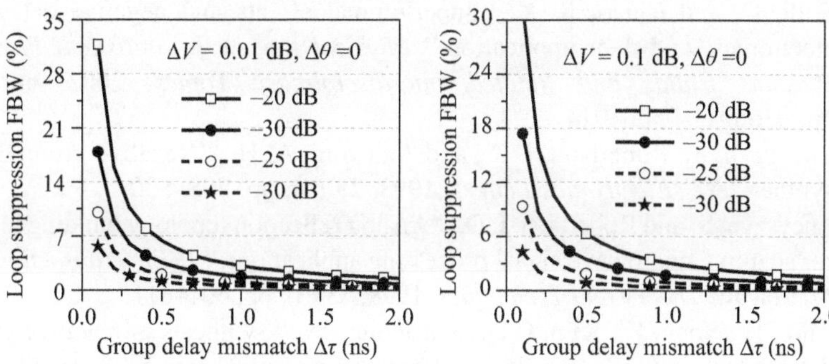

Figure 1.32 Calculated loop suppression FBW according to the GD mismatch

Table 1.1 Measured results of bandwidth-enhanced feedback amplifier using NGD circuit

Frequency (GHz)	2.14	2.14
Output power P_{1dB} (dBm)	43	43
Test signal	2-tone	WCDMA
Cancellation BW (MHz)	>50	>50
Linearity (dBc)	−51	−52
Improvement (dB)	22	25

by EPA, and then injected into the input port (*D*) of the MPA in phase to generate the error-free RF_{OUT}. By introducing the NGD circuit into the feedback loop, the feedback path delay can be minimized to increase the cancellation bandwidth.

Figure 1.32 illustrates the calculated loop suppression FBW variation with respect to the GD mismatch. In case of the 0.3 ns GD mismatch, it is expected that the FBW for 20 dB cancellation will be 10%. When the mismatch is 1.0 ns, the bandwidth is considerably reduced to 3.2%. The cancellation bandwidth is reduced to 30% of the bandwidth for a 0.3 ns mismatch. The measurement results of feedback amplifier employing NGD circuit are summarized in Table 1.1 where cancellation bandwidth greatly improved as compared to the conventional ones.

References

[1] Brillouin, L., *Wave Propagation and group velocity*, New York: Academic Press, 1960.

[2] Hymel, C. M., Skolnick, M. H, Stubbers, R. A, and Brandt, M. E., "Temporally advanced signal detection: a review of the technology and potential applications," *IEEE Circuits and Systems Magazine*, 2011, **11** (3), pp.10–25.

[3] Mitchell, M. W. and Chia, R. Y. "Causality and negative group delay in a simple bandpass amplifier," *American Journal of Physics*, 1998, **66** (1), pp. 14–19.

[4] Kitano, M., Nakanishi, T., and Sugiyama, K., "Negative group delay and superluminal propagation: an electronic circuit approach," *IEEE Journal of Selected Topics in Quantum Electronics*, 2003, **9** (1), pp. 43–51.

[5] Solli, D. and Chiao, R. Y. "Superluminal effects and negative delays in electronics, and their application," *Physical Review E, Statistical Physics, Plasmas Fluids and Related Interdisciplinary Topics*, 2002, **66** (6), pp. 0566011-05661014.

[6] Lucyszyn, S., Robertson, I. D., and Aghvami, A. H., "Negative group delay synthesizer," *Electronics Letters*, 1993, **29** (9), pp. 798–800.

[7] Lucyszyn, S. and Robertson, I. D., "Analog reflection topology building blocks for adaptive microwave signal processing applications," *IEEE Transactions on Microwave Theory and Techniques*, 1998, **43** (3), pp. 601–611.

[8] Choi, H., Song, K., Kim, C.D., and Jeong, Y., "Synthesis of negative group delay time circuit," *Asia Pacific Microwave Conference*, 2008, pp. 05–08.

[9] Choi, H., Kim, Y, Jeong, Y., and Kim, C. D., "Synthesis of reflection type negative group delay circuit using transmission line resonator," *Proceedings of 39th European Microwave Conference*, 2009, pp. 902–605.

[10] Choi, H., Kim, Y., Jeong, Y., and Lim, J., "A design of size-reduced negative group delay circuit using a stepped impedance resonator," *Proceedings of Asia-Pacific Microwave Conference*, 2010, pp. 118–1121.

[11] Ravelo, B., Perennec, P., and Roy, M. L, "Synthesis of broadband negative group delay active circuits," *IEEE MTT-S, International Microwave Symposium Digest*, 2007, pp. 2177–2180.

[12] Ravelo, B., Perennec, P., and Roy, M. L, "Active microwave circuit with negative group delay," *IEEE Microwave Wireless Component Letters*, 2007, **17** (12), 861–863.

[13] Ravelo, B., Perennec, A., and Roy, M. Le., "Synthesis of broadband negative group delay active circuits," *IEEE International Microwave Symposium Digest*, 2007, pp. 2177–2180.

[14] Kandic, M. and Bridges, G. E., "Asymptotic limit of negative group delay in active resonator-based distributed circuits," *IEEE Transactions on Circuit System-I*, 2011, **58** (8), pp.1727–1735.

[15] Broomfield, C. D. and Everard, J. K. A., "Broadband negative group delay networks for compensation of microwave oscillators and filters," *Electronics Letters*, 2000, **9** (23), pp. 1931–1932.

[16] Kandic, M. and Bridges, G. E., "Bilateral Gain-compensated negative group delay circuit," *IEEE Microwave Wireless Component Letters*, 2011, **21** (6), pp. 308–310.

[17] Chaudhary, G. and Jeong, Y., "Transmission-type negative group delay networks using coupled line doublet structure," *IET Microwave, Antenna and Propagation*, 2015, **9** (8), pp. 748–784.

[18] Choi, H., Chaudhary, G., Moon, T., Jeong, Y., Lim, J., and Kim, C. D., "A design of composite negative group delay circuit with lower signal attenuation for performance improvement of power amplifier linearization techniques," *IEEE International Microwave Symposium Digest*, 2011, pp. 1–4.

[19] Chaudhary, G. and Jeong, Y., "Distributed transmission line negative group delay circuit improved signal attenuation," *IEEE Microwave and Wireless Component Letters*, 2014, **24** (1), pp. 20–22.

[20] Chaudhary, G. and Jeong, Y., "Transmission-line negative group delay networks with improved signal attenuation," *IEEE Antenna and Wireless Propagation Letters*, 2014, **13**, pp. 1039–1042.

[21] Chaudhary, G. and Jeong, Y., "Low signal attenuation negative group network topologies using coupled lines," *IEEE Transactions on Microwave Theory and Techniques*, 2014, **62** (10), pp. 2316–2324.

[22] Chaudhary, G., Jeong, Y., and Lim, J., "Miniaturized negative group delay circuit using defected microstrip structure and lumped elements," *IEEE MTT-S International Microwave Symposium*, 2013, pp. 1–3.

[23] Chaudhary, G., Jeong, J, Kim, P., Jeong, Y., and Lim, J., "Compact negative group delay circuit using defected group structure," *IEEE Asia Pacific Microwave Conference*, 2013, pp. 22–24.

[24] Chaudhary, G., Jeong, Y., and Lim, J. "Realization of negative group delay network using defected microstrip structure," *International Journal of Antenna and Propagation*, 2014, **2014**, Article ID 836950, pp. 1–5.

[25] Chaudhary, G., Kim, P, Jeong, J., Jeong, Y., and Lim, J, "Dual-band negative group delay circuit using defected microstrip structure," *IEEE Radio and Wireless Symposium*, 2015, pp. 129–131.

[26] Chaudhary, G., Jeong, Y., and Lim, J., "Miniatured dual-band negative group delay circuit dual-plane defected structures," *IEEE Microwave Wireless Component Letters*, 2014, **24** (8), pp. 521–523.

[27] Chaudhary, G. and Jeong, Y., "A design of compact wideband negative group delay network using cross coupling," *Microwave and Optical Technology Letters*, 2014, **56** (11), pp. 2494–2497.

[28] Wu, C. T. M. and Itoh, T., "Maximally flat negative group delay circuit: a microwave transversal filter approach," *IEEE Transactions on Microwave Theory and Techniques*, 2014, **62** (6), pp. 1330–1342.

[29] Chaudhary, G., Jeong, Y., and Lim, J., "Microstrip line negative group delay filters for microwave circuits," *IEEE Transactions on Microwave Theory and Techniques*, 2014, **62** (2), pp. 234–243.

[30] Chaudhary, G. and Joeng, Y., "Negative group delay phenomenon analysis using finite unloaded quality factor resonators," *Progress in Electromagnetic Research*, 2016, **156**, pp. 55–62.
[31] Chaudhary, G. and Jeong, Y., "Tunable center frequency negative group delay filter using coupling matrix approach," *IEEE Microwave Wireless Component Letters*, 2017, **27** (1), pp. 37–39.
[32] Chaudhary, G. and Jeong, Y., "A finite unloaded quality factor resonators based negative group delay circuit and its application to design power divider," *Microwave and Optical Technology Letters*, 2016, **58** (12), pp. 2918–2921.
[33] Ravelo, B., Roy, M. L., and Perennec, A., "Application of negative group delay active circuits to the design of broadband and constant phase shifters," *Microwave Optical Technology Letters*, 2008, **50** (12), pp. 3078–3080.
[34] Jeong, Y., Choi, H., and Kim, C. D., "Experimental verification for time advancement of negative group delay in RF electronics circuits," *IET Electronics Letters*, 2010, **46** (4), pp. 306–307.
[35] Choi, H., Jeong, Y., Kim, C. D., and Kenney, J. S., "Efficiency enhancement of feedforward amplifiers by employing a negative group delay circuit," *IEEE Transactions on Microwave Theory and Techniques*, 2010, **58** (5), pp. 1116–1125.
[36] Choi, H., Jeong, Y., Kim, C. D., and Kenney, J. S., "Bandwidth enhancement of an analog feedback amplifier by employing a negative group delay circuit," *Progress in Electromagnetic Research*, 2010, **105**, pp. 253–272.
[37] Noto, H., Yamauchi, K., Nakayama, M., and Isota, Y. "Negative group delay circuit for feed forward amplifier," *IEEE International Microwave Symposium Digest*, 2007, pp. 1103–1106.
[38] Oh, S. S. and Shafai, L., "Compensated circuit with characteristics of lossless double negative materials and its application to array antennas," *IET Microwave, Antennas & Propagation*, 2007, **1** (1), pp. 29–38.
[39] Mirzaei, H. and Eleftheriades, G. V., "Realizing non-Foster reactive elements using negative group delay networks," *IEEE Transactions on Microwave Theory and Techniques*, 2013, **61** (12), pp. 4322–4332.
[40] Mirzaei, H. and Eleftheriades, G. V., "Arbitrary-angle squint-free beamforming in series-fed antenna arrays using non-Foster elements synthesized by negative group delay networks," *IEEE Transactions on Antennas and Propagation*, 2015, **63** (5), pp. 1997–2010.
[41] Larose, C. L. and Ghannouchi, F. M., "Optimization of feedforward amplifier power efficiency on the basis of drive statistics," *IEEE Transactions on Microwave Theory and Techniques*, 2003, **51** (1), pp. 41–54.
[42] Cripps, S. C., *Advanced techniques in RF power amplifier design*, 2002, Norwood, MA: Artech house.
[43] Konstantinou, K. and Paul, D. K., "Analysis and design of broadband high efficiency feedforward amplifier," *IEEE MTT-S International Microwave Symposium*, 1996, pp. 867–870.
[44] Seidel, H. and Warren, N. J., "Reentrant signal feedback amplifier," *U.S. Patent* 3,624,532; issued Nov. 30, 1971.

[45] Seidel, H. and Warren, N. J., "Feedback amplifier," *U.S. Patent* 3,656,831; issued Apr. 18, 1972.
[46] McRory, J. G. and Johnston, R. H. "An RF amplifier for low intermodulation distortion," in *IEEE MTT-S International Microwave Symposium Digest*, 1994, pp. 1741–1744.
[47] Kim, Y., Yang, Y., Kang, S., and Kim, B. "Linearization of 1.85 GHz amplifier using feedback predistortion loop," in *IEEE MTT-S International Microwave Symposium Digest*, 1998, pp. 1675–1678.
[48] Qiang, L., Ying, Z. Z., and Wei, G., "Design of a feedback predistortion linear power amplifier," *Microwave Journal*, 2005, **48** (5), pp. 232–241.
[49] Ezzeddine, A. K., Hung, H., and Huang, H., "An MMAC C-band FET feedback power amplifier," *IEEE Transactions on Microwave Theory and Techniques*, 1990, **38** (4), pp. 350–357.
[50] Faulkner, M., Contos, D., and Johansson, M. "Linearization of power amplifiers using RF feedback," *Electronics Letters*, 1995, **31** (23), pp. 2023–2024.

Chapter 2

Negative group delay power dividers

Girdhari Chaudhary[1] and Yongchae Jeong[1]

Abstract

Power dividers have been widely adopted as basic building components in microwave circuits and systems and extensively studied over the past decades including the unequal and tunable power division ratio, circuit miniaturization, and multiband operation. Because the group delay (GD) can influence the performances of RF/electronic circuits and systems, understanding the effect of GD has become critical for communication systems. However, GD analysis of power divider has rarely been considered in previous works and the positive GDs have only been provided through different transmission paths. In this chapter, several topologies of power divider and coupler with the predefined NGD characteristics will be presented including equal and unequal power division ratios. Similarly, coupling matrix synthesis with finite unloaded-quality resonators will be also presented in this chapter to design the power divider with the predefined NGD.

Keywords

Arbitrary power division ratio, negative group delay, positive group delay, power divider, rat-race coupler.

2.1 Context and introduction

Power dividers have been widely utilized as basic building components in microwave circuits and systems. Power dividers have extensively studied over the past decades including the unequal and tunable power division ratio, circuit miniaturization, and multiband operation [1–9]. Because the group delay (GD) can influence the performances of RF/electronic circuits and systems, the GD effect of power divider in RF/electronics systems has become critical for communication systems. However, the GD analysis of the conventional power divider has rarely

[1]Division of Electronics and Information Engineering, Chonbuk National University, Republic of Korea

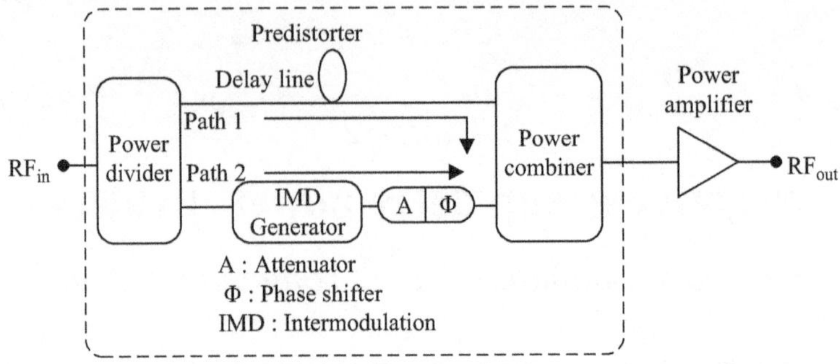

Figure 2.1 Block diagram of the analog predistortion amplifier [11]

been considered. Moreover, the conventional power dividers exhibit the positive GDs through different transmission paths [10].

In modern RF wireless communication systems, complex modulation and demodulation techniques are employed to meet the demands of the high data transmission rate. Therefore, these complex modulation techniques require a highly linear high-power amplifier. A predistortion method is one of the low-cost linearization techniques, which have advantages of low-power consumption and simple circuit configuration [11,12]. In this technique, the predistortion circuit having inverse distortion characteristics of the power amplifier is connected in front of the power amplifier as shown in Figure 2.1. To improve the linearity, the GDs, magnitudes, and phases of two path signals should be matched in the predistorter. For this purpose, the delay element, attenuator, and phase shifter are used in the predistorter. Therefore, the research that can demonstrate power divider/combiner with the negative group delay (NGD) will be beneficial in such systems to compensate the positive GD and help to enhance linearity performance as well as the elimination of GD element and attenuator.

In this chapter, several topologies of power divider with the predefined NGD characteristics will be presented including equal and unequal power division ratios. Similarly, coupling matrix synthesis with finite unloaded-quality resonators will be also presented to design the power divider with the predefined positive and negative GDs.

2.2 Wilkinson power divider with arbitrarily prescribed NGD

Figure 2.2 depicts the schematic of Wilkinson power divider with arbitrarily prescribed NGD [13,14]. This structure comprises $\lambda/4$ series transmission lines loaded with shunt resistor-connected short-circuited coupled lines. The characteristic impedance of transmission line, even- and odd-mode impedances of the coupled line are denoted as Z_1, Z_{0_e}, and Z_{0_o}, respectively.

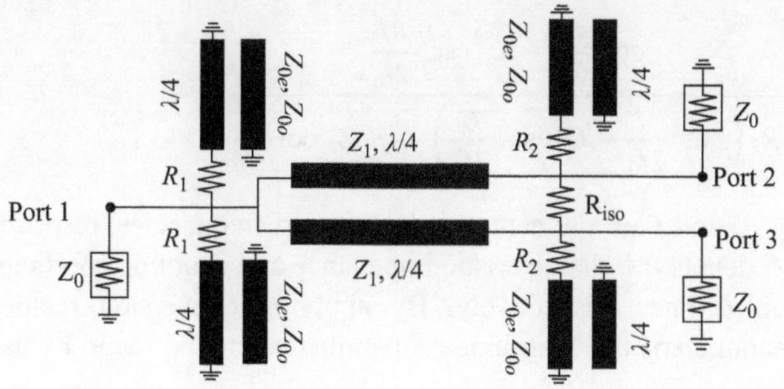

Figure 2.2 Schematic of Wilkinson power divider with prescribed NGD [13]

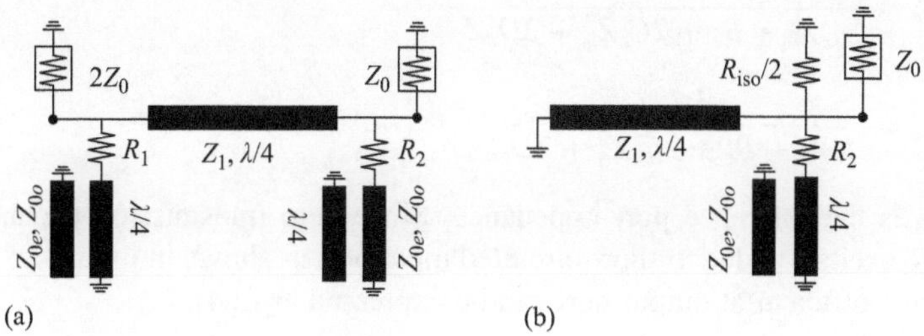

Figure 2.3 (a) Even- and (b) odd-mode equivalent subcircuits of the NGD power divider

2.2.1 Theoretical analysis

Figure 2.3 depicts the equivalent subcircuits of the NGD power divider under even- and odd-mode excitations. Total even-mode ABCD-parameters of sub-circuit shown in Figure 2.3(a) is determined by using (2.1):

$$\begin{bmatrix} A_e & B_e \\ C_e & D_e \end{bmatrix} = \begin{bmatrix} 1 & 0 \\ Y_{\text{in}_1} & 1 \end{bmatrix} \begin{bmatrix} \cos\dfrac{\pi f}{2f_0} & jZ_1 \sin\dfrac{\pi f}{2f_0} \\ j\dfrac{1}{Z_1}\sin\dfrac{\pi f}{2f_0} & \cos\dfrac{\pi f}{2f_0} \end{bmatrix} \begin{bmatrix} 1 & 0 \\ Y_{\text{in}_2} & 1 \end{bmatrix}, \tag{2.1}$$

where

$$Y_{\text{in}_1} = \frac{\cot^2\dfrac{\pi f}{2f_0} - C^2\csc^2\dfrac{\pi f}{2f_0}}{R_1\left(\cot^2\dfrac{\pi f}{2f_0} - C^2\csc^2\dfrac{\pi f}{2f_0}\right) + jCZ_c\cot\dfrac{\pi f}{2f_0}} \tag{2.2a}$$

$$Y_{in_2} = \frac{\cot^2\frac{\pi f}{2f_0} - C^2\csc^2\frac{\pi f}{2f_0}}{R_2\left(\cot^2\frac{\pi f}{2f_0} - C^2\csc^2\frac{\pi f}{2f_0}\right) + jCZ_c\cot\frac{\pi f}{2f_0}}. \tag{2.2b}$$

Similarly, f_0 and f are the center and operating frequencies, respectively. Likewise, Z_c and C denote the characteristic impedance and coupling coefficient of short-circuited coupled lines, respectively. By applying conversion relation between ABCD to S-parameters, the even-mode S-parameters can be found by using (2.3):

$$S_{11e} = \frac{A_eZ_0 + B_e - 2C_eZ_0^2 - 2D_eZ_0}{A_eZ_0 + B_e + 2C_eZ_0^2 + 2D_eZ_0} \tag{2.3a}$$

$$S_{22e} = \frac{-A_eZ_0 + B_e - 2C_eZ_0^2 + 2D_eZ_0}{A_eZ_0 + B_e + 2C_eZ_0^2 + 2D_eZ_0} \tag{2.3b}$$

$$S_{21e} = \frac{2\sqrt{2}Z_0}{A_eZ_0 + B_e + 2C_eZ_0^2 + 2D_eZ_0}, \tag{2.3c}$$

where Z_0 is the reference port impedance. Since zero transmission occurs under odd-mode excitation due to short-circuited input port as shown in Figure 2.3(b), the reflection coefficient at output port can be expressed as (2.4):

$$S_{22o} = \frac{1 - Z_0\left(\frac{2}{R_{iso}} + Y_{in_2} - j\frac{1}{Z_1}\cot\frac{\pi f}{2f_0}\right)}{1 + Z_0\left(\frac{2}{R_{iso}} + Y_{in_2} - j\frac{1}{Z_1}\cot\frac{\pi f}{2f_0}\right)} \tag{2.4}$$

where R_{iso} is an isolation resistor. Finally, the overall S-parameters of the power divider can be determined by combining the even- and odd-mode S-parameters as (2.5):

$$S_{11}(f) = S_{11e}(f) \tag{2.5a}$$

$$S_{21}(f) = S_{12}(f) = S_{31}(f) = S_{13}(f) = \frac{S_{21e}(f)}{\sqrt{2}} \tag{2.5b}$$

$$S_{22}(f) = S_{33}(f) = \frac{S_{22e}(f) + S_{22o}(f)}{2} \tag{2.5c}$$

$$S_{23}(f) = S_{32}(f) = \frac{S_{22e}(f) - S_{22o}(f)}{2}. \tag{2.5d}$$

For zero reflection from input and output ports ($S_{ii} = 0$) and infinite isolation between ports 2 and 3 ($S_{23} = 0$) at f_0, the condition (2.6) should be satisfied.

$$R_1 = 2R_2 \tag{2.6a}$$

$$Z_1 = \sqrt{\frac{2R_1R_2Z_0^2}{(R_1 - 2R_2)Z_0 + R_1R_2 - 2Z_0^2}} \tag{2.6b}$$

$$R_{\text{iso}} = \frac{(R_1 + 2Z_0)Z_1^2}{Z_0 R_1} \tag{2.6c}$$

Additionally, the transmission coefficient magnitudes and GDs through different transmission paths at f_0 are determined as (2.7) by using (2.3) and (2.5) for $R_1 = 2R$ and $R_2 = R$.

$$S_{21}\Big|_{f=f_0} = S_{31}\Big|_{f=f_0} = \left|\frac{2Z_0 R^2 Z_1}{(2RZ_0 + R^2 + Z_0^2)Z_1^2 + 2R^2 Z_0^2}\right|, \tag{2.7a}$$

$$\tau_{21}\Big|_{f=f_0} = \tau_{31}\Big|_{f=f_0} = -\frac{3}{4f_0}\left\{\frac{Z_c}{2RC} - \frac{(R+Z_0)(2RCZ_0 + Z_1 Z_c)Z_1 + 2RZ_0^2 Z_c}{C\left[2R(2Z_0 + R)Z_1^2 + 2Z_0^2(Z_1^2 + 2R^2)\right]}\right\}. \tag{2.7b}$$

where

$$Z_c = \frac{2Z_{0_e}}{Z_{0_e}/Z_{0_o} - 1} = Z_{0_e}\frac{1-C}{C} = Z_{0_o}\frac{1+C}{C} \tag{2.8}$$

As observed from (2.8), very high Z_c can be acquired if the ratio of Z_{0_e} to Z_{0_o} is close to unity or C becomes very small. For explanatory demonstration of analytical expressions, the computed normalized GDs ($\tau_{21} \times f_0$) are shown in Figure 2.4.

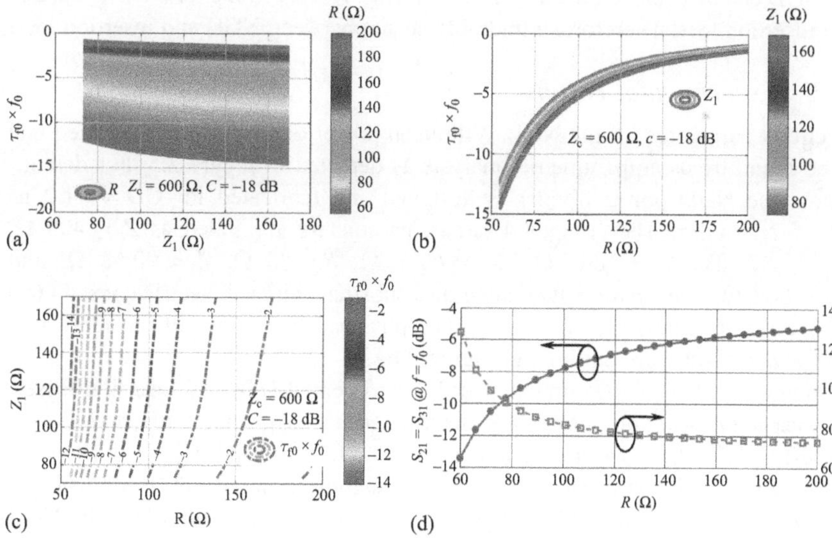

Figure 2.4 Calculated GD and magnitude $|S_{21}| = |S_{31}|$ for different Z_1 and R for matched input/output ports and infinite isolation with $Z_c = 600\ \Omega$ and $C = -18$ dB: (a) GD according to Z_1; (b) GD according to R; (c) relation between R and Z_1; and (d) magnitude of S_{21}, S_{31}

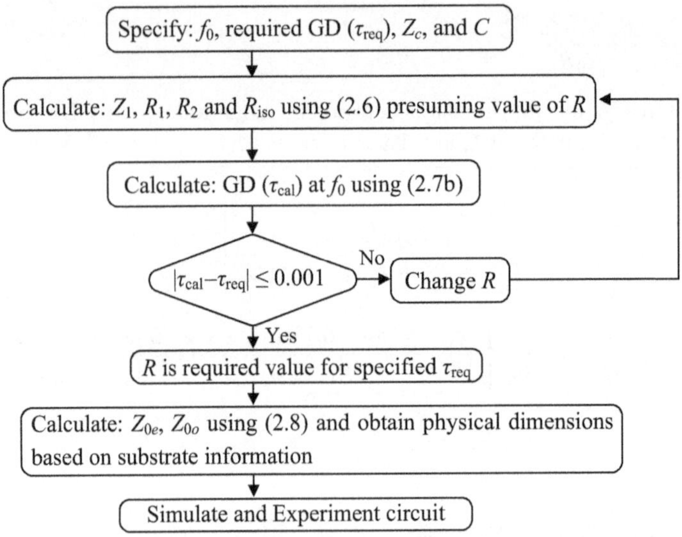

Figure 2.5 Flowchart to calculate circuit parameters of NGD Wilkinson power divider

In this computation, the Z_c and C of the short-circuited coupled lines are presumed as 600 Ω and −18 dB, respectively. From these graphs, it can be concluded that high NGD can be achieved for high Z_1 and low R. However, higher NGD generates more insertion loss. Therefore, a trade-off occurs between NGD and insertion loss.

2.2.2 Experimental results

The circuit parameters of the NGD Wilkinson power divider for the specified GD can be found by using parametric analysis as depicted in Figure 2.5. For demonstration, the NGD power divider is designed and fabricated for GD = −1.2 ns at f_0 of 2.14 GHz. The computed circuit parameters are stated as $Z_c = 400\ \Omega$, $C = -13.57$ dB, $Z_{0_e} = 106.1\ \Omega$, $Z_{0_o} = 69.33\ \Omega$, $R = 80\ \Omega$, $Z_1 = 90.58\ \Omega$, and $R_{iso} = 266.6\ \Omega$. The circuit is fabricated on a substrate with a dielectric constant (ε_r) of 2.2 and a thickness (h) of 31 mils. The physical dimensions of the fabricated NGD power divider are outlined in Figure 2.6.

Figure 2.7 depicts the simulated and experimented GDs and insertion losses. The measured insertion losses are determined as $|S_{21}| = 9.29$ dB and $|S_{31}| = 9.30$ dB, while the GDs are found as $\tau_{21} = -1.16$ ns and $\tau_{31} = -1.17$ ns at $f_0 = 2.143$ GHz. Likewise, 3-dB transmission magnitude bandwidth (BW), which is defined as the 3-dB variation from the center frequency S_{21}, is 150 MHz. The NGD BW and magnitude flatness of power divider can be enhanced by cascading the number of sections with slightly different center frequencies [14]. Likewise, Figures 2.8 and 2.9 show return losses/isolation and amplitude/phase imbalances characteristics of NGD power divider, respectively.

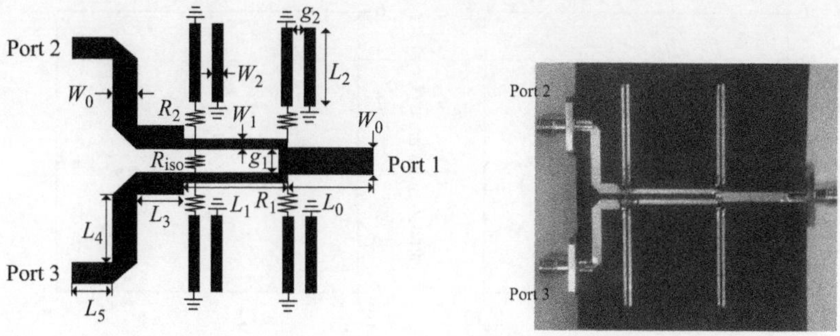

Figure 2.6 (a) Simulation schematic and (b) a photograph of the fabricated circuit. Physical dimensions: $W_0 = 2.4$, $L_0 = 24$, $W_1 = 0.94$, $L_1 = 27.4$, $g_1 = 1.8$, $W_2 = 0.8$, $L_2 = 26.3$, $g_2 = 0.76$, $L_3 = 5$, $L_4 = 14.1$, and $L_5 = 4$ (Unit: mm)

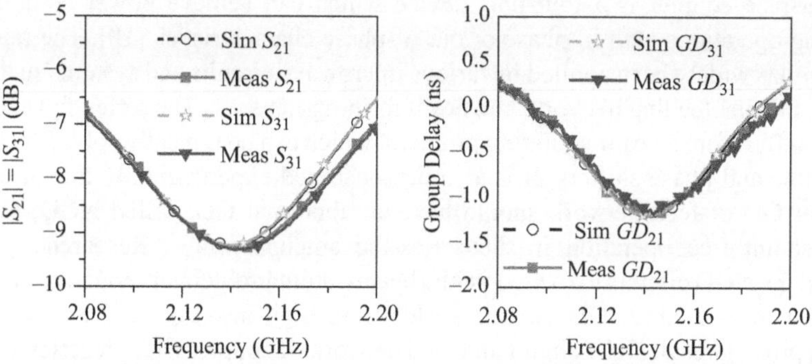

Figure 2.7 Simulated and measured S-parameter and GD characteristics

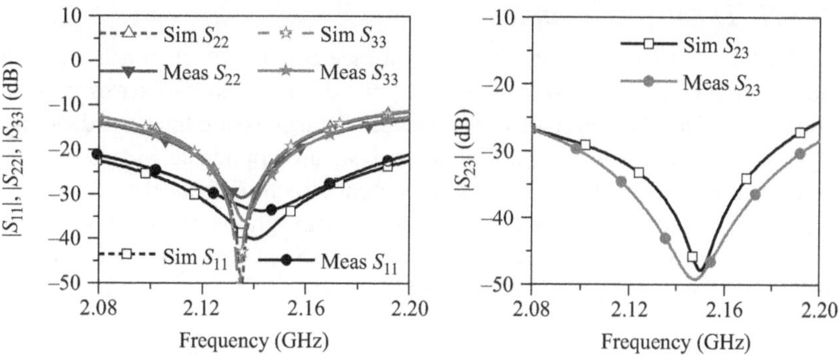

Figure 2.8 Simulated and measured return losses and isolation characteristics

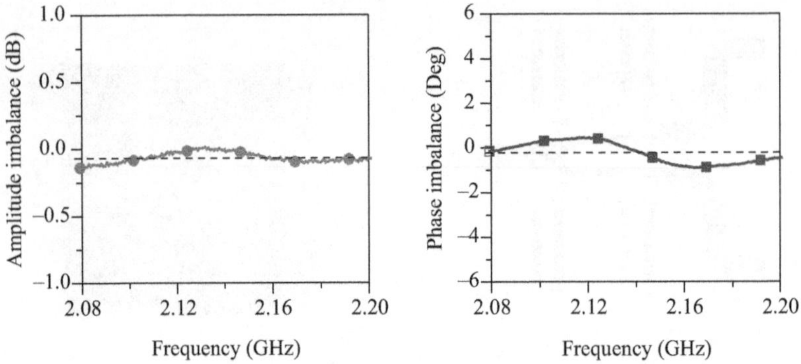

Figure 2.9 Measured amplitude and phase imbalance characteristics

2.3 Rat-race coupler with arbitrary prescribed NGD

The rat-race coupler is a four-port device which can achieve power combining/dividing operations with in-phase or out-of-phase characteristics [10]. The rat-race coupler has widely been applied in various microwave circuits and systems such as a mixer, antenna feeding network, and beam forming-network. The series-fed antenna arrays suffer from a beam-squinting problem, which can be typically solved by using delay line and phase shifters. It is also demonstrated experimentally that an inter-element GD of feed network should have an abnormal GD, called NGD, for the beam-squint free operation in the series-fed antenna arrays. Researchers have applied the NGD circuits to overcome the beam-squint problem in series-fed antenna arrays. However, the NGD circuits suffer from high insertion loss and separate design procedures of NGD circuit and feed network. If coupler with prescribed NGD and low insertion loss is possible to design, it will be more beneficial to design these systems [15,16].

2.3.1 Analytical analysis of arbitrary power division ratio NGD rat-race coupler

Figure 2.10 depicts the schematic of NGD rat-race coupler which consists of shunt resistors R connected to $\lambda/4$ short-circuited coupled line, and two series transmission lines [16]. The characteristics impedances of transmission lines are denoted as Z_2, Z_3, and Z_4. The even- and odd-mode analysis are applicable in this circuit to find the S-parameters because of the symmetrical structure about the XX'-axis.

Figure 2.11 shows even- and odd-mode decompositions of the NGD rat-race coupler where the overall even- and odd-mode ABCD-parameters can be obtained through matrix multiplication as illustrated by (2.9).

$$\begin{bmatrix} A_e & B_e \\ C_e & D_e \end{bmatrix} = \begin{bmatrix} a & jZ_4 b \\ j\dfrac{b}{Z_4} & a \end{bmatrix} \begin{bmatrix} 1 & 0 \\ Y_{in_3} & 1 \end{bmatrix} \begin{bmatrix} a & jZ_2 b \\ j\dfrac{b}{Z_2} & a \end{bmatrix} \begin{bmatrix} 1 & 0 \\ Y_{in_4} & 1 \end{bmatrix} \begin{bmatrix} a & jZ_4 b \\ j\dfrac{b}{Z_4} & a \end{bmatrix} \tag{2.9a}$$

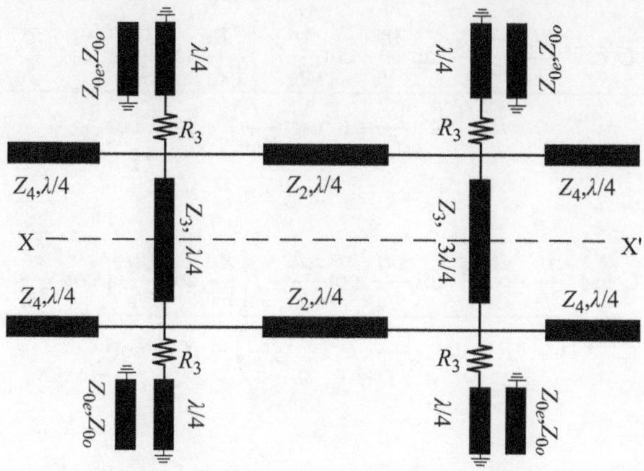

Figure 2.10 Schematic of arbitrary power division NGD rat-race coupler

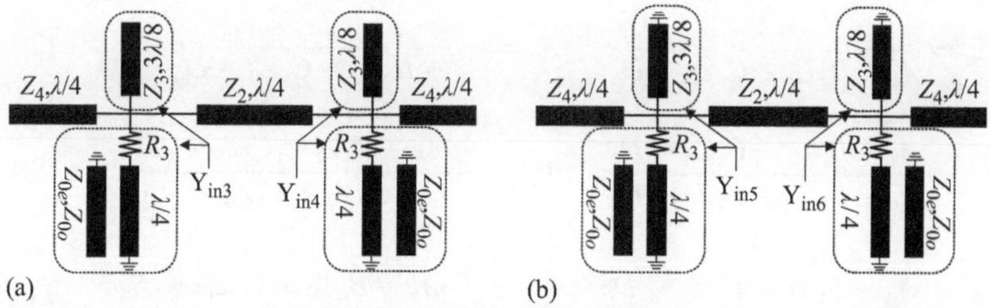

Figure 2.11 Equivalent circuits of NGD rat-race coupler under: (a) even-mode excitation and (b) odd-mode excitation

$$\begin{bmatrix} A_o & B_o \\ C_o & D_o \end{bmatrix} = \begin{bmatrix} a & jZ_4 b \\ j\dfrac{b}{Z_4} & a \end{bmatrix} \begin{bmatrix} 1 & 0 \\ Y_{\text{in}_5} & 1 \end{bmatrix} \begin{bmatrix} a & jZ_2 b \\ j\dfrac{b}{Z_2} & a \end{bmatrix} \begin{bmatrix} 1 & 0 \\ Y_{\text{in}_6} & 1 \end{bmatrix} \begin{bmatrix} a & jZ_4 b \\ j\dfrac{b}{Z_4} & a \end{bmatrix} \tag{2.9b}$$

where

$$a = \cos\frac{\pi f}{2f_0},\ b = \sin\frac{\pi f}{2f_0} \tag{2.10a}$$

$$Y_{\text{in}_3} = \frac{\cot^2\dfrac{\pi f}{2f_0} - C^2\csc^2\dfrac{\pi f}{2f_0} - \dfrac{CZ_c}{Z_3}\tan\dfrac{\pi f}{4f_0}\cot\dfrac{\pi f}{2f_0} + j\left\{\dfrac{R_3}{Z_3}\tan\dfrac{\pi f}{4f_0}\left(\cot^2\dfrac{\pi f}{2f_0} - C^2\csc^2\dfrac{\pi f}{2f_0}\right)\right\}}{R_3\left(\cot^2\dfrac{\pi f}{2f_0} - C^2\csc^2\dfrac{\pi f}{2f_0}\right) + jCZ_c\cot\dfrac{\pi f}{2f_0}} \tag{2.10b}$$

$$Y_{\text{in}_4} = \frac{\cot^2\dfrac{\pi f}{2f_0} - C^2\csc^2\dfrac{\pi f}{2f_0} - \dfrac{CZ_c}{Z_3}\tan\dfrac{3\pi f}{4f_0}\cot\dfrac{\pi f}{2f_0} + j\left\{\dfrac{R_3}{Z_3}\tan\dfrac{3\pi f}{4f_0}\left(\cot^2\dfrac{\pi f}{2f_0} - C^2\csc^2\dfrac{\pi f}{2f_0}\right)\right\}}{R_3\left(\cot^2\dfrac{\pi f}{2f_0} - C^2\csc^2\dfrac{\pi f}{2f_0}\right) + jCZ_c\cot\dfrac{\pi f}{2f_0}} \tag{2.10c}$$

$$Y_{\text{in}_6} = \frac{\cot^2\dfrac{\pi f}{2f_0} - C^2\csc^2\dfrac{\pi f}{2f_0} + \dfrac{CZ_c}{Z_3}\cot\dfrac{3\pi f}{4f_0}\cot\dfrac{\pi f}{2f_0} - j\left\{\dfrac{R_3}{Z_3}\cot\dfrac{3\pi f}{4f_0}\left(\cot^2\dfrac{\pi f}{2f_0} - C^2\csc^2\dfrac{\pi f}{2f_0}\right)\right\}}{R_3\left(\cot^2\dfrac{\pi f}{2f_0} - C^2\csc^2\dfrac{\pi f}{2f_0}\right) + jCZ_c\cot\dfrac{\pi f}{2f_0}} \tag{2.10e}$$

The relation between C and Z_c can be expressed by (2.8). Assuming reference port impedance $Z_0 = 1/Y_0$, the even- and odd-mode S-parameters can be directly obtained from ABCD-parameters with the following expressions (2.11):

$$S_e = \frac{2}{A_e + B_e Y_0 + C_e Z_0 + D_e}, \quad S_o = \frac{2}{A_o + B_o Y_0 + C_o Z_0 + D_o} \tag{2.11a}$$

$$\Gamma_{e_1} = \frac{A_e + B_e Y_0 - C_e Z_0 - D_e}{A_e + B_e Y_0 + C_e Z_0 + D_e}, \quad \Gamma_{e2} = \frac{-A_e + B_e Y_0 - C_e Z_0 + D_e}{A_e + B_e Y_0 + C_e Z_0 + D_e} \tag{2.11b}$$

$$\Gamma_{o_1} = \frac{A_o + B_o Y_0 - C_o Z_0 - D_o}{A_o + B_o Y_0 + C_o Z_0 + D_o}, \quad \Gamma_{o_2} = \frac{-A_o + B_o Y_0 - C_o Z_0 + D_o}{A_o + B_o Y_0 + C_o Z_0 + D_o} \tag{2.11c}$$

Subsequently, S-parameters of overall NGD coupler can be obtained using (2.12):

$$S_{11} = \frac{\Gamma_{e_1} + \Gamma_{o_1}}{2}, \quad S_{22} = \frac{\Gamma_{e_2} + \Gamma_{o_2}}{2} \tag{2.12a}$$

$$S_{21} = S_{43} = \frac{S_e + S_o}{2}, \quad S_{31} = S_{42} = \frac{S_e - S_o}{2} \tag{2.12b}$$

$$S_{41} = \frac{\Gamma_{e_1} - \Gamma_{o_1}}{2}, \quad S_{23} = \frac{\Gamma_{e_2} - \Gamma_{o_2}}{2} \tag{2.12c}$$

For the matched all ports ($S_{ii} = 0$) with reference impedance Z_0 at f_0, the relation among Z_2, Z_3, Z_4, and R_3 is found as (2.13):

$$Z_4 = \left\{\frac{Z_2^2 Z_0^2}{1 + Z_2^2/R^2 + Z_3^2/Z_3^2}\right\}^{1/4} \tag{2.13}$$

Additionally, the magnitude of *S*-parameters through different transmission paths at f_0 are determined by using (2.14):

$$|S_{21}|_{f=f_0} = |S_{43}|_{f=f_0} = \frac{R_3^2}{2R_3Z_2 + R_3^2(Z_4^2/Z_0Z_2 + Z_2Z_0/Z_4^2) + Z_2Z_4^2/Z_0(1 + R_3^2/Z_3^2)} \tag{2.14a}$$

$$|S_{41}|_{f=f_0} = -|S_{23}|_{f=f_0} = \frac{R_3^2Z_2}{Z_3\{2R_3Z_2 + R_3^2(Z_4^2/Z_0Z_2 + Z_2Z_0/Z_4^2) + Z_2Z_4^2/Z_0(1 + R_3^2/Z_3^2)\}} \tag{2.14b}$$

$$|S_{31}|_{f=f_0} = |S_{42}|_{f=f_0} = 0 \tag{2.14c}$$

The isolation between output ports is independent of design parameters as observed from (2.14c). Likewise, the power division ratio *k* of the NGD coupler at f_0 is independent of R_3, Z_c, and *C* and simply evaluated by Z_2 and Z_3 as shown in (2.15).

$$\left.\frac{|S_{41}|}{|S_{21}|}\right|_{f=f_0} = \left.\frac{|S_{23}|}{|S_{43}|}\right|_{f=f_0} = \frac{Z_2}{Z_3} = k \tag{2.15}$$

Subsequently, the GDs through different transmission path at f_0 are simplified as (2.16)

$$\tau_{21}\Big|_{f=f_0} = \tau_{43}\Big|_{f=f_0} = -\frac{1}{4f_0}\left(\frac{2Z_c}{R_3C} - \frac{b_1 + b_2/Z_0 + b_3Z_0}{2R_3Z_2 + R_3^2(Z_4^2/Z_0Z_2 + Z_2Z_0/Z_4^2) + Z_2Z_4^2/Z_0(1 + R_3^2/Z_3^2)}\right) \tag{2.16a}$$

$$\tau_{41}\Big|_{f=f_0} = \tau_{23}\Big|_{f=f_0} = \tau_{21}\Big|_{f=f_0} - \frac{Z_4^2}{4f_0Z_0Z_3} = \tau_{43}\Big|_{f=f_0} - \frac{Z_4^2}{4f_0Z_0Z_3} \tag{2.16b}$$

where

$$b_1 = 2(1 + Z_4/Z_2 + Z_2/Z_4)R_3^2 + 2Z_4(Z_2 + kR_3^2/Z_3) + 4kR_3^2 + 2Z_2Z_c/C \tag{2.17a}$$

$$b_2 = 2R_3\{kZ_4^2(2 + Z_c/Z_3C) + Z_4(Z_4 + Z_2) + Z_4^2Z_c/Z_2C\} \tag{2.17b}$$

$$b_3 = 2R_3Z_2/Z_4(1 + Z_c/Z_4C) \tag{2.17c}$$

For a graphical illustration of analytical design expressions, the computed *k*, GDs, and magnitudes of S_{21} and S_{31} as function of Z_c, *C*, and R_3 are depicted in

Figure 2.12. As can be seen in Figure 2.12, the practically realizable k will be in between 0.14 and 7 if characteristics impedances of transmission lines are in the range of 20–140 Ω. Likewise, the GDs will become more negative if Z_c increases and R_3 decreases, however, the insertion losses increase for low R. Similarly, if the GDs at f_0 is fixed, the NGD BW (BW of GD when GD < 0) will be affected by Z_c and C.

To evaluate the effect of Z_c and C on the NGD BW, the computed results are shown in Figure 2.13 by fixing GDs at $f = f_0$. As observed from these results, NGD BW will decrease if Z_c extends toward high value. However, insertion loss of transmission paths outdoes which necessitates a trade-off between the NGD BW and insertion loss. Based on these analytical expressions, the design method of arbitrary k NGD rat-race coupler is summarized with the flowchart shown in Figure 2.14.

For illustration, two types of NGD rat-race couplers with $k = 0.52$ and $k = 1$ are designed and fabricated at 2.14 GHz using a substrate with dielectric constant (ε_r) of 2.2 and thickness (h) of 0.787 mm.

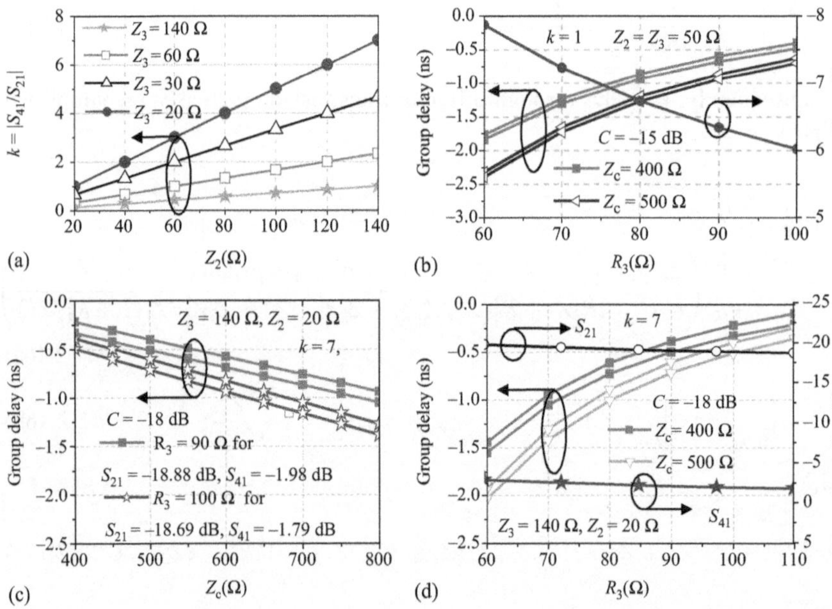

Figure 2.12 Power division ratio k and magnitudes and GDs at $f_0 = 2.14$ GHz: (a) k as function of Z_2 and Z_3; (b) GD and magnitudes of S_{21} and S_{31} according R_3 for fixed Z_c, C, and k = 1; (c) GD according to Z_c with fixed R_3, S_{21}, S_{31}, C, and k = 7; and (d) GD and magnitude of S_{21} and S_{31} according to R_3 for fixed Z_c, C, and k = 7

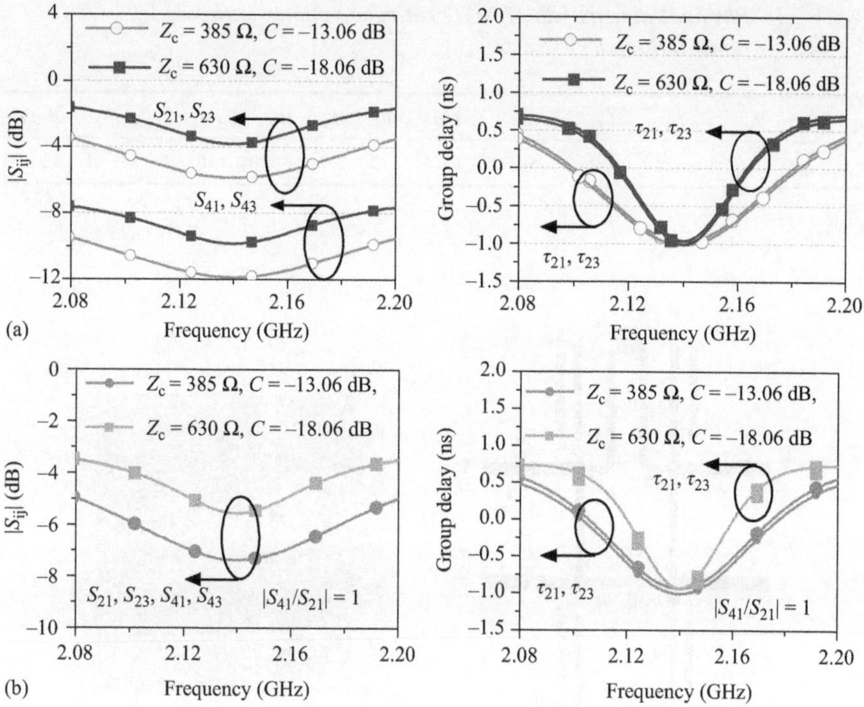

Figure 2.13 Computed magnitudes and GDs of the proposed rat-race coupler for different values of Z_c and k: (a) k = 0.5 and (b) k = 1

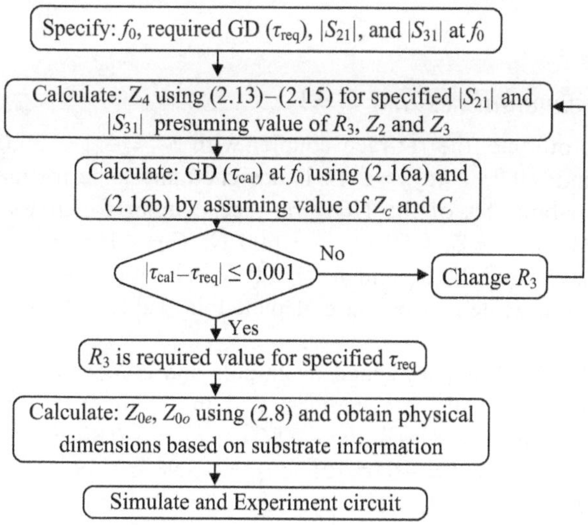

Figure 2.14 Design flowchart of NGD rat-race coupler with arbitrary prescribed GD

Table 2.1 Physical dimensions of NGD rat-race coupler with k = 1 (unit: millimeter)

L_1	L_2	L_3	L_4	L_5	L_6	L_7	W_1	W_2	W_3	W_4	g_1	R (Ω)
25.6	1 6	2 25.8	3 8.2	4 25.2	5 27.5	6 26	7 2	8 2	9 1.04	10 3.2	11 1.1	12 100

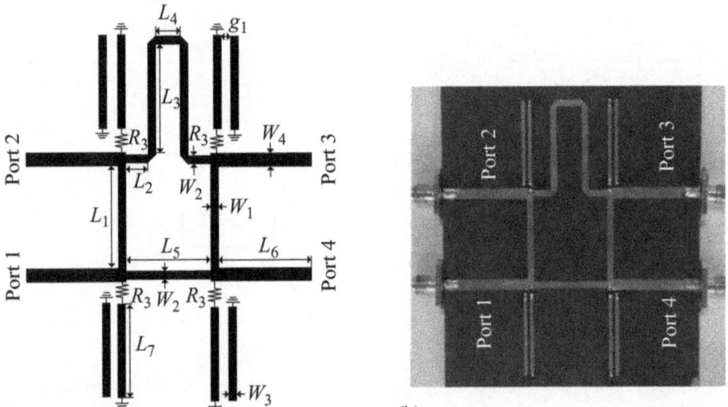

(a) (b)

Figure 2.15 (a) EM simulation layout with physical dimensions with k = 1 and (b) photograph of fabricated circuit

2.3.2 Experimental results: $k = 1$

For the first prototype, the rat-race coupler with $S_{21} = S_{41} = -6.02$ dB ($k = 1$) and GDs around −0.9 ns at $f_0 = 2.14$ is designed and experimented. The computed circuit parameters of the NGD rat-race coupler for given specification are determined as $Z_2 = Z_3 = 50$ Ω, $Z_c = 600$ Ω, $C = -15$ dB, $Z_{0_e} = 129.77$ Ω, $Z_{0_o} = 90.59$ Ω, $Z_4 = 40.82$ Ω, and $R_3 = 100$ Ω. The physical dimension and a photograph of fabricated circuit are depicted in Table 2.1 and Figure 2.15, respectively.

The simulated and experimented magnitude and GD characteristics between different transmission paths are demonstrated in Figure 2.16 where GDs and magnitudes are summarized as $\tau_{21} = -0.86$ ns, $\tau_{41} = -0.80$ ns, $\tau_{43} = -0.77$ ns, and $\tau_{23} = -0.71$ ns, $|S_{21}| = -6.05$ dB, $|S_{41}| = -5.98$ dB, $|S_{43}| = -5.98$ dB, and $|S_{23}| = -6.04$ dB at $f_0 = 2.143$ GHz.

Likewise, the return loss/isolation and phase imbalance characteristics are depicted in Figure 2.17. The return losses are acquired higher than 18.11 dB at $f_0 = 2.143$ GHz. In addition, the isolation and phase imbalance characteristics are

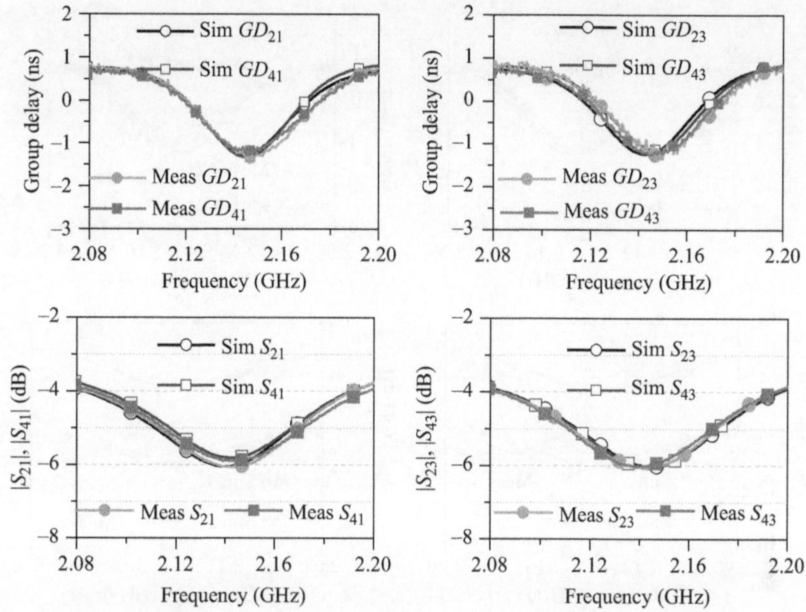

Figure 2.16 Simulated and experimented GDs and magnitudes of NGD rat-coupler with k = 1

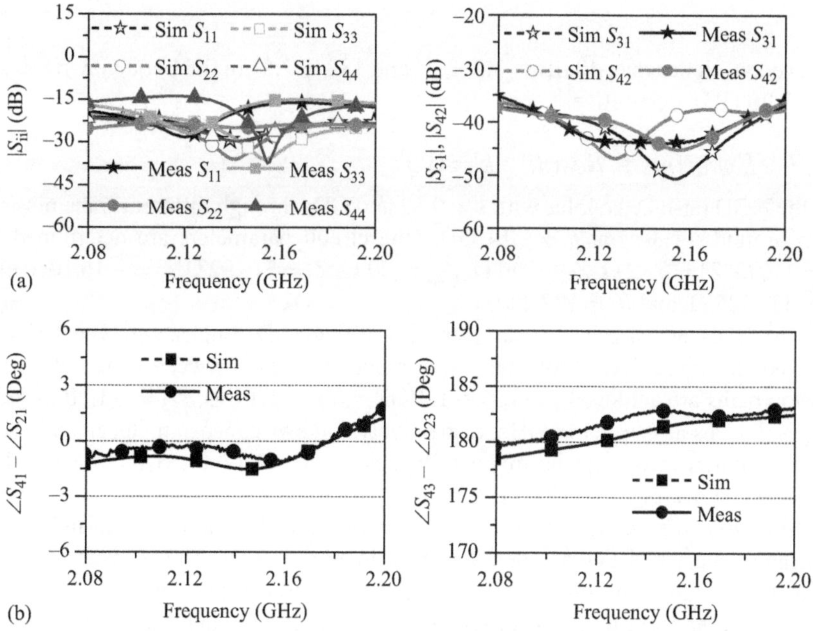

Figure 2.17 Simulated and experimented results of the NGD rat-race coupler with k = 1: (a) return losses/isolation and (b) phase imbalance

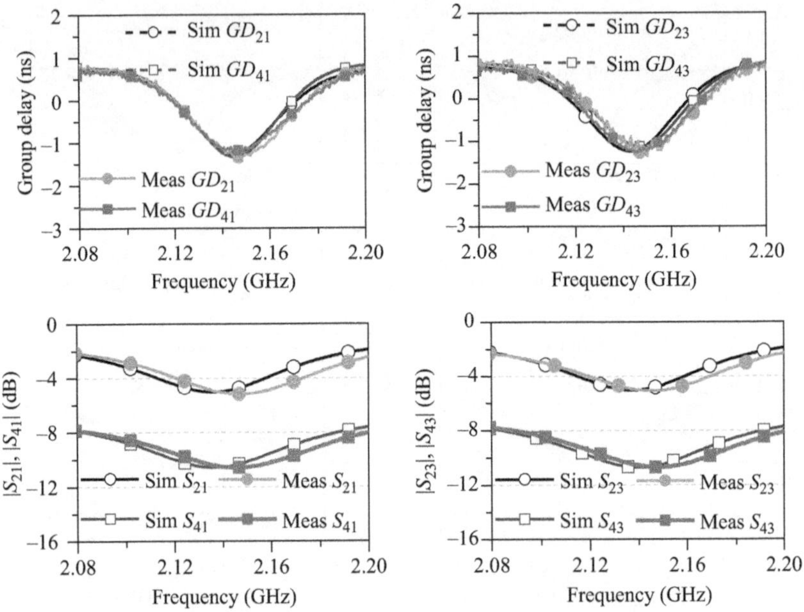

Figure 2.18 Simulated and experimented GDs and magnitudes of the NGD rat-race coupler with $k = 0.52$

achieved as higher than 40 dB, $-1.2 \pm 2°$ and $182 \pm 2.5°$ in the frequency range of 2.08–2.20 GHz, respectively.

2.3.3 Experiment Results: $k = 0.52$

For the NGD rat-race coupler with $k = 0.52$ and GDs through different transmission paths around -1.3 ns at $f_0 = 2.14$ GHz, the circuit parameters are determined as $Z_2 = 110\,\Omega$, $Z_3 = 57.2\,\Omega$, $Z_{0_e} = 96\,\Omega$, $Z_{0_o} = 70\,\Omega$, $Z_c = 516.92\,\Omega$, $C = -16.1026$ dB, $Z_4 = 47.6725\,\Omega$, and $R_3 = 102.2\,\Omega$.

The GDs and magnitude characteristics of the NGD coupler with $k = 0.52$ are depicted in Figure 2.18. From the experiment, the GDs between different transmission paths are achieved as $\tau_{21} = -1.26$ ns, $\tau_{41} = -1.19$ ns, $\tau_{43} = -1.20$ ns, and $\tau_{23} = -1.16$ ns at $f_0 = 2.144$ GHz, respectively. Likewise, *S*-parameter magnitudes of different transmission paths are quantified as $|S_{21}| = -5.17$ dB, $|S_{41}| = -10.59$ dB, $|S_{43}| = -5.06$ dB, $|S_{23}| = -10.67$ dB at $f_0 = 2.14$ GHz.

Similarly, the simulated and experimented return losses, isolation and phase imbalance characteristics are demonstrated in Figure 2.19. The measured return losses and isolation are quantified as >15 dB and 40 dB, respectively. Similarly, the in-phase and out-of-phase differences are $1.1 \pm 2.5°$ and $182 \pm 2.6°$ in the frequency range of 2.08–2.20 GHz.

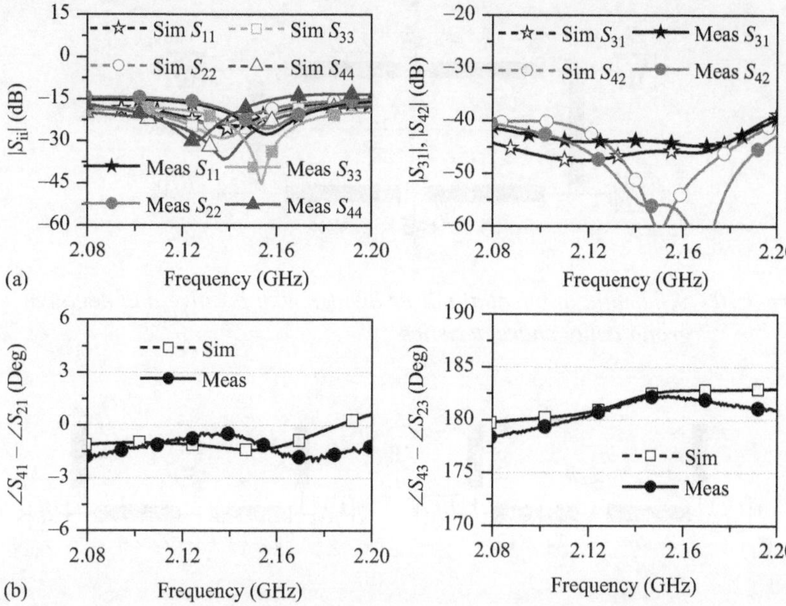

Figure 2.19 Simulated and experimented results of the NGD rat-race coupler with $k = 0.52$: (a) return loss/isolation and (b) phase imbalance

2.4 Arbitrary power division ratio power divider with positive GD and NGD

2.4.1 Analytical analysis

Figure 2.20 depicts the structure of the unequal power divider with positive GD and NGD [17]. The structure is composed of vertical transmission lines with a characteristic impedance of Z_2 and electrical lengths of $\lambda/4$ and $3\lambda/4$ and horizontal transmission lines with characteristic impedance Z_1 and the electrical length of $\lambda/4$. The shunt resistors R_4 are connected between the two horizontal transmission lines Z_1.

Under even- and odd-mode excitations, the circuit is decomposed into equivalent subcircuits depicted in Figure 2.21 where the overall even- and odd-mode ABCD-parameters are found as (2.18). In these expressions, subscripts e and o represent even- and odd-modes, respectively.

$$A_e = \cos^2\frac{\pi f}{f_0} - \frac{Z_5}{Z_6}\tan\frac{3\pi f}{4f_0}\sin\frac{\pi f}{f_0} - j\left(\frac{Z_5^2}{Z_6 R_4}\tan\frac{3\pi f}{4f_0}\sin^2\frac{\pi f}{2f_0} - \frac{Z_5}{2R_4}\sin\frac{\pi f}{f_0}\right) \tag{2.18a}$$

$$B_e = B_o = -\frac{Z_5^2}{R_4}\sin^2\frac{\pi f}{2f_0} + jZ_5\sin\frac{\pi f}{f_0} \tag{2.18b}$$

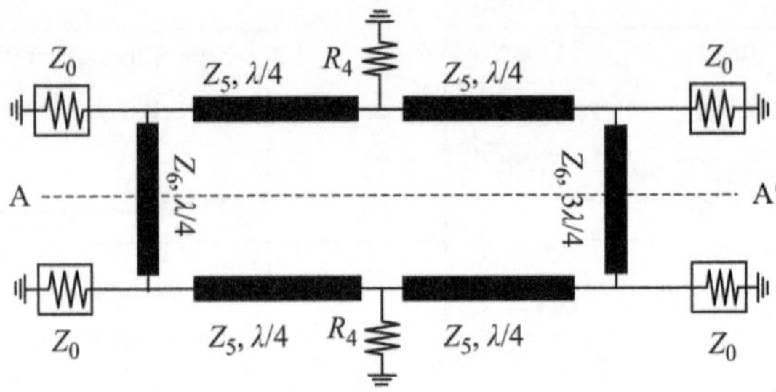

Figure 2.20 Schematic of unequal power divider with positive and negative group delay characteristics

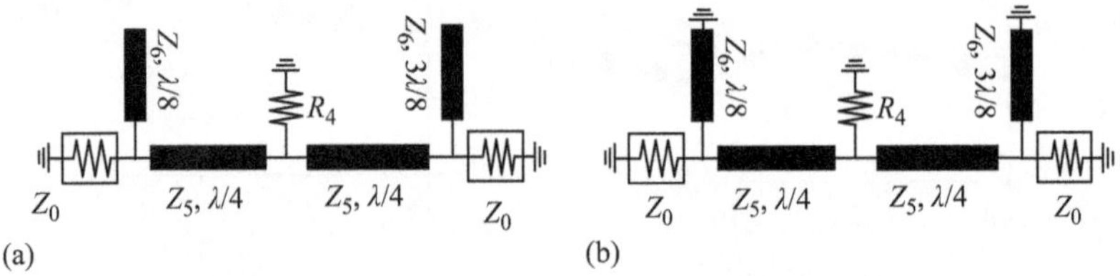

Figure 2.21 Equivalent circuit of unequal power divider with positive and negative group delays under (a) even- and (b) odd-mode excitations

$$C_e = \frac{1}{R_4}\cos^2\frac{\pi f}{2f_0} + \frac{Z_5^2}{R_4 Z_6^2}\tan\frac{\pi f}{4f_0}\tan\frac{3\pi f}{4f_0}\sin^2\frac{\pi f}{2f_0} - \frac{Z_5}{2R_4 Z_6}\left(\tan\frac{\pi f}{4f_0} + \tan\frac{3\pi f}{4f_0}\right)\sin\frac{\pi f}{f_0}$$
$$+ j\left\{\frac{1}{Z_6}\left(\tan\frac{\pi f}{4f_0} + \tan\frac{3\pi f}{4f_0}\right)\cos\frac{\pi f}{f_0} + \sin\frac{\pi f}{f_0}\left(\frac{1}{Z_5} - \frac{Z_5}{Z_6^2}\tan\frac{\pi f}{4f_0}\tan\frac{3\pi f}{4f_0}\right)\right\} \tag{2.18c}$$

$$D_e = \cos\frac{\pi f}{f_0} - \frac{Z_5}{Z_6}\tan\frac{\pi f}{4f_0}\sin\frac{\pi f}{f_0} - j\left(\frac{Z_5^2}{R_4 Z_5}\tan\frac{\pi f}{4f_0}\sin^2\frac{\pi f}{2f_0} - \frac{Z_5}{2R_4}\sin\frac{\pi f}{f_0}\right) \tag{2.18d}$$

$$A_o = \cos\frac{\pi f}{f_0} + \frac{Z_5}{Z_6}\cot\frac{3\pi f}{4f_0}\sin\frac{\pi f}{f_0} + j\left(\frac{Z_5^2}{R_4 Z_6}\cot\frac{3\pi f}{4f_0}\sin^2\frac{\pi f}{2f_0} + \frac{Z_5}{2R_4}\sin\frac{\pi f}{f_0}\right) \tag{2.18e}$$

$$C_o = \frac{1}{R_4}\cos^2\frac{\pi f}{2f_0} + \frac{Z_5^2}{Z_6^2 R_4}\cot\frac{3\pi f}{4f_0}\cot\frac{\pi f}{4f_0}\sin^2\frac{\pi f}{2f_0} + \frac{Z_5}{2R_4 Z_6}\left(\cot\frac{\pi f}{4f_0} + \cot\frac{3\pi f}{4f_0}\right)\sin\frac{\pi f}{f_0}$$
$$- j\left\{\frac{1}{Z_6}\cos\frac{\pi f}{f_0}\left(\cot\frac{\pi f}{4f_0} + \cot\frac{3\pi f}{4f_0}\right) - \sin\frac{\pi f}{f_0}\left(\frac{1}{Z_5} - \frac{Z_5}{Z_6^2}\cot\frac{\pi f}{4f_0}\cot\frac{3\pi f}{4f_0}\right)\right\} \tag{2.18f}$$

$$D_o = \cos\frac{\pi f}{f_0} + \frac{Z_5}{Z_6}\cot\frac{\pi f}{4f_0}\sin\frac{\pi f}{f_0} + j\left(\frac{Z_5^2}{R_4 Z_6}\cot\frac{\pi f}{4f_0}\sin^2\frac{\pi f}{2f_0} + \frac{Z_5}{2R_4}\sin\frac{\pi f}{f_0}\right) \quad (2.18g)$$

The resulting *S*-parameters are determined by combining even- and odd-mode *S*-parameters as described in (2.11) and (2.12). For all ports matched, the value of Z_6 is determined as (2.19) by setting the input/output reflection coefficients (S_{11}, S_{22}, S_{33}, and S_{44}) equal to zero at $f = f_0$.

$$Z_6 = Z_0 \quad (2.19)$$

Subsequently, the power division ratio k of power divider is expressed as (2.20) for all matched reflection coefficients. For fixed Z_5, k can be enhanced by simply increasing R_4.

$$k = \left|\frac{S_{21}}{S_{31}}\right|_{f=f_0} = \left|\frac{S_{43}}{S_{42}}\right|_{f=f_0} = \frac{R_4 Z_0}{Z_5^2} \quad (2.20)$$

Additionally, the magnitudes of transmission coefficients and isolation between output ports at $f = f_0$ are ascertained as (2.21):

$$|S_{31}|_{f=f_0} = |S_{42}|_{f=f_0} = \frac{Z_5^2}{R_4 Z_0 + Z_5^2} = \frac{1}{k+1} \quad (2.21b)$$

$$|S_{32}|_{f=f_0} = |S_{41}|_{f=f_0} = 0 \quad (2.21c)$$

As can be seen from (2.21c), the isolation between output ports is perfectly zero and independent of Z_5, Z_6, and R_4. However, the magnitude of transmission coefficients at f_0 is a function of R_4 and Z_5. Subsequently, the GDs through different transmission paths at $f = f_0$ can be obtained as (2.22) using the phase of transmission coefficients.

$$\tau_{21}\Big|_{f=f_0} = -\frac{1}{2\pi}\frac{d\angle S_{21}}{df}\Big|_{f=f_0} = \frac{2k^2 Z_5^2 + kZ_0(k+1)(Z_0 + 2Z_5)}{4f_0 Z_0 k Z_5 (k+1)} \quad (2.22a)$$

$$\tau_{31}\Big|_{f=f_0} = -\frac{1}{2\pi}\frac{d\angle S_{31}}{df}\Big|_{f=f_0} = -\frac{Z_0 + 2kZ_5}{4f_0 Z_0} + \tau_{21}\Big|_{f=f_0} \quad (2.22b)$$

From (2.22), it can be noted that the GD between transmission path 2 and 1 is always positive, whereas the GD through path 3 and 1 can be negative by appropriately selecting Z_5 and R_4. In addition, the GD will move toward higher negative value if R_4 increases for fixed Z_5.

To better understanding these analytical design equations graphically, the computed magnitudes of *S*-parameters through different transmission paths are

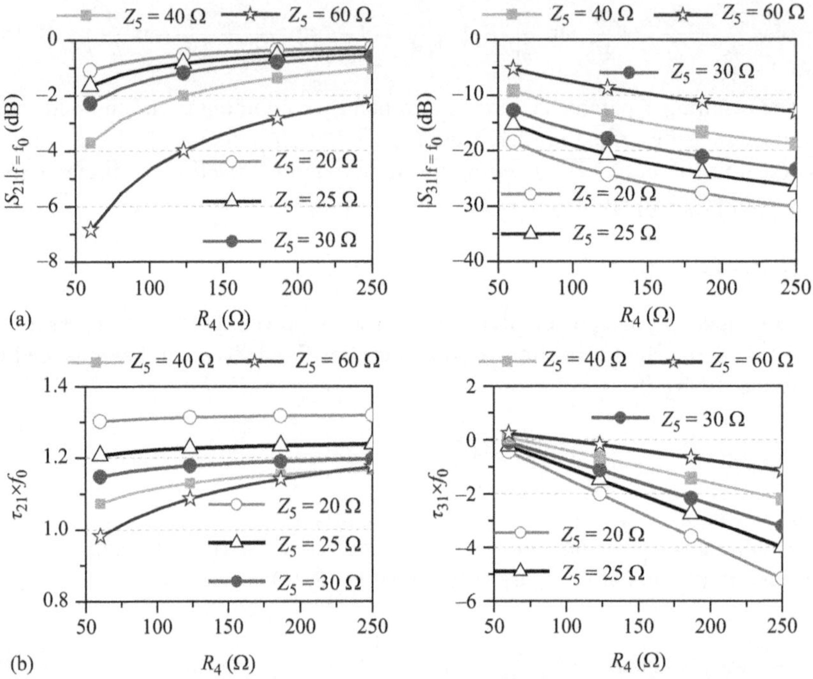

Figure 2.22 Calculated magnitudes and normalized GDs of different transmission paths as a function of R and Z_1: (a) magnitudes and (b) GDs

depicted in Figure 2.22(a). As can be noted from this figure, if R_4 increases, the magnitude of S_{21} will be increased, whereas the magnitude of S_{31} will be reduced, resulting in high k. Likewise, the normalized GDs (maximum achievable GD multiplied by f_0) of different transmission paths are shown Figure 2.22(b). As seen from this figure, it is observed that the GD between transmission path 2 and 1 is increased toward higher positive value by increasing R_4 and decreasing Z_5. Moreover, the positive GD is almost saturated for low Z_5. Similarly, the GD between transmission path 3 and 1 is increased toward higher negative value by increasing R_4 and decreasing Z_5. Therefore, a low Z_5 is preferable for high k and NGD.

To scrutinize the effect of Z_5 in the GD BW, the computed magnitudes and GDs of unequal power dividers are depicted in Figure 2.23. In the first case, the GD through path 3 and 1 is retained constant while in the second case, the magnitude is retained constant at f_0. As observed from this figure, the GD BW is slightly wider in case of lower Z_5, however, the insertion loss is increased. Similarly, magnitude flatness is enhanced in case of lower Z_5, but the NGD is decreased. Therefore, the lower Z_5 is preferable for wide GD and magnitude flatness BW by maintaining the trade-off among GD BW, magnitude flatness, and insertion loss.

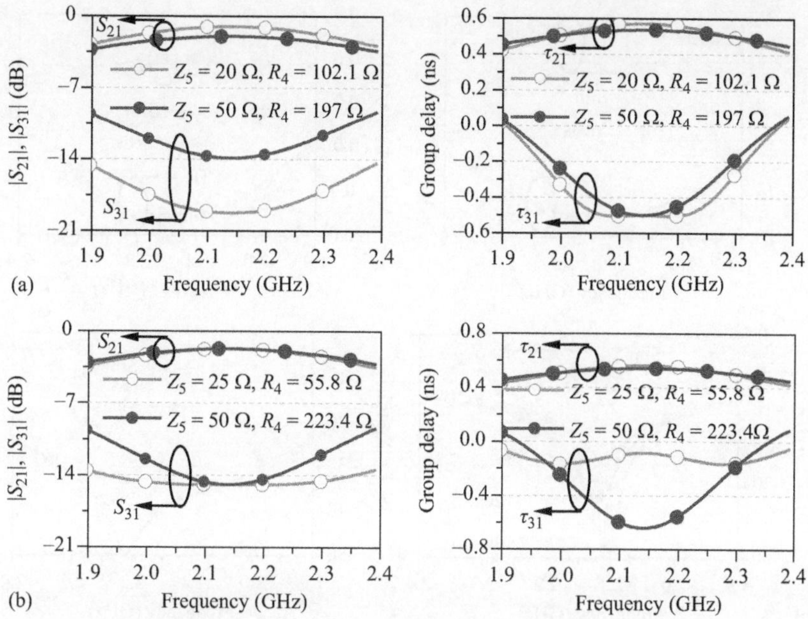

Figure 2.23 Synthesized magnitudes and GDs of unequal power divider for illustrating the effect of Z_5: (a) fixed GD and (b) fixed magnitude

2.4.2 Implementation and experimental performances: $k = 16.9$ dB

For experimental justification of the power divider, two prototypes are fabricated at f_0 of 2.14 GHz using a substrate with a dielectric constant (ε_r) of 2.2 and thickness (h) of 0.787 mm.

The first prototype circuit is designed for $k = 16.9$ dB and GD of -0.5 ns through path 3 and 1. The computed circuit parameters for above specification are obtained as $Z_5 = 30\ \Omega$, $Z_6 = 50\ \Omega$, and $R_4 = 126\ \Omega$. The simulated and experimented results of GDs and magnitudes are depicted in Figure 2.24. From the experiment, the magnitude transmission coefficients are determined as $|S_{21}| = -1.19$ dB and $|S_{31}| = -18.1$ dB at 2.14 GHz providing k is 16.92 dB. In addition, the GDs through different transmission paths are determined as $\tau_{21} = 0.58$ ns and $\tau_{31} = -0.48$ ns at $f_0 = 2.14$ GHz. Similarly, the return losses and isolation between output ports are >15 dB and 30 dB in the frequency range of 2–2.22 GHz, respectively.

2.4.3 Implementation and experiment performances: $k = 20$ dB

For the second prototype, a goal is to accomplish $k = 20$ dB and -0.7 ns GD through path 3 and 1 at $f_0 = 2.14$ GHz. Based on this specification, the computed circuit parameters are acquired as $Z_5 = 25\ \Omega$, $Z_6 = 50\ \Omega$, and $R_4 = 125\ \Omega$.

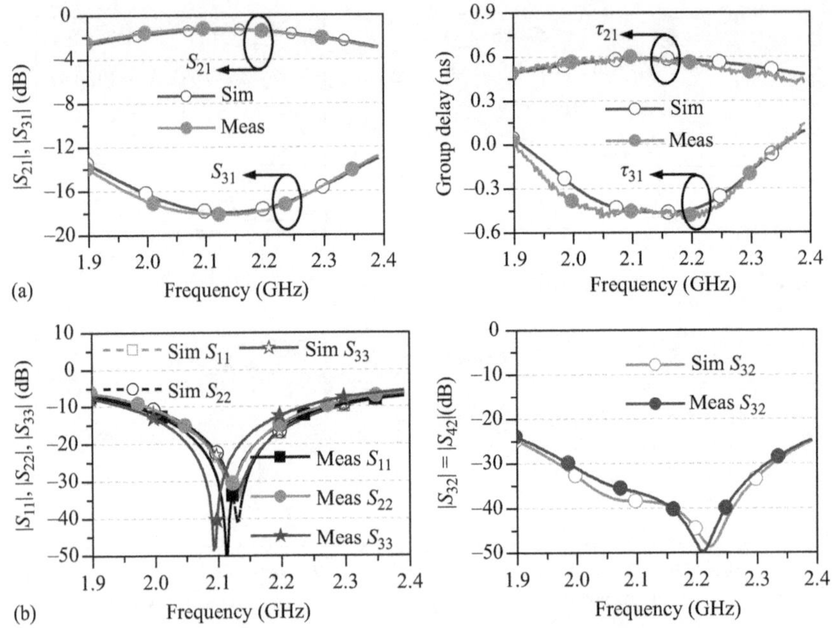

Figure 2.24 Simulated and experimented results of unequal power divider with $k = 16.9$ dB: (a) GD/magnitude and (b) return losses/isolation

The simulated and experimented results of GD and magnitudes are depicted in Figure 2.25. The measured magnitude of transmission coefficients are determined as $|S_{21}| = -0.95$ dB and $|S_{31}| = -21.01$ dB at f_0 resulting $k = 20.06$ dB. Similarly, the measured GDs through different transmission paths are summarized as $\tau_{21} = 0.58$ ns and $\tau_{31} = -0.48$ ns at f_0. Similarly, the return losses and isolation between output ports are >15 dB and 28 dB in the frequency range of 2–2.22 GHz, respectively.

2.5 Power divider with positive GD and NGD: coupling matrix approach

The coupling matrix method is widely adapted to synthesize narrow filters and filtering power divider with distinct features including tunable center frequency as well as BW, switchable filter from bandstop to all-pass filter, and the bandstop to bandpass switchable filter [18–20]. In this section, the coupling matrix method is applied to design the power divider with arbitrarily prescribed GDs and power division ratio [21–23]. The structure is composed of the microstrip line resonators with finite unloaded quality factors Q_u. The main advantage of this structure is that it does not need any of lumped resistor to exhibit NGD, otherwise the NGD is mainly determined by input/output external couplings, Q_u, and inter-resonator couplings between resonators.

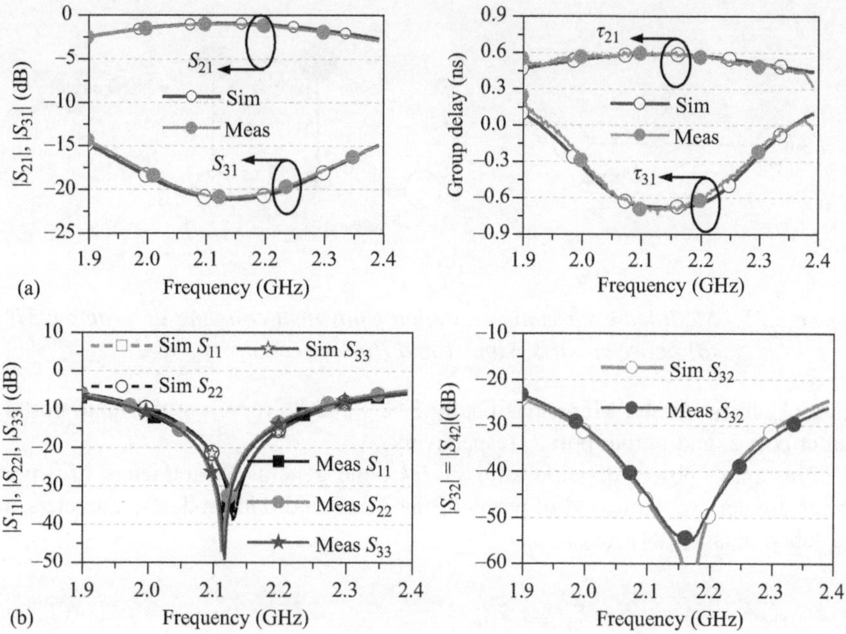

Figure 2.25 Simulated and experimented results of unequal power divider with $k = 20$ dB: (a) GD/magnitude and (b) return losses/isolation

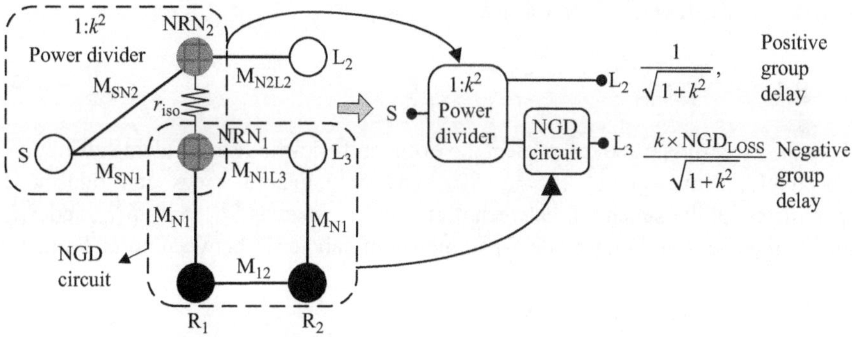

Figure 2.26 Coupling diagram of NGD power divider with finite unloaded quality factor resonators

2.5.1 Mathematical analysis

Figure 2.26 depicts the coupling diagram of the power divider where 1:k^2 power divider arm contains the NGD circuit [23]. In this coupling diagram, nodes S, NRN, L, R_1, and R_2, present the source, nonresonating node, load, and the resonators with

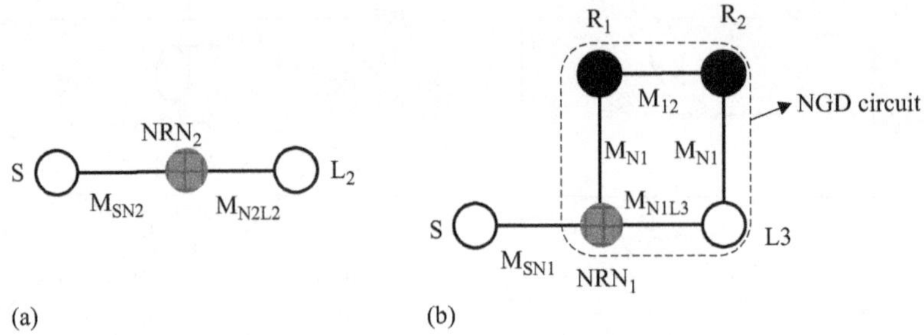

Figure 2.27 Modified even-mode excitation equivalent coupling diagrams [23]: (a) between ports 2 and 1 and (b) between ports 3 and 1

finite Q_u, respectively. More specifically S, L_2, and L_3 represent the input port 1, output port 2, and output port 3, respectively.

Presuming power division ratio of $1{:}k^2$, the general *S*-parameters of 3-port power divider are expressed in terms of the even- and odd-mode *S*-parameters of the subcircuits as [24]:

$$S_{11} = \frac{S_{11e}^{21} + S_{11e}^{31}}{2} = S_{11e}^{21} = S_{11e}^{31} \tag{2.23a}$$

$$S_{22} = \frac{k^2 S_{22e}^{21} + S_{22o}^{21}}{1+k^2}, \quad S_{33} = \frac{k^2 S_{33e}^{31} + S_{33o}^{31}}{1+k^2} \tag{2.23b}$$

$$S_{21} = \frac{S_{21e}^{21}}{\sqrt{1+k^2}}, \quad S_{31} = \frac{k S_{31e}^{31}}{\sqrt{1+k^2}} \tag{2.23c}$$

$$S_{23} = \frac{k\left(S_{22e}^{21} - S_{33o}^{31}\right)}{1+k^2}, \tag{2.23d}$$

where the subscripts e and o represent the even- and odd-mode equivalent subcircuits, respectively. Similarly, S_{11e}^{21}, S_{22e}^{21}, S_{21e}^{21}, and S_{22o}^{21} denote even- and odd-mode *S*-parameters of the subcircuit between port 2 and 1, whereas S_{11e}^{31}, S_{33e}^{31}, S_{31e}^{31}, and S_{33o}^{31} indicate the even- and odd-mode *S*-parameters of subcircuit between ports 3 and 1.

2.5.2 Even-mode analysis

The equivalent even-mode coupling diagram between ports 2 and 1 is depicted in Figure 2.27(a) where M_{SN2} and M_{N2L2} indicate the coupling between S and NRN_2, and between L_2 and NRN_2, respectively. The equivalent normalized port reference impedances of the source and load between port 2 and 1 are expressed as (2.24):

$$r_{S_e}^{21} = 1 + k^2, \; r_{L2_e}^{21} = 1 \tag{2.24}$$

The coupling between the source/load and NRNs can be implemented by using a $\lambda/4$ transmission line such that the normalized characteristic impedance is identical to the inverse of coupling the coefficient. In addition, the even-mode *S*-parameters of subcircuit can be determined by using the equivalent even-mode coupling diagram between port 2 and 1 as depicted in Figure 2.26(a). The normalized coupling between the source/load and NRN_2 is expressed as (2.25) for the zero input/output reflection coefficients ($S_{11e}^{21} = S_{22e}^{21} = 0$) at f_0.

$$M_{SN2} = \frac{1}{k}\frac{1}{\sqrt{1+k^2}} \tag{2.25a}$$

$$M_{N2L2} = \frac{1}{k} \tag{2.25b}$$

Subsequently, the magnitude of transmission coefficient and corresponding GDs between ports 2 and 1 of the power divider at f_0 are found as (2.26) for zero input/output reflection coefficients.

$$S_{21p}\Big|_{f=f_0} = \frac{S_{21e}^{21}}{\sqrt{1+k^2}}\Bigg|_{f=f_0} = \frac{2\sqrt{r_{se}^{21}}M_{N2L2}M_{SN2}}{\sqrt{1+k_1^2}\left(r_{se}^{21}M_{SN2}^2 + M_{N2L2}^2\right)} = \frac{1}{\sqrt{1+k^2}} \tag{2.26a}$$

$$\tau_{21p}\Big|_{f=f_0} = \frac{1+2k^2+\sqrt{(1+k^2)}+k^2\sqrt{(1+k^2)}}{8f_0\sqrt{k^2(1+k^2)}} \tag{2.26b}$$

Similarly, Figure 2.26(b) depicts the coupling diagram between ports 3 and 1 under even-mode excitation where M_{SN1} denotes the coupling between S and NRN_1. Similar to the previous case, the normalized port reference impedance between ports 3 and 1 can be expressed as (2.27).

$$r_{Se}^{31} = \left(1+k^2\right)/k^2, r_{L3e}^{31} = 1 \tag{2.27}$$

The normalized source impedance shown in (2.27) will be converted into "1" by coupling M_{SN1}. In addition, NRN_1, L_3, R_1, and R_2 represent the second-order NGD filter, of which the $(N+2)\times(N+2)$ normalized coupling matrix is expressed by (2.28)

$$M_{NGD} = \begin{bmatrix} 0 & M_{N1} & 0 & M_{N1L3} \\ M_{N1} & \dfrac{-j}{Q_u\Delta} & M_{12} & 0 \\ 0 & M_{12} & \dfrac{-j}{Q_u\Delta} & M_{N1} \\ M_{N1L3} & 0 & M_{N1} & 0 \end{bmatrix}, \tag{2.28}$$

where the subscripts N1, L3, 1, and 2 represent the NRN_1 as an input port, output port 3, the first resonator, and the second resonator, respectively. Similarly, a 3-dB FBW is denoted by Δ. Utilizing equivalent even-mode sub-circuit depicted in Figure 2.25(b), the coupling between the source (S) and NRN_1 can be determined as (2.29) for zero reflection coefficient ($S_{11e}^{31} = 0$) at f_0.

$$M_{SN1} = \frac{k}{\sqrt{1+k^2}} \tag{2.29}$$

Similarly between port 2 and 1, the magnitude of transmission coefficient between ports 3 and 1 is found as (2.30) and (2.31) by assuming $M_{12} = aM^2_{\mathrm{N1}}$

$$S_{31\mathrm{p}}\Big|_{f=f_0} = \frac{kS^{31}_{31e}}{\sqrt{1+k^2}}\Bigg|_{f=f_0} = \frac{k \times S_{S-N_1}\Big|_{f=f_0} \times S_{\mathrm{NGD}}\Big|_{f=f_0}}{\sqrt{1+k^2}}, \tag{2.30}$$

where

$$S_{S-N_1}\Big|_{f=f_0} = \frac{2\sqrt{r^{31}_{se}}M_{\mathrm{SN1}}}{\left(r^{31}_{se}M^2_{\mathrm{SN1}}+1\right)} = 1, \tag{2.31a}$$

$$S_{\mathrm{NGD}}\Big|_{f=f_0} = \frac{2\left(\dfrac{M_{\mathrm{N1L3}}}{Q_u^2\Delta^2} + a^2M_{\mathrm{N1L3}}M^4_{\mathrm{N1}} - aM^4_{\mathrm{N1}}\right)}{\left(M^2_{\mathrm{N1L3}}+1\right)\left(\dfrac{1}{Q_u^2\Delta^2} + a^2M^4_{\mathrm{N1}}\right) + \dfrac{2M^2_{\mathrm{N1}}}{Q_u\Delta} + M^4_{\mathrm{N1}}\left(1 - 2aM_{\mathrm{N1L3}}\right)}. \tag{2.31b}$$

As noted from (2.30) and (2.31), the magnitude of transmission coefficient between ports 3 and 1 depends on the transmission coefficient magnitudes of source and NRN_1, and NGD circuit. The insertion loss between port 3 and 1 increases as compared to the conventional power divider due to the additional insertion loss of NGD circuit. Likewise, the GD between ports 3 and 1 can be found by using phase of $S_{31\mathrm{p}}$ as (2.32):

$$\tau_{31\mathrm{p}}\Big|_{f=f_0} = \frac{1}{4f_0}\left\{\frac{1+2k^2}{2k\sqrt{1+k^2}} + \frac{8Q_\mathrm{u}M_{\mathrm{N1L3}}}{\pi aM^4_{\mathrm{N1}}(1-aM_{\mathrm{N1L3}})Q_\mathrm{u}^2\Delta^2 - \pi M_{\mathrm{N1L3}}}\right\}$$
$$+\frac{1}{\pi f_0}\left\{\frac{M^2_{\mathrm{N1}}Q_\mathrm{u}^2\Delta + Q_\mathrm{u}M^2_{\mathrm{N1L3}} + Q_\mathrm{u}}{\left[0.5\left(M^2_{\mathrm{N1L3}}+1\right)\left(1+Q_\mathrm{u}^2\Delta^2a^2M^4_{\mathrm{N1}}\right) + M^2_{\mathrm{N1}}Q_\mathrm{u}\Delta - Q_\mathrm{u}^2\Delta^2M^4_{\mathrm{N1}}\left(aM_{\mathrm{N1L3}}-0.5\right)\right]}\right\} \tag{2.32}$$

where

$$M_{\mathrm{N1L3}} = \frac{aM^4_{\mathrm{N1}} \mp \sqrt{a^2M^8_{\mathrm{N1}} + \left(a^2M^4_{\mathrm{N1}} + \dfrac{1}{Q_\mathrm{u}^2\Delta^2}\right)\left\{M^4_{\mathrm{N1}}(a^2-1) + \dfrac{1}{Q_\mathrm{u}^2\Delta^2}\right\}}}{a^2\mathrm{M}^4_{\mathrm{N1}} + \dfrac{1}{Q_\mathrm{u}^2\Delta^2}} \tag{2.33}$$

In (2.33), the choice of positive sign and negative sign depends on the value of M_{N1}. If $M_{\mathrm{N1}} > 0$, a positive sign is preferred while the negative sign is assigned for $M_{\mathrm{N1}} < 0$. As noted from (2.26b), the GD between paths 2 and 1 is always positive, whereas the GD of paths 3 and 1 can become negative subjecting to the coupling coefficient and Q_u of the resonators.

2.5.3 Odd-mode analysis

Figure 2.28 depicts the equivalent sub-circuit under odd-mode excitation where normalized termination impedances of source between ports 2 and 1 (r_{So}^{21}) and ports 3 and 1 (r_{So}^{31}) can be expressed as (2.34):

$$r_{So}^{21} = r_{So}^{31} = 0 \tag{2.34}$$

No signal transmission occurs under odd-mode excitation because of short-circuited input port where the normalized isolation resistors are determined as (2.35):

$$r_{iso}^{21} = k^2, \quad r_{iso}^{31} = 1 \tag{2.35}$$

For the ease of mathematical simplicity, the termination impedances of NRN_2 and NRN_1 are presumed same as $r^{21}{}_{iso}$ and $r^{31}{}_{iso}$, which can change to load impedance at port 2 and 3 by M_{N2L2} and NGD circuit, respectively. Similarly, for perfect isolation ($S_{23} = 0$) between output ports, the value of the normalized isolation resistor (r_{iso}) can be determined as (2.36):

$$r_{iso} = 1 + k^2 \tag{2.36}$$

To demonstrate the validation of the analytical equations of the power divider, the computed results are depicted in Figure 2.29. Table 2.2 shows the computed coupling matrix for GD of −1 ns at 2.14 GHz through ports 3 and 1. As can be noted from the computed results, the magnitude of transmission coefficients is affected with k^2. In addition, since the NGD circuit can be obtained under insertion loss condition, the magnitude of S_{31} in the NGD power divider is decreased than the conventional power divider as given in (2.31) and (2.32). However, the GDs of the different transmission paths (τ_{21} and τ_{31}) are almost constant and independent of k^2.

To investigate the effect of Q_u on the transmission coefficients magnitudes and associated GDs, the computed results are depicted in Figure 2.30. In this computation, the circuit parameters are presumed as $f_0 = 2.14$ GHz, $k^2 = 2$, $M_{N1} = 0.70$,

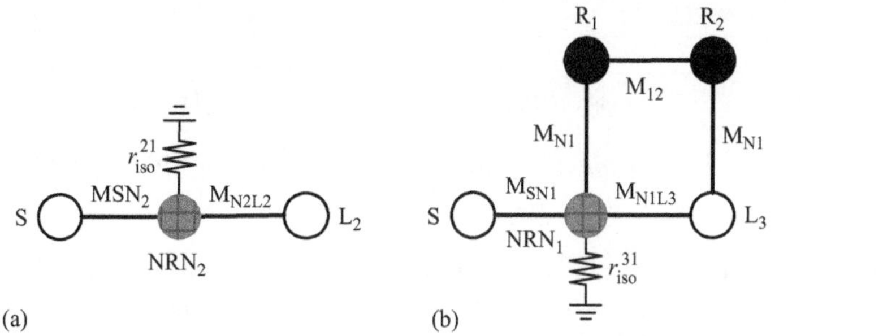

Figure 2.28 Modified odd-mode excitation equivalent coupling diagrams: (a) between ports 2 and 1 and (b) between ports 3 and 1

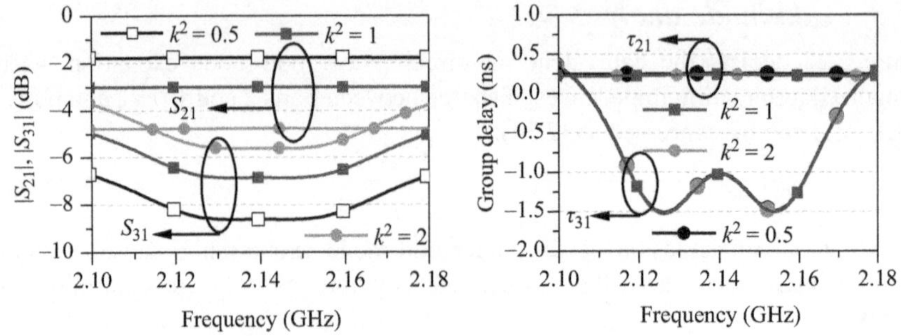

Figure 2.29 Analytically computed magnitude/GD of power divider with different k^2 and $Q_u = 50$, $\Delta = 2\%$, and $f_0 = 2.14$ GHz

Table 2.2 Circuit parameters and S-parameter magnitudes for power divider with NGD = −1 ns with $Q_u = 50$ and $\Delta = 2\%$

k^2	M_{SN1}	M_{SN2}	$M_{N21.2}$	M_{N1}	M_{NIL3}	a	r_{iso}	S_{31c}	S_{NGD}	S_{31p}
0.5	1.1547	0.5773	1.4142	0.7	−0.734	0.86	1.5	−4.77	−3.84	−8.61
1	0.7070	0.7070	1	0.7	−0.734	0.86	2	−3.01	−3.84	−6.85
2	0.4082	0.8165	1	0.7	−0.734	0.86	3	−1.76	−3.84	−5.60

S-parameter magnitudes at f_0
S_{31c}: conventional power divider, S_{31p}: proposed power divider,
$S_{21c} = S_{21p}$

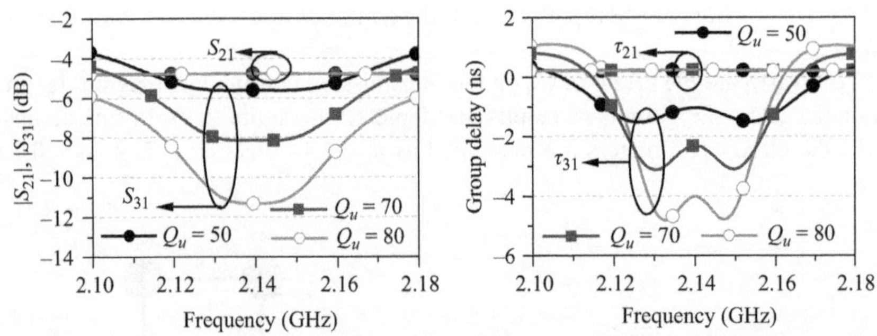

Figure 2.30 The calculated magnitude and GD responses of transmission path 3 and 1 with $k^2 = 2$ and different Q_u

and $M_{12} = aM_{N1}^2$. As can be seen from this figure, the magnitude and GD of S_{31} will decrease if Q_u alters from 50 to 86. Similarly, the NGD BW (which is defined as the BW of GD < 0) slightly will reduce if NGD increases. This necessitates a trade-off among NGD, the insertion loss, and the NGD BW. In contrary, the magnitude and GD of S_{21} are nearly invariable for every Q_u.

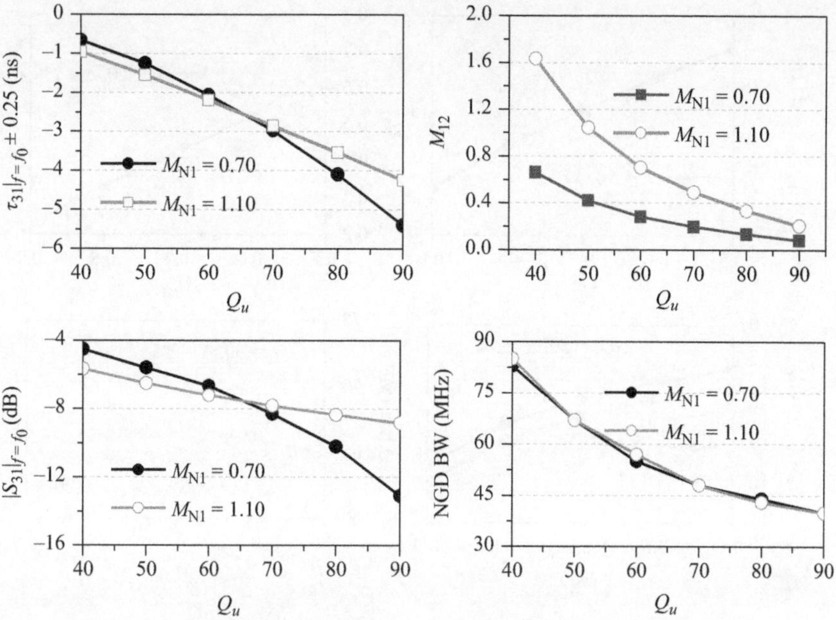

Figure 2.31 GD, magnitude, M_{12} and NGD BW variations as a function of Q_u with $k^2 = 2$ and $\Delta = 2\%$

For thoroughly exploring the effect of Q_u, the computed magnitude, GD of S_{31}, the NGD BW, and the inter-resonator couplings are depicted in Figure 2.31 for distinct Q_u. The value of M_{N1} is presumed as 0.70 and 1.10 while the GD of S_{31} was retained as $\tau_{31}|_{f0} \pm 0.25$ ns in this computation. As can be noted from these graphs, if the value Q_u is adjusted from 40 to 90, NGD of S_{31} increases, the magnitude of S_{31} and the inter-resonator coupling coefficient will decrease. However, the NGD BW will reduce if the GD becomes more and more negative. It can be also noted that the NGD BW is nearly invariable in both cases ($M_{N1} < 1$ and $M_{N1} > 1$).

Likewise, for thoroughly scrutinizing the effect of the coupling coefficient (M_{N1}) between NRN_1 and the resonator, the computed magnitude and GD of S_{31} at $f_0 = 2.14$ GHz and the coupling coefficient (M_{12}) between the resonators and the NGD BW are depicted in Figure 2.32. As observed from these graphs, the NGD BW is enhanced from 65 to 74 MHz, when M_{N1} is increased toward a higher value. However, NGD BW remains constant when $M_{N1} > 1$. Similarly, when GD is moved toward higher negative value, M_{12} is reduced in case of $M_{N1} < 1$ and enlarged in the case of $M_{N1} > 1$. However, the $|S_{31}|$ at f_0 in case $M_{N1} < 1$ is smaller than in case of $M_{N1} > 1$. It can be also noted from Figures 2.29 and 2.30 that the NGD and BW are invariable with k^2 and fully controlled by parameters of the NGD circuit such as M_{N1} and Q_u.

Based on this analysis, the design method of NGD power divider is summarized with the flowchart shown in Figure 2.33.

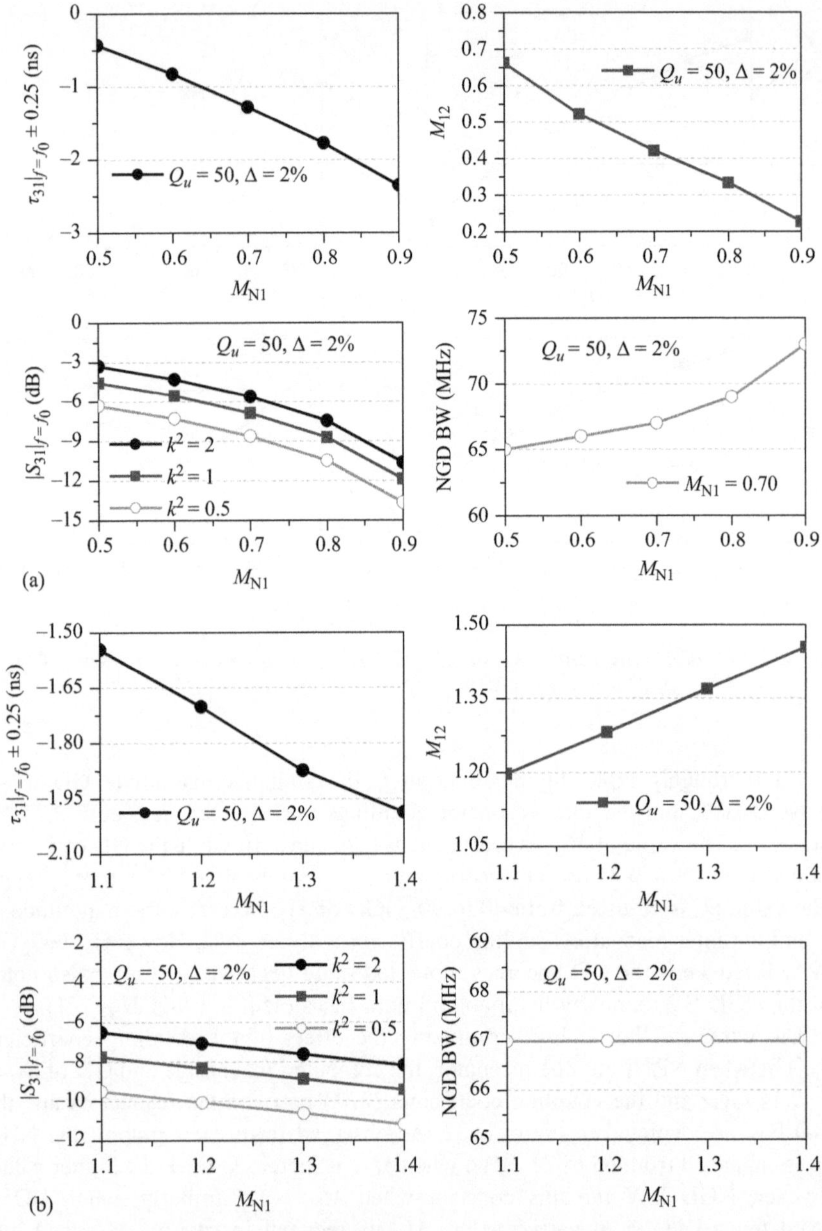

Figure 2.32 Calculated magnitude/GD and NGD BW variations according to M_{N1} (a) $M_{N1} < 1$ and (b) $M_{N1} > 1$

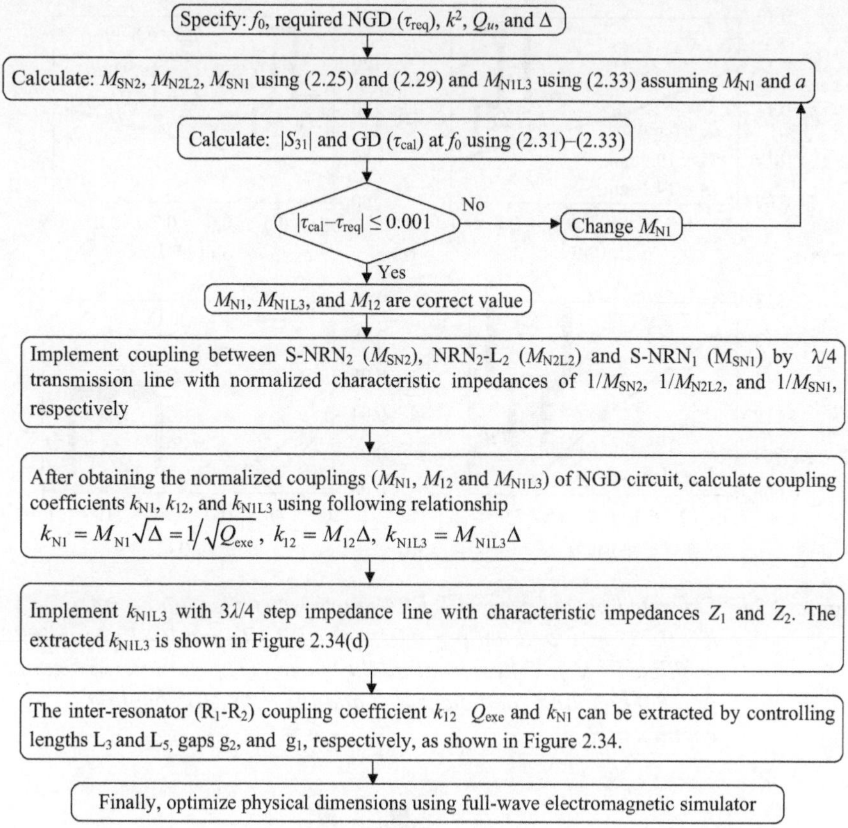

Figure 2.33 Design flowchart of NGD power divider using coupling matrix

For experimental validation, the NGD power divider with $k^2 = 1$ and GD $= -1$ ns is designed and measured at $f_0 = 2.14$ GHz using FR4 epoxy substrate with a dielectric constant (ε_r) of 4.48, thickness (h) of 0.787 mm, and loss tangent of 0.02. The extracted coupling parameters for the above specification are expressed as $M_{SN2} = 0.7071$, $M_{N2L2} = 1$, $M_{SN1} = 0.7071$, $M_{N1} = 1.04$, $M_{12} = 1.1897$, $M_{N1L3} = 1.4269$ $Q_u = 50$, and $\Delta = 2\%$. The finite Q_u-resonators (R$_1$–R$_2$) are implemented using $\lambda/2$ resonators. The extracted coupling coefficients are depicted in Figure 2.34.

Figure 2.35 depicts the physical dimensions of the fabricated NGD power divider. The simulated and experimented results of NGD power divider are demonstrated in Figure 2.36. From the experiment, the S-parameters are ascertained as $|S_{21}| = -3.21$ dB, $|S_{31}| = -8.701$ dB, $|S_{11}| = -19.92$ dB, $|S_{22}| = -18.26$ dB, $|S_{33}| = -17.79$ dB, and $|S_{23}| = -33.97$ dB at $f_0 = 2.14$ GHz. Likewise, the GDs through different

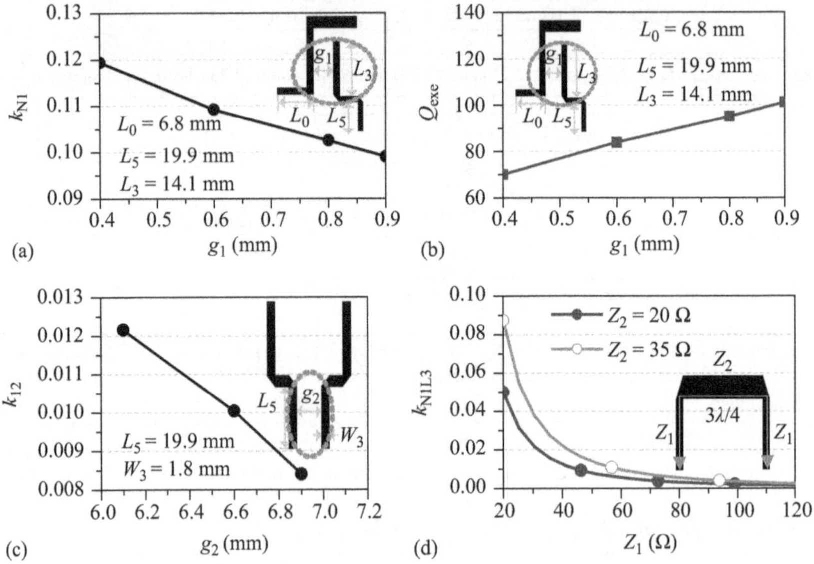

Figure 2.34 Extracted parameters with FR4-epoxy substrate with $\varepsilon_r = 4.4$, $h = 0.787$ mm, and tan $\delta = 0.02$: (a) NRR$_1$-R$_1$ and R$_2$–L$_3$ coupling coefficients k_{N1}; (b) external quality factor Q_{exe}; (c) inter-resonator (R$_1$–R$_2$) coupling coefficient k_{12}; and (d) NRN$_1$-L$_3$ coupling coefficient k_{N1L3}

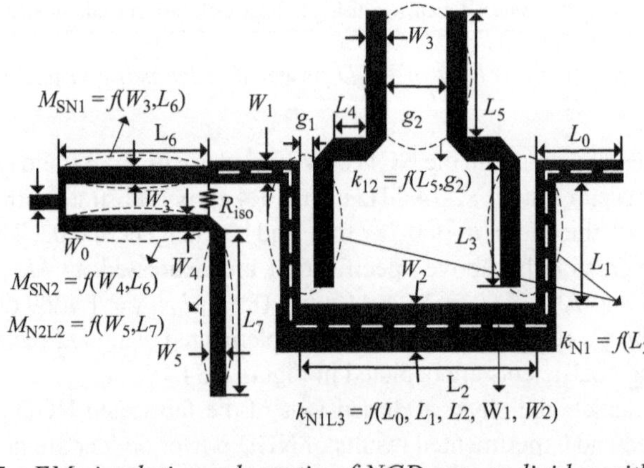

Figure 2.35 EM-simulation schematic of NGD power divider with $k^2 = 1$. Physical dimensions: $L_0 = 15.8$, $L_1 = 12.2$, $L_2 = 18.1$, $L_3 = 14$, $L_4 = 1.5$, $L_5 = 20$, $L_6 = 19$, $L_7 = 19$, $W_0 = 1.5$, $W_1 = 1.8$, $W_2 = 3.70$, $W_3 = 0.76$, $W_4 = 0.76$, $W_5 = 1.5$, $g_1 = 0.5$, and $g_2 = 6.9$ (unit: mm)

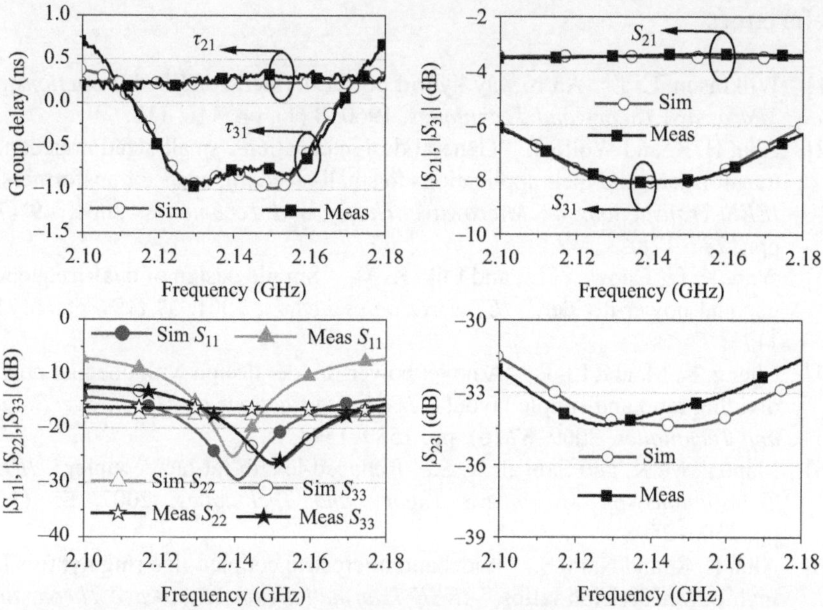

Figure 2.36 Simulated and measured results of NGD power divider with $k^2 = 1$

Table 2.3 S-parameters comparison of the conventional and the NGD power dividers

k^2	Conventional PD*		Measured results of NGD PD		
	S_{21c} (dB)	S_{31c} (dB)	S_{21p} (dB)	S_{31p} (dB)	S_{NGD} (dB)
0.6	−2.04	−4.26	−2.26	−9.96	−5.70
1	−3.01	−3.01	−3.21	−8.70	−5.69
2	−4.77	−1.76	−4.98	−7.48	−5.72

PD : Power divider
* : Ideal calculated results for conventional PD

transmission paths at $f_0 = 2.14$ GHz are determined as $\tau_{21} = 0.3079$ ns and $\tau_{31} = -0.8281$ ns. Similarly, the measured NGD BW is estimated as 60 MHz.

In Table 2.3, the magnitudes of transmission coefficients of NGD power divider are compared with the conventional ones. As noted from this comparison table, the insertion losses between ports 3 and 1 of the NGD power divider are higher than the conventional because of the additional insertion of NGD circuit in this path.

References

[1] Wilkinson, E. J., "An N-way hybrid power divider," *IEEE Transactions on Microwave Theory and Techniques*, 1960, **8** (1), pp. 116–118.

[2] Ahn, H. R. and Wolff, I., "General design equations, small-sized impedance transformers, and their applications to small-sized impedance transformers," *IEEE Transactions on Microwave Theory and Techniques*, 2001, **49** (7), pp. 1277–1288.

[3] Wan, K. L., Chow, Y. L., and Luk, K. M., "Simple design of dual-frequency unequal power-divider," *IET Electronics Letters*, 2001, **37** (19), pp. 1171–1173.

[4] Cheng, K. M. and Li, P., "A novel power-divider design with unequal power dividing ratio and simple layout," *IEEE Transactions on Microwave Theory and Techniques*, 2009, **57** (6), pp. 1589–1594.

[5] Mandal, M. K. and Sanyal, X. S., "Reduced-length rat-race coupler," *IEEE Transactions on Microwave Theory and Techniques*, 2007, **55** (12), pp. 2593–2598.

[6] Ahn, H. R. and Nam, S., "Wideband microstrip coupled-line ring hybrids for high power division ratios," *IEEE Transactions on Microwave Theory and Techniques*, 2013, **61** (5), pp. 1768–1780.

[7] Ahn, H. R., "Compact CVT-/CCT-unequal power dividers for high-power division ratios and design methods for arbitrary phase differences," *IEEE Transactions on Microwave Theory and Techniques*, 2014, **62** (12), pp. 2954–2964.

[8] Ahn, H. R., Kim, Y., and Kim, B., "Planar 10:1 unequal three-port dividers using general design equations," *IET Electronics Letters*, 2012, **48** (15), pp. 934–935.

[9] Li, B., Wu, X., and Wu, W., "A 10:1 unequal Wilkinson power divider using coupled lines with two shorts," *IEEE Microwave Wireless Component Letters*, 2009, **19** (12), pp. 789–791.

[10] Chik, M. J. and Cheng, K. K. M., "Group delay investigation of rat-race coupler design with tunable power dividing ratio," *IEEE Microwave Wireless Component Letters*, 2014, **24** (5), pp. 324–326.

[11] Yi, J., Yang, Y., Park, M., Kang, W., and Kim, B., "Analog predistortion linearizer for high power RF amplifiers," *IEEE Transactions on Microwave Theory and Techniques*, 2000, **48** (12), pp. 2709–2913.

[12] Kim, Y., Chang, I. S., and Jeong, Y., "An analog predistortion linearizer design," *Microwave Journal*, 2005, **48** (2), pp. 118–226.

[13] Chaudhary, G. and Jeong, Y., "A design of power divider with negative group delay characteristics," *IEEE Microwave Wireless Component Letters*, 2015, **25** (6), pp. 394–396.

[14] Chaudhary, G., Jeong, S., Kim, P., and Jeong, Y., "Coupled line power divider with multiple-pole negative group delay characteristics," *Progress in Electromagnetic Symposium*, 2016, pp. 2033.

[15] Chaudhary, G., Han, J., Jeong, Y., and Lim, J., "Arbitrary power division ratio power divider with prescribed wideband negative group delay," *European Microwave Conference*, 2017, pp. 13–16.

[16] Chaudhary, G. and Jeong, Y., "Arbitrary power division ratio rat-race coupler with negative group delay characteristics," *IEEE Microwave Wireless Component Letters*, 2016, **26** (8), pp. 565–567.

[17] Chaudhary, G., Kim, P., Jeong, J., and Jeong, Y., "A power divider with positive and negative group delay characteristics," *URSI Asia-pacific Radio Science Conference*, 2016, pp. 1202–1204.

[18] Naglich, E. J., Lee, J., Peroulis, D., and Chappell, W. J., "Switchless tunable bandstop to all-pass reconfigurable filter," *IEEE Transactions on Microwave Theory and Techniques*, 2012, **60** (5), pp. 1258–1265.

[19] Lee, J., Naglich, E. J., Sigmarsson, H. H., Peroulis, D., and Chappell, W. J., "New bandstop filter circuit topology and its application to design of a bandstop to bandpass switchable filter," *IEEE Transactions on Microwave Theory and Techniques*, 2013, **61** (3), pp. 1114–1123.

[20] Chen, C. F., Huang, T. Y., Shen, T. M., and Wu, R. B., "Design of miniaturized filter power dividers for system-in-a package," *IEEE Transactions on Component, Packaging and Manufacturing Technology*, 2013, **3** (10), pp. 1663–1672.

[21] Wu, Y. and Liu, Y., "A unequal coupled-line Wilkinson power divider for arbitrary terminated impedances," *Progress in Electromagnetics Research*, 2011, **117**, pp. 181–194.

[22] Chaudhary, G., Kim, P., Jeong, J., and Jeong, Y., "A design of negative group delay power divider: Coupling matrix approach with finite unloaded-Q_u resonators," *IEEE International Microwave Symposium*, 2016, pp. 1–3.

[23] Chaudhary, G. and Jeong, Y., "A finite unloaded quality-factor resonators based negative group delay circuit and its application to design power divider," *Microwave and Optical Technology Letters*, 2016, **58** (12), pp. 2918–2921.

[24] Chaudhary, G., and Jeong, Y., "Negative group delay phenomenon analysis in power divider: coupling matrix approach," *IEEE Transactions on Components, Packaging and Manufacturing Technology*, 2017, **7** (9), pp. 1543–1551.

Chapter 3

Delay-induced negative group delay

Henning U. Voss[1]

Abstract

Real-time prediction of signals is one of the most fundamental tasks encountered in control problems, in machine learning and artificial intelligence applications, as well as by living systems. It is known that the phenomenon of negative group delay (NGD) can be utilized to perform real-time prediction of smooth input signals. In the present contribution, it is first shown that a particular system property that can cause an NGD is time-delayed feedback. Based on this insight, very simple linear signal predictors that use past predicted, or feedback, values rather than past signal values for prediction are presented. These *delay-induced negative group delay* (*DINGD*) predictors exist both for discrete- and continuous-time signals and afford real-time prediction of complex signals without the need for a specific signal model. The theory of DINGD predictors is outlined and demonstrated on various types of broadband input data simulations. As an example for a technical application, an algorithm is presented that predicts, in real time, smooth voluntary manual movements of a computer mouse or stylus. In general, manufactured or natural systems utilizing the DINGD predictor would not require a memory of input signal values but only of already predicted, internalized states. Along this line, there is evidence for the significance of DINGD prediction in biological systems, including the human motor system. Finally, the DINGD predictor as a special type of prediction by anticipatory coupling is discussed.

Keywords

Prediction, forecasting, negative group delay, human motor control, man-machine interfacing.

3.1 Introduction

Signal prediction is one of the most basic tasks of signal analysis. In the majority of cases, signal prediction relies on a prior analysis of the signal to be predicted, and a subsequent model fit. The fitting procedure can be off-line, such as in some neural

[1]Weill Cornell Medical College, Citigroup Biomedical Imaging Center, USA

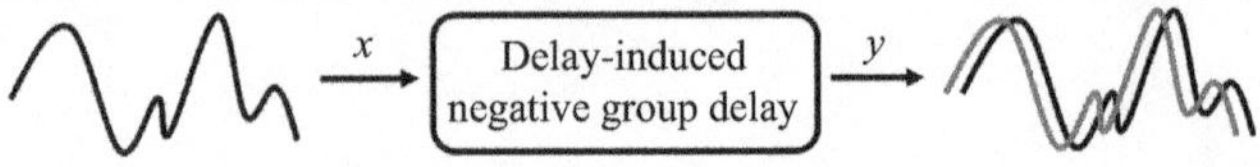

Figure 3.1 Schematics of real-time signal prediction with a DINGD predictor

network approaches, which use a learning and a test dataset, or in real time, such as in adaptive filtering. It would be highly advantageous for many applications if the modeling step could be simplified considerably or even omitted as much as possible.

In general, negative group delay (NGD) of an input/output system causes the output signal to anticipate or predict characteristics of the input signal. This allows for the real-time prediction of simple, smooth signals, independent of the specific system that generated the signal. For example, in the pioneering work of Mitchell and Chiao [1], it is shown how the NGD of a simple electronic circuit can predict the peak of a Gaussian waveform. Therefore, some systems with NGD can be viewed as real-time signal predictors.

Only recently it has been shown that NGD can also occur in linear input/output systems with time-delayed feedback, or mathematically, non-autonomous delay-differential equations [2]. This type of NGD is called *delay-induced negative group delay*, or *DINGD*. The group delay (GD) of a DINGD filter defines the prediction horizon, i.e., the time the system output $y(t)$, here called a predictor, predicts the system input $x(t)$ ahead of time. This is schematically shown in Figure 3.1. Specifically, a DINGD predictor does not need to model the signal generating system. Rather, its coefficients only depend on the desired prediction horizon and basic spectral properties of the input signal. Generally, DINGD prediction works for band-limited signals.

In the following, the continuous-time DINGD predictor is reviewed, before this concept will be extended to discrete-time DINGD predictors. As an example, it will be shown how DINGD prediction can be used to predict smooth voluntary manual movements of a cursor on a computer screen in real time. Time delays are a typical component of biological neuronal networks, and it is reasonable to hypothesize a possible relevance of DINGD prediction in neuronal computations [3] involved, for example, in human motor control [4]. These applications, as well as how prediction by DINGD fits into the more general concept of anticipatory coupling of a system to other systems or signals, will be discussed at the end of this chapter.

3.2 The DINGD predictor

DINGD predictors exist for discrete- and continuous-time signals. In the following, both cases will be described and applied to artificial and real-world data sets.

3.2.1 Continuous-time signal prediction with DINGD

The continuous-time DINGD predictor is modeled as a non-autonomous linear delay-differential equation in the variable y with a single input parameter x,

$$\dot{y}(t) = -a\,y(t) + b\,x(t) - c\,y(t-\tau). \tag{3.1}$$

Time t is a continuous variable, and the time delay τ is assumed to be larger than zero. The parameters are the relaxation coefficient $a \geq 0$, the input scaling b, and the feedback gain $c \geq 0$. The input signal $x(t)$, which is the signal to be predicted by the output $y(t)$, is assumed to be bounded with finite variance and zero mean, and to be generated by a stationary, not necessarily known, process. The DINGD predictor output is supposed to predict the signal in the following sense:

$$y(t) \approx x(t + \delta), \tag{3.2}$$

where $\delta > 0$ is the *prediction horizon*, to be properly defined below. In other words, the output of the filter is a real-time estimate of future signal values. In addition, it is assumed that the parameters are chosen in a way that this model is stable, i.e., it only has bounded solutions. The specification of stability conditions will be postponed until later. Briefly, it is assumed that for $b = 0$, the parameters a and c are chosen such that the fixed point $y = 0$ does not become unstable.

Autonomous linear delay-differential equations have been studied for quite a while, in particular, with respect to their stability and oscillatory solutions [5–11]. Here, we are concerned with the *nonautonomous* case (3.1). Only recently it has been shown that (3.1) can generate an NGD between input and output signals, which can be used for predicting the input signal [2]. This will be reviewed below. In contrast to the well-studied case of conventional NGD filters without time delay [1,12–14], the GD in (3.1) is caused by the delayed-feedback term; it is a DINGD. It can be positive or negative, depending on the model parameters.

There are immediate applications of this model. For example, it is evident that this model constitutes nothing but a leaky integrator with time-delayed feedback, which can be implemented easily by electronic circuits [15,16]. In the neurosciences, this equation models a building block of a neuron [3], the leaky integrator. The leaky integrator consists of a trans-membrane capacitance, the integrator, in parallel to the membrane resistance, the leak [17,18]. Add recurrent inhibition, modeled as a linear time-delayed feedback term with negative gain [19], and (3.1) arises.

Prediction performance of the DINGD predictor will be quantified *a-posteriori* by the cross-correlation function (CCF) between filter output and time-shifted signal values

$$C_{xy}(\tau) = \frac{E\left[\left(x_\tau - \mu_{x_\tau}\right)\left(y - \mu_y\right)\right]}{\sigma_{x_\tau}\sigma_y}, \tag{3.3}$$

where E, μ, and σ denote the expectation value, mean, and standard deviation, respectively, and $x_\tau(t)$ is defined as $x(t - \tau)$. For the sake of notational simplicity, here and in the following the same symbols are used for random variables and actual samples of them, and for parameters and actual estimates of them. The prediction horizon δ is properly defined as that argument that maximizes the CCF: $\delta = \arg\max C_{xy}(-\tau)(\tau > 0)$. Should the CCF have a global maximum of its magnitude for positive delays, there is no prediction but lag. This convention will become useful later on when the argument of the maximum of the CCF is identified with the GD of the filter, which is negative when the filter is predictive. For real-time prediction, no future time points with respect to the reference time t can be used. Whereas the DINGD predictor has been written already in causal terms, it remains to rewrite the prediction equation (3.2). Due to signal stationarity, one can perform a time shift on both sides of (3.2) to obtain the equivalent causal prediction expression

$$y(t-\delta) \approx x(t). \tag{3.4}$$

In this causal expression, the filter output is predicting or anticipating the filter input, too, but no reference to future time points has been made.

Equation (3.1) is linear and thus can be described by its frequency response function [2,20,21]

$$H(\omega) = \frac{b}{a + \mathrm{i}\omega + c\mathrm{e}^{-\mathrm{i}\omega\tau}} = \frac{b}{\beta}(\beta_1 + \mathrm{i}\beta_2), \tag{3.5}$$

with

$$\beta_1 = a + c\cos(\omega\tau), \quad \beta_2 = c\sin(\omega\tau) - \omega, \quad \beta = \beta_1^2 + \beta_2^2. \tag{3.6}$$

The imaginary unit i is defined as usual as the solution of $i^2 = -1$. This frequency response function defines the steady-state input/output relationship between x and y in Fourier space as

$$Y(\omega) = H(\omega)X(\omega), \tag{3.7}$$

where $\omega = 2\pi f$ and f is the frequency, and

$$x(t) = \int X(\omega)\mathrm{e}^{\mathrm{i}\omega t}d\omega, \quad y(t) = \int Y(\omega)\mathrm{e}^{\mathrm{i}\omega t}d\omega. \tag{3.8}$$

If the frequency response function is written as

$$H(\omega) = G(\omega)\mathrm{e}^{\mathrm{i}\Phi(\omega)}, \tag{3.9}$$

its gain is

$$G(\omega) = \frac{|b|}{\sqrt{\beta}}, \tag{3.10}$$

and its phase

$$\Phi(\omega) = \arg(\beta_1 + \mathrm{i}\beta_2). \tag{3.11}$$

The expression for the gain can be used to compute the magnitude of the input scaling parameter b. In order to achieve a correctly scaled prediction, the goal would be to set the gain to approximately unity for the range of signal frequency components to be predicted. The GD is frequency-dependent and given by

$$\delta(\omega) = -\frac{d\Phi(\omega)}{d\omega}. \tag{3.12}$$

Specifically, the GD of (3.1) follows as

$$\delta(\omega) = \frac{c\cos(\omega\tau) - c^2\tau - a(c\tau\cos(\omega\tau) - 1) + c\tau\omega\sin(\omega\tau)}{\beta}. \tag{3.13}$$

The latter expression is properly defined outside of the poles of the frequency response function only. This GD can be positive, zero, or negative. NGD in general means a group advance, or real-time prediction of the input signal [1,12]. In order to characterize the GD for low-frequency signal components, δ_0, the GD $\delta(\omega)$ is expanded for small ω. It turns out that there are no linear terms in ω. Neglecting quadratic and higher-order terms in the counter and denominator of the expansion, it follows

$$\delta_0 = \delta(\omega \approx 0) = \frac{1 - c\tau}{a + c}. \tag{3.14}$$

This expression for low-frequency input signal components, i.e., for small ω, has four important consequences:

1. Equation (3.1) allows both for positive GD and NGD. The GD is positive for $c\tau < 1$. This means that the output signal will lag the input. For $c\tau > 1$, the GD δ_0 is negative, i.e., the output signal predicts the input signal.
2. The GD δ_0 is independent of ω. This is desirable for limiting distortions in the predicted signal [12,14].
3. For unconstrained feedback delay times, maximization of the NGD requires $a = c$. This follows from the fact that the delay-independent stability of (3.1) with $b = 0$ requires $a > c$ [22,23]. Therefore, the maximum prediction horizon is achieved next to the edge of the parameter regime that guarantees stability for all possible, including arbitrarily large, feedback delays.
4. The maximum NGD, or prediction horizon, is achieved for $a = c$ in the limit of infinite a. It is

$$\delta_{0,\max} = -\tau/2. \tag{3.15}$$

This value provides a universal upper bound for the prediction horizon of the stable continuous-time DINGD predictor, (3.1). The predictor (3.1) with $a < c$ including the conservative case with $a = 0$ still can have an NGD, but its stability depends on the feedback delay. A more detailed analysis reveals that the limit obtained in (3.15) cannot be overcome.

These theoretical results are demonstrated in Figure 3.2, along with an application. Signal prediction was performed by numerically solving (3.1), i.e., its solution $y(t)$ was used as a predictor for the signal $x(t)$. Equation (3.1) was solved with a Runge–Kutta scheme with $\tau = 1.0$ s and $y(t) = 0$ for $t \in [-\tau, 0]$. The other parameters were $a = 100$, $b = c = 90$. All computations in this chapter were performed with MATLAB® R2017a (The MathWorks, Inc., Natick, MA, USA).

The real-world signal used in this example consists of 45 s of the horizontal coordinate of a mouse pointer sampled with a MATLAB script at an average time

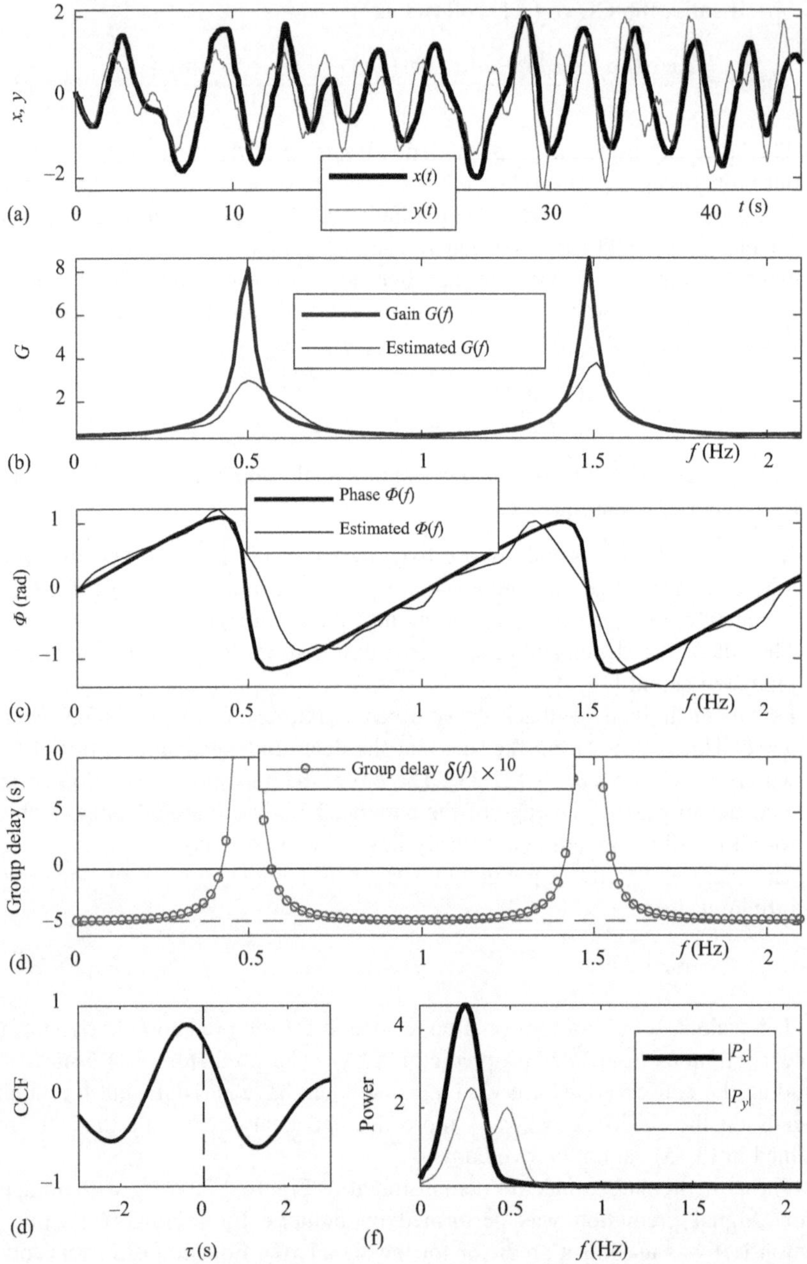

Figure 3.2 Example application: DINGD prediction of voluntary manual computer mouse movements. All graphs are linear (nonlogarithmic) plots. Please refer to the text for details

interval of 14.7 ms. The mouse pointer was manually moved smoothly on a personal computer screen in two directions by the author, arbitrarily describing circles, back-and-forth movements, and figure-eight shapes. There was no conscious systematics behind the movements. In this sense, the signal can be considered as a pseudo-random signal. The application to simulated smoothed Gaussian white noise signals has been described previously, with similar results as presented here [2].

Figure 3.2(a) shows the input signal in bold and the output signal in normal line thickness (please refer to legend). On average, the output predicts the input. This is verified in Figure 3.2(e), which shows the CCF between input and output. Its maximum is attained for negative values of the delay; specifically, at the maximum, CCF $(-328 \text{ ms}) = 0.8$. On the other hand, for the given parameters a, b, c, and τ, the theoretical zero-frequency limit GD is $\delta_0 = -468$ ms, from (3.14). The discrepancy between these two values has at least three contributions: (i) transients at the beginning, (ii) the estimation error of the CCF due to the finite dataset sampled, and (iii) the fact that (3.14) is the zero-frequency limit, whereas the signal has significant frequency components of up to 0.5 Hz (Figure 3.2(f), bold). These differences cause signal distortions that affect the CCF in two ways: the argument of the maximum CCF value differs from δ_0, and the maximum itself becomes less than 1.0. Theoretical results pertaining to the exact relationship between the CCF and the frequency content of a signal are provided in Ref. [24]. Of note, the theoretically expected GD is smaller than the maximally possible GD of 500 ms, given by (3.15), due to the finite parameters a, b, and c.

Figure 3.2(b) shows the analytic frequency response function gain G in bold blue. Figure 3.2(c) shows the corresponding phase Φ in bold, derived from (3.10) and (3.11). The GD given by (3.13) is shown as connected circles in Figure 3.2(d). The dashed line indicates the theoretically expected GD for zero frequency as per (3.14). The GD is negative for the interval from 0 to 0.42 Hz. Then, around the first resonance at 0.5 Hz, it turns positive and starts alternating between positive and negative values. Note that only the very low frequency range is shown here, as the Nyquist frequency would be 34 Hz. If the whole range were plotted, 34 resonances would be visible. Figure 3.2(b) and (c) also shows, as thin lines, numerical estimates of the gain and phase obtained from a Fourier analysis of the input and output signals.

In order for the prediction via DINGD to work, as in this example, it is essential that the band-limited signal has most of its power restricted to the first NGD interval, which here extends from 0 to 0.42 Hz. The signal, however, has significant power of up to 0.5 Hz, reaching the resonance of the chosen DINGD predictor. This is the main reason for the distortions in the predicted signal. Those higher frequency components are amplified, as visible in Figure 3.2(f).

This example application constitutes an off-line prediction of cursor movements that have been recorded beforehand. Whereas the continuous-time DINGD predictor might have its merits in certain fields, for most implementations a discrete-time version would be advantageous. Therefore, in the next section, discrete-time DING predictors will be discussed. After having gained a basic understanding of how prediction via DINGD works, the discrete-time case will enable us also to gain further insights into how band-limited high-frequency signals can be predicted, and will enable us to provide an even simpler algorithm for the real-time prediction of mouse pointer movements.

3.2.2 Discrete-time time signal prediction with DINGD

Rather than treating only the case of a single feedback delay as in the continuous-time case, for the discrete time case, multiple feedback delays will be considered. This will provide us with more flexibility of filter designs. For the sake of simplicity, the discrete sampling time Δt will be set to unity and omitted in the following. Given a stationary, zero-mean, variance- and band-limited signal $x(t)(t = ..., -1, 0, 1, 2, ...)$, the DINGD filter is given by

$$y(t) = bx(t) - \sum_{k=0}^{\tau-1} c_k y(t - (\tau - k)). \tag{3.16}$$

The filter order τ equals the largest delay used in this filter. Here, the frequency is defined up to the Nyquist frequency, i.e., $\omega \in [0, \pi]$ for the angular frequency and $f \in [0, 1/2]$ for the frequency $f = \omega/2\pi$. The filter coefficients b and c_k need to be chosen to make the filter predictive, equivalently to the continuous-time case before. There are probably many possible configurations for the coefficients, and here only few possibilities will be worked out.

The DINGD filter will be analyzed by its frequency response function

$$H(\omega) = \frac{b}{1 + \sum_{k=0}^{\tau-1} c_k e^{-i(\tau-k)\omega}}. \tag{3.17}$$

The stability of the filter is determined by the location of the roots of the denominator with respect to the complex unit circle [25,26]. Due to the infeasibility to compute all the roots for general sets of coefficients, here a less general but sufficient condition for stability is used. The DINGD filter (3.16) is stable when the real part of the denominator of the frequency response function (3.17) is strictly positive [27–29]. This can be simplified to the following condition on the *stability spectrum S(ω)*:

$$S(\omega) = 1 + \sum_{k=0}^{\tau-1} c_k \cos((\tau - k)\omega) > 0 \,\forall\, \omega. \tag{3.18}$$

Again, this condition is a sufficient but not necessary condition for stability. The validity of this stability criterion is plausible for a filter of order $\tau = 1$. In this case, the filter is $y(t) = bx(t) - cy(t - 1)$, and it is well known that the filter is stable for any $|c| < 1$. From (3.18) follows $S(\omega) = 1 + c\cos(\omega) > 0 \,\forall\, \omega$, and since the cosine is limited to the interval $[-1,1]$, the same condition $|c| < 1$ follows. This stability criterion is easy to verify numerically for any filter order and set of coefficients.

3.2.3 The simplest DINGD predictor

The simplest DINGD predictor is defined as

$$y(t) = bx(t) - cy(t-\tau). \tag{3.19}$$

As in this expression, unnecessary indexes on coefficients will be dropped in the following. The frequency response function of the discrete-time DINGD predictor (3.19) is

$$H(\omega) = \frac{b}{1+ce^{-i\omega\tau}} = \frac{b}{\beta(\omega)}(1 + c\cos(\omega\tau) + i\,c\sin(\omega\tau)) \tag{3.20}$$

with

$$\beta(\omega) = 1 + 2\,c\cos(\omega\tau) + c^2. \tag{3.21}$$

Its gain is

$$G(\omega) = \frac{|b|}{\sqrt{\beta(\omega)}}. \tag{3.22}$$

This expression can be used to compute the magnitude of the input scaling parameter b, again by setting the gain to approximately unity for relevant frequencies. The GD follows as

$$\delta(\omega) = -\frac{d\Phi(\omega)}{d\omega} = -\frac{c\tau(c+\cos(\omega\tau))}{\beta(\omega)}, \tag{3.23}$$

an expression that is properly defined only outside of the poles of the frequency response function.

In order to make predictions one sample step or more ahead, we are seeking to obtain an integer GD $\delta \leq -1$ for $\tau \geq 1$ (causal system) and $0 < c \leq 1$ (to avoid instability). We consider first the zero frequency case ($\omega = 0$) and from there proceed to the GD for general frequencies. In other words, we are now interested again in the prediction of low-frequency band-limited signals, which are defined as signals with significant power only in the baseband $[0, f_0]$. The GD for zero frequencies is

$$\delta(0) = -\frac{c\tau}{1+c}. \tag{3.24}$$

Specific cases for the time delay τ are considered:

- $\tau = 1$: There is no *integer* solution for $\delta(0)$ for any $0 < c \leq 1$. (This is a special case of the elementary NGD IIR filter introduced by Ravelo [30] ($b = 0$ in (3.1) there, $T_s = 1$).)
- $\tau = 2$: For $c = 1$ the GD at zero frequency is $\delta(0) = -1$. Furthermore, the GD equals -1 for all frequencies where it is defined, i.e., outside of the poles of

the frequency response function. The frequency response function has poles defined by the zeros of its denominator. The poles are located at $\omega = (2n + 1)\pi/2$ ($n = 0, 1, \ldots$). The Nyquist frequency is $\omega_N = \pi$, so there is one pole, at $f_1 = \omega_1/2\pi = ¼$. This pole separates two frequency bands with qualitatively different properties: Due to a phase jump at f_1 the second band causes a sign reversal of the output and thus cannot be used together with the first band for prediction.

Figure 3.3(a) shows the frequency response function expressed through its gain and phase, as well as the GD over frequency for this case. It is clear that the DINGD predictor depends, as all NGD-based prediction, on the frequency content of the signal to be predicted. Note that although the NGD is constant over a wide

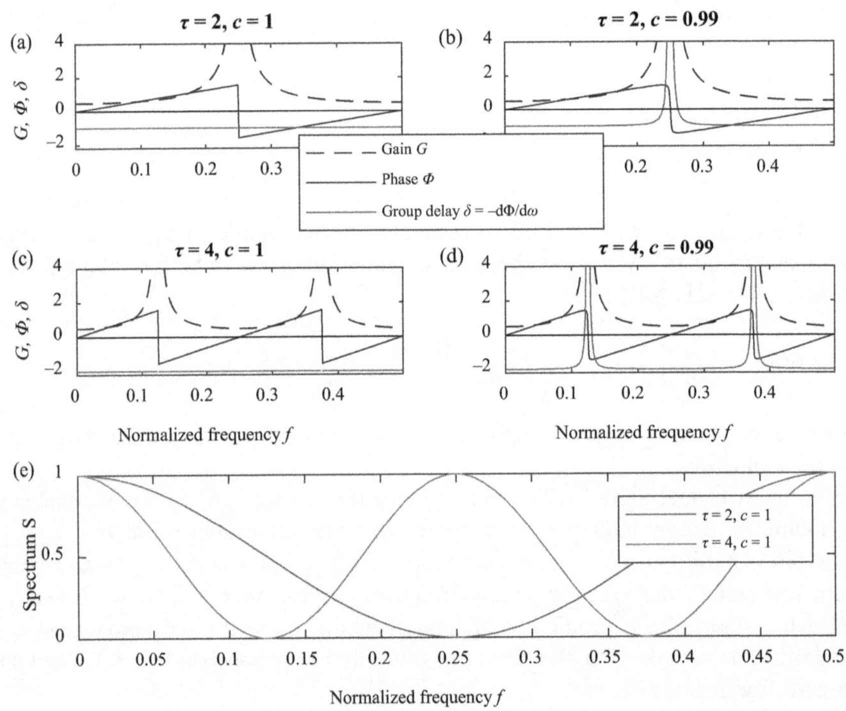

Figure 3.3 Theory I: (a) to (d) Gain and phase of frequency response function (3.20) as well as corresponding GD (3.23) for feedback delays $\tau = 2$ ((a), (b)) and $\tau = 4$ ((c), (d)), for feedback gain $c = 1$ ((a), (c)) and $c = 0.99$ ((b), (d)). The meaning of the graphs is shown in the legend; (e) Stability spectrum for $\tau = 2$ and $\tau = 4$ for $c = 1$. The spectrum touches the zero line, indicating marginal instability. The spectrum for the cases with $c = 0.99$ is visually similar but does not touch the zero line, which indicates stability

frequency range, the filter has poles, thus restricting applicability of the DINGD predictor to baseband signals.

In order to keep prediction stable and also in physical implementations, a value of $c = 1$ might not be feasible and it is more useful to proceed with $c = 1 - \varepsilon$, with $0 < \varepsilon \ll 1$. This has the additional advantage that the poles of the frequency response function are resolving into mere resonances (i.e., the gain is not diverging). Figure 3.3(b) shows the corresponding frequency response function as well as the GD for $c = 0.99$. The instability for the case of $c = 1$ and stability for $c = 0.99$ is verified via the stability spectrum $S(\omega)$ in Figure 3.3(e).

- $\tau = 4$: For $c = 1$, the GD at zero frequency is $\delta(0) = -2$. Furthermore, the GD is -2 for all frequencies for which it is defined. The poles of the frequency response function are located at $\omega = (2n + 1)\pi/4$ ($n = 0, 1, \ldots$). There are two poles, at $f_1 = \omega_1/2\pi = 1/8$ and $f_2 = \omega_2/2\pi = 3/8$. These two poles separate three frequency bands with qualitatively different properties. The first band and the third band can be used combined. The second band causes prediction with reversed sign and cannot be combined with the other two bands. However, this band could be useful, too. To use it, one can set $b < 0$ in (3.19) to compensate for the sign reversal.
- Larger, even τ: It is straightforward to generalize to larger delays as long as $c = 1$. In general, as the delay increases, there will be more poles. This means the frequency bands will be split up into more sections. In addition, in general the prediction horizon will always be half of the predictor feedback delay τ as per (3.24).

3.2.4 Application examples

The performance of the discrete-time DINGD predictor will be illustrated with the help of three examples, Figures 3.4–3.6. The examples are described in the figure captions. In all examples, $c = 1 - \varepsilon$ ($0 < \varepsilon \ll 1$) is used. The time unit is 1.

3.2.5 Cascading

As it has been shown above, one way to increase the NGD is to increase the feedback delay of the DINGD predictor. However, there is another way: feeding the output of the predictor into another predictor. This way, it is possible to increase the NGD without increasing the feedback delay. This has the advantage that one can work with a delay of $\tau = 1$; although the NGD without cascading would be 0.5, as we have seen before, with cascading one can again obtain integer values. Cascading is a special case of DINGD filters with multiple delays, which will be described further below. However, it will be described in detail here because it provides one way of understanding why multiple delays can enhance prediction with DINGD.

Cascading DINGD predictors might increase anticipation times at the cost of prediction accuracy or system stability, which is a general problem of cascading NGD systems [12–14]. However, if the filter is stable for all frequencies, as can be

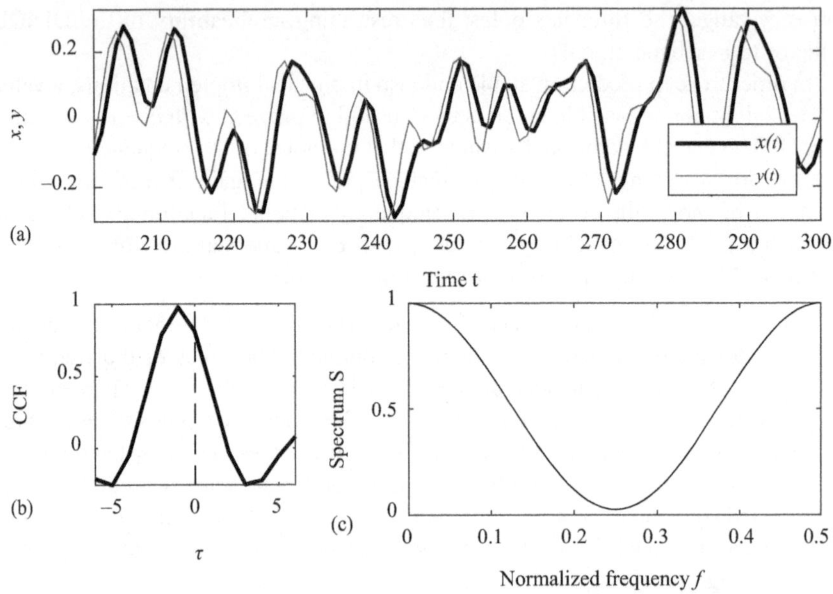

Figure 3.4 Band-limited noise I: (a) A band-limited noise signal x(t) (bold, see legend) and its prediction signal y(t), the output of the predictor (3.19). Out of 1,000 simulated time points, 100 are shown; (b) the CCF peaks at a lag of −1. Its peak value is CCF(−1) = 0.98. This shows that y(t − 1) ≈ x(t), or, equivalently, y(t) ≈ x(t + 1). Therefore, y(t) is a predictor of x(t); (c) the stability spectrum S(ω) indicates stability of this filter; (d) power spectral density function estimates for x(t) (black, thick line) and y(t) (red). One can see that frequency components of Y(ω) that are closer to the resonance, where the gain increases (see (e)), are amplified more, which causes the more jittery appearance of the predictor time series y(t) as compared with the signal time series x(t) in the first row; (e) the frequency response function and GD. Shown are analytic values as well as gain and phase values estimated from the data, see legend. The large estimation errors of gain and phase for frequencies > 0.25 are due to the lack of input signal power for those frequencies. Signal and prediction parameters: The signal x(t) consists of uniformly distributed white noise, low-pass filtered with a seventh-order Butterworth filter with cutoff frequency of 0.15. The parameters of (3.19) are b = 1.50, c = 0.95, and τ = 2

checked with the stability criterion (3.18), feeding the filter output back into the same filter, again results in a stable combined filter. One reason for reduced accuracy is that out-of-band frequency components are amplified in the first filtering step and when fed into the next filter, become even more amplified.

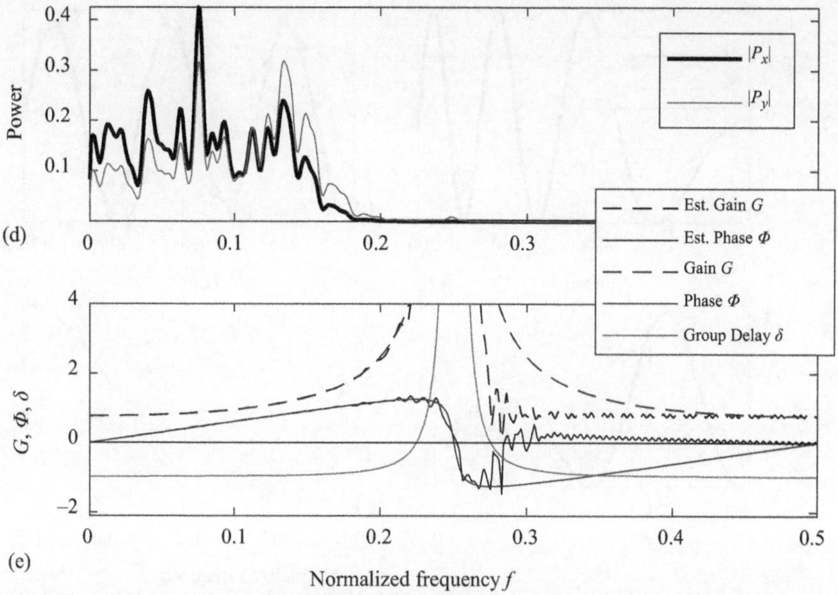

Figure 3.4 (*Continued*)

Cascading $m > 1$ DINGD predictors in the time domain means that the output $y(t) = y_m(t)$ relates to the input $x(t)$ via

$$\begin{aligned} y_1(t) &= bx(t) - cy_1(t-\tau), \\ y_2(t) &= by_1(t) - cy_2(t-\tau), \\ &\dots \\ y_m(t) &= by_{m-1}(t) - cy_m(t-\tau). \end{aligned} \tag{3.25}$$

The frequency response function is

$$H_m(\omega) = H(\omega)^m = |H(\omega)|^m e^{im\Phi(\omega)}. \tag{3.26}$$

Its gain is

$$G_m(\omega) = \frac{|b|^m}{\beta(\omega)^{m/2}}, \tag{3.27}$$

and its GD

$$\delta(\omega) = -\frac{d\Phi_m(\omega)}{d\omega} = m\delta(\omega), \tag{3.28}$$

defined wherever $\delta(\omega)$ is defined.

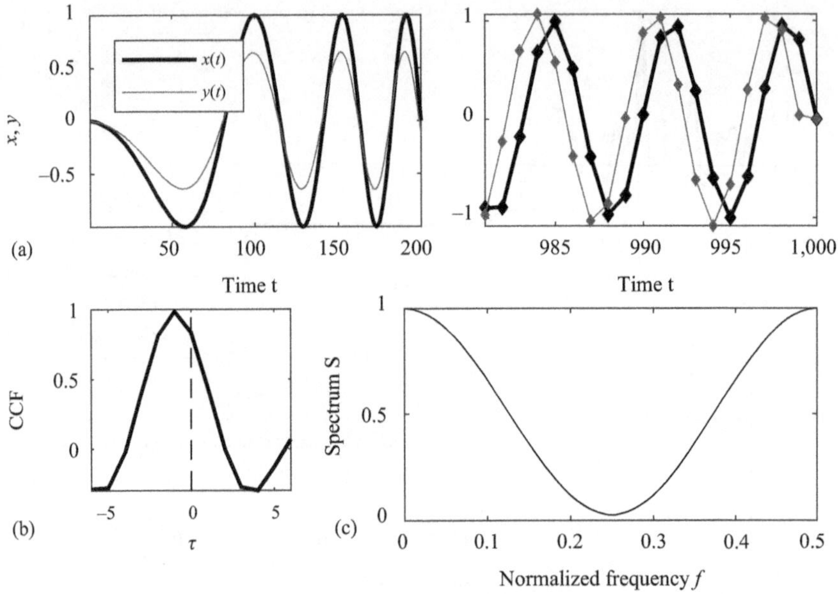

Figure 3.5 *Chirp signal I: (a) The left plot shows the first 200 data points of a chirp signal (a frequency-swept sine function) x(t) and its prediction signal y(t). The right plot shows the last 20 data points. Whereas an enlargement of the left plot would show prediction (with reduced amplitude), too, prediction is more evident in the right plot with the higher frequency oscillations. The DINGD predictor predicts the signal with the same GD of −1, for both very low and high frequencies. This is an example for a non-stationary signal that still can be predicted in real time; (b) the CCF has a peak value of CCF (−1) = 0.99. This shows that* $y(t-1) \approx x(t)$*, or, equivalently,* $y(t) \approx x(t+1)$*. Therefore, y(t) is a predictor of x(t); (c) the stability spectrum S(ω) indicates stability of this filter; (d) power spectral density function estimates for x(t) (bold) and y(t). It is evident that frequency components of the signal that are closer to the resonance of the frequency response function (e), where its gain increases, are amplified more relative to lower frequency components. This explains the relatively smaller amplitude of the predictor for low frequencies. In other words, the parameter b has been set to provide correct amplitudes for high frequencies only. The large estimation errors of gain and phase for frequencies > 0.15 are due to the lack of input signal power for those frequencies; (e) the frequency response function and GD. Shown are analytic values, as well as gain and phase values estimated from the data, see the legend. Signal and prediction parameters: The signal x(t) consists of 1,000 samples of a linear chirp signal with cutoff frequency of 0.15. The parameters of (3.19) are* $b = 1.25$*,* $c = 0.95$*, and* $\tau = 2$

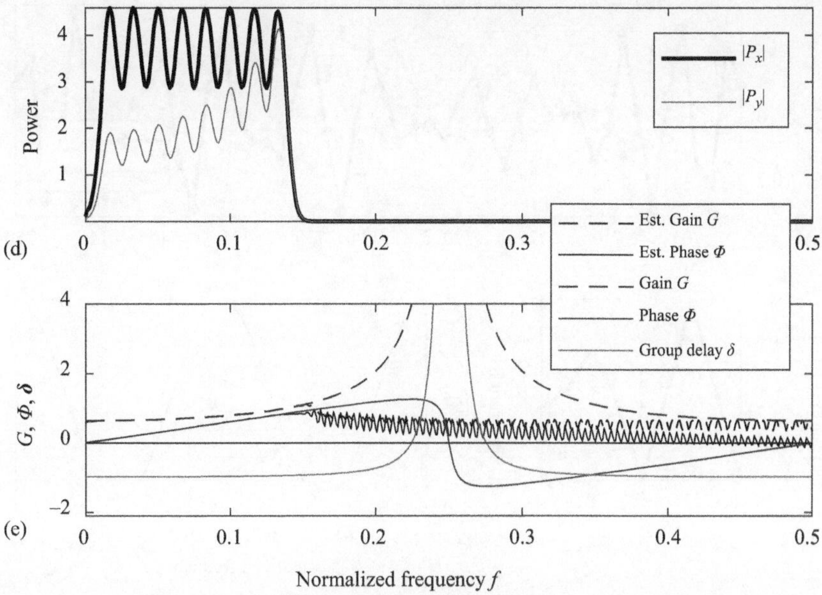

Figure 3.5 (Continued)

We consider the special case of $m = 2$, $\tau = 1$ first. Again, this is a special case of the IIR filter considered in Ref. [30], this time with cascading. As before, in order to make predictions one sample step or more ahead, we need to obtain an integer GD $\delta \leq -1$. Since the GD for zero frequencies is just m times the GD for the case of no cascading, for $c = 1$ the GD at zero frequency is $\delta_2(0) = -1$. Furthermore, for $c = 1$, the GD is constant -1 for all frequencies where it is defined. The poles of the frequency response function are located at $\omega = (2n + 1)\pi$ $(n = 0, 1, \ldots)$. Therefore, there is only one pole at $f_1 = \omega_1/2\pi = ½$, at the edge of the frequency range. In summary, for $\tau = 1$ and $m = 2$, the GD is negative throughout the entire frequency range. This means that cascading allows for using a delay of $\tau = 1$ without a split of frequency bands, and the NGD can, in the ideal case, extend over the entire frequency band. Still, the variable gain of the frequency response function prevents the prediction of signals with certain frequencies. It is straightforward to derive cases with $\tau > 1$ and $m > 2$, etc. However, the case of $\tau = 1$ stands out as the NGD bands are not split up into separate bands.

Figure 3.7 shows the frequency response for $m = 2$ and $m = 4$, again for $c = 1$ and $c = 1 - \varepsilon$, with $0 < \varepsilon \ll 1$. Figure 3.8 provides a numerical example. As before, the time unit is 1.

3.2.6 *Multiple delays*

Cascading two DINGD predictors corresponds to a total delay that is larger than the delays of the subsystems. Similarly, one could try to shape the frequency response function by using more than one delay in a single, noncascaded, predictor. With respect to the possible application to human motor control, the advantage of having

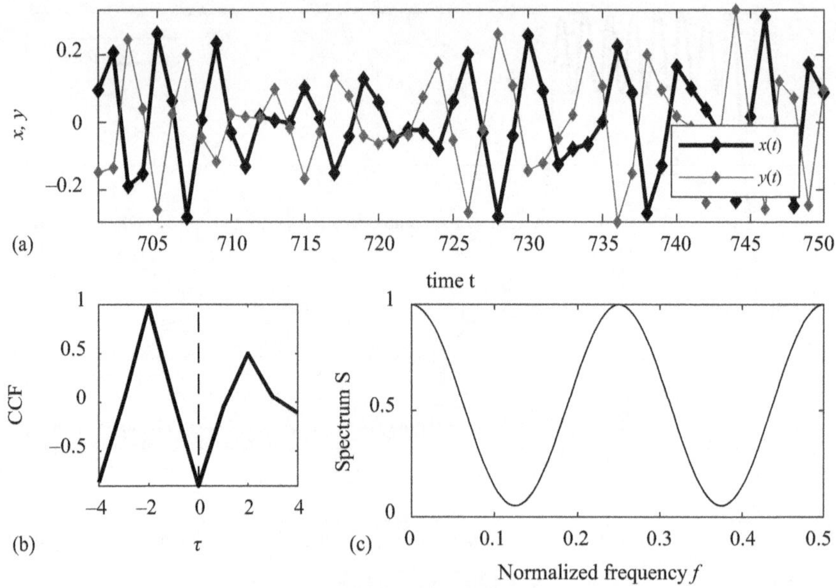

Figure 3.6 Band-limited noise II—prediction with the second band and a GD of −2: (a) A band-limited noise signal x(t) (bold) and its prediction signal y(t), the output of the predictor (3.19). Out of 1,000 simulated time points, 50 are shown. It is evident that the signal is predicted two time steps ahead. It is also evident that, although it is not smooth, the signal is predicted with high accuracy, including patterns of data points that are not formed by the envelope of an oscillatory signal; (b) the CCF between x(t) and y(t) peaks at a lag of −2. Its peak value is CCF(−2) = 0.98. This shows that $y(t-2) \approx x(t)$, or, equivalently, $y(t) \approx x(t+2)$. Therefore, y(t) is a predictor of x(t); (c) the stability spectrum S(ω) indicates stability of this filter; (d) power spectral density function estimates for x(t) (black, thick line) and y(t) (red); (e) the frequency response function as well as corresponding GD. Shown are analytic values as well as gain and phase values estimated from the data, see legend. Again, outside of the signal frequency band the estimates naturally have a large error. Signal and prediction parameters: The signal x(t) consists of uniformly distributed white noise, band-pass filtered with a seventh-order Butterworth filter with cutoff frequencies of 0.18 and 0.32. The parameters of (3.19) are $b = -1.50$, $c = 0.90$, and $\tau = 4$. Note that the phase behavior of the second NGD band relative to the first band requires a negative value for the parameter b

multiple delays in predictive schemes has recently been emphasized by Stepp and Turvey [31]. Multiple feedback delays also play an important role in artificial neural networks [32] but their potential for prediction in the sense as described here has not been widely used yet (an exception is Ref. [33]).

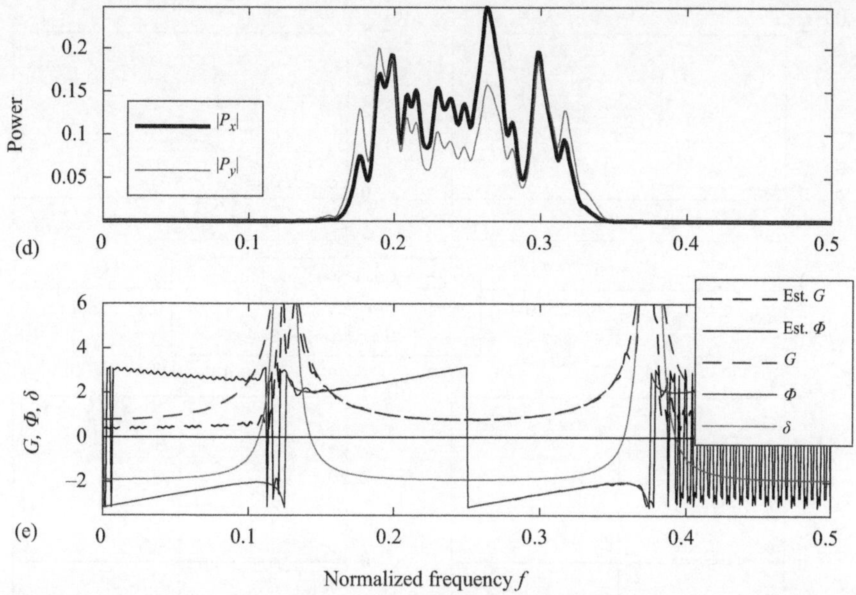

Figure 3.6 (Continued)

There are many possibilities for designing DINGD predictors with multiple delays, and here only three special cases with two delays and useful prediction properties are presented: a highpass, a lowpass, and a bandpass predictor with NGD. Another, more complex case with an arbitrary number of delays is described in Ref. [24]. The three cases presented here are given by two-delay systems with a maximum delay of two and four. Adopting the notation of (3.16), the DINGD predictor with a maximum delay of two is given by

$$y(t) = bx(t) - c_0 y(t-2) - c_1 y(t-1), \tag{3.29}$$

and the one with maximum delay of four by

$$y(t) = bx(t) - c_0 y(t-4) - c_1 y(t-3) - c_2 y(t-2) - c_3 y(t-1). \tag{4.30}$$

However, since only two delays are considered, these two cases can be summarized in one expression, which will also allow stating the results in a simpler way. With the two feedback delays τ_1 and τ_2, the DINGD filter (3.16) can be written as

$$y(t) = bx(t) - d_1 y(t-\tau_1) - d_2 y(t-\tau_2). \tag{3.31}$$

The two-delay frequency response function is

$$H(\omega) = \frac{b}{1 + d_1 e^{-i\omega\tau_1} + d_2 e^{-i\omega\tau_2}}. \tag{3.32}$$

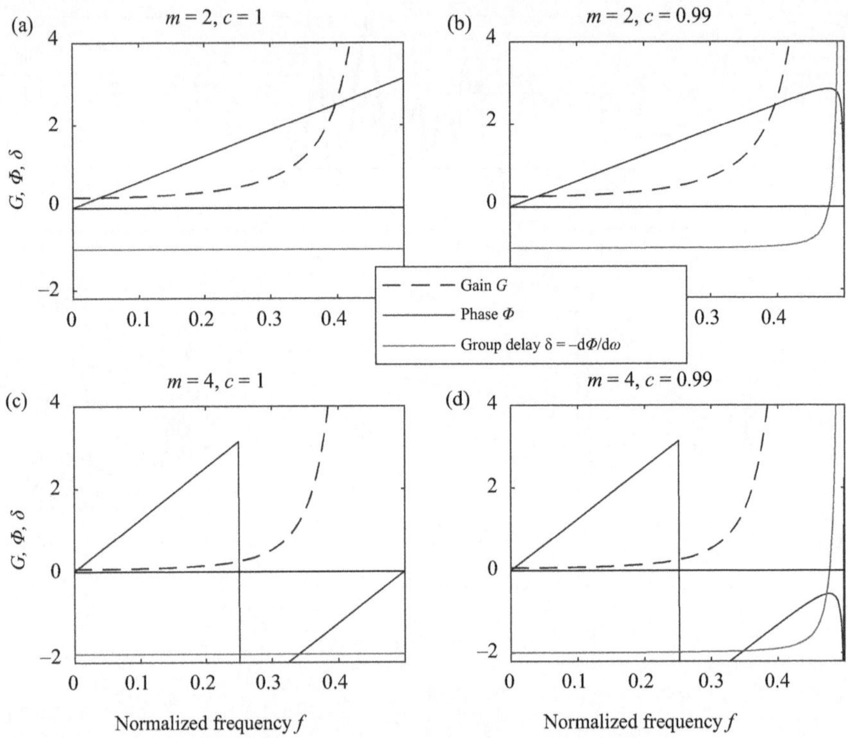

Figure 3.7 Theory II—cascading: The frequency response function (3.26) as well as corresponding GD (3.28) for the cascaded predictor with τ = 1 and cascading level m = 2 ((a), (b)) and m = 4 ((c), (d)), for feedback gain c = 1 ((a), (c)) and c = 0.99 ((b), (d)). Note that although the NGD is constant over a wide frequency range, the gain increases for high frequencies, restricting applicability of the DINGD predictor to signals containing only lower frequencies. This is demonstrated in the example of Figure 3.8

With

$$\begin{aligned}\beta(\omega) &= 1 + d_1^2 + d_2^2 + 2d_1d_2\cos(\omega(\tau_1 - \tau_2)) \\ &\quad + 2d_1\cos(\omega\tau_1) + 2d_2\cos(\omega\tau_2)\end{aligned} \tag{3.33}$$

it follows

$$G(\omega) = \frac{|b|}{\sqrt{\beta(\omega)}} \tag{3.34}$$

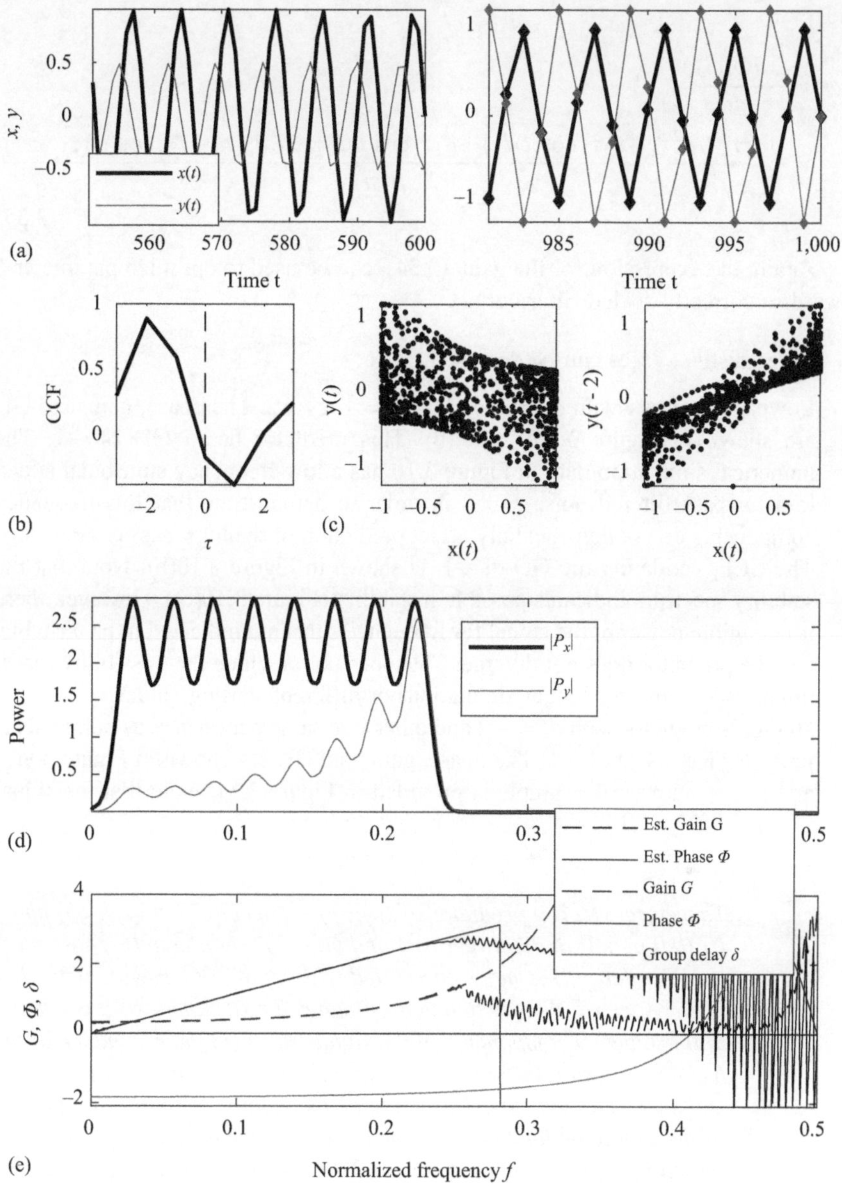

Figure 3.8 Chirp signal II—cascading four predictors: (a) Left panel: 50 data points of a chirp signal x(t) and its prediction signal y(t). Right panel: Same for the last 20 data points. The DINGD predictor predicts the signal with the same GD of −2, for both low and high frequencies. This is an example for a non-stationary signal that still can be predicted in real time; (b) the CCF has a peak value of CCF(−2) = 0.89. This shows that y(t − 2) ≈ x(t), or, equivalently, y(t) ≈ x(t + 2).

and

$$\delta(\omega) = -\frac{d\Phi(\omega)}{d\omega}$$
$$= -\frac{d_1^2\tau_1 + d_2^2\tau_2 + d_1\tau_1\cos(\omega\tau_1) + d_2\tau_2\cos(\omega\tau_2) + d_1 d_2(\tau_1+\tau_2)\cos(\omega(\tau_1-\tau_2))}{\beta(\omega)}. \quad (3.35)$$

Again, the expression for the gain, (3.34), can be used to optimize parameter b towards a correctly scaled filter output.

Three distinct cases can be designed:

- Lowpass predictor with $d_1 = 2$, $d_2 = 1$, $\tau_1 = 1$, $\tau_2 = 2$. The phase, gain, and GD are shown in Figure 3.9(a) and (b). This predictor has a GD of -1. The numerical simulation data in Figure 3.10 has a low-frequency sinusoidal added to a lowpass filtered noise signal in order to demonstrate that low-frequency components do not detrimentally affect prediction of the high-pass components. The CCF, validating the GD of -1, is shown in Figure 3.10(b). Note that the stability spectrum indicates possible instability (Figure 3.10(c)). However, there is negligible power of the signal for frequencies that are indicated to be unstable, and the predictor does not diverge. This does not preclude the possibility that it might "blow up" for longer simulations or different starting values.
- Highpass predictor with $d_1 = -2$ and otherwise same parameters as before. This predictor has a GD of -1. The phase, gain, and GD are shown in Figure 3.9(c) and (d). A numerical example is provided in Figure 3.11. Note that the signal

Figure 3.8 (Continued)

Therefore, y(t) is a predictor of x(t); (c) scatterplots between x(t) and y (t) (left panel) and between x(t) and y(t – 2) (right panel). These plots confirm that y(t) is more correlated with x(t + 2) than with x(t); (d) power spectral density function estimates for x(t) and y(t). It is evident that frequency components of the signal closer to the resonance of the frequency response function are amplified more relative to lower frequency components. This explains the relatively smaller amplitude of the predictor for low frequencies; (e) the frequency response function as well as corresponding GD. Shown are analytic values as well as gain and phase values estimated from the data, see legend. The large estimation errors of gain and phase for frequencies > 0.25 are due to the lack of input signal power for those frequencies. Signal and prediction parameters: The signal x(t) consists of 1,000 samples of a linear chirp signal with cutoff frequency of 0.25. Note that the chirp frequency sweeps a larger range than in Figure 3.5, made possible by the move of the resonance to higher frequencies. The parameters of (3.25) are m = 4, b = 1.40, c = 0.85, and τ = 1

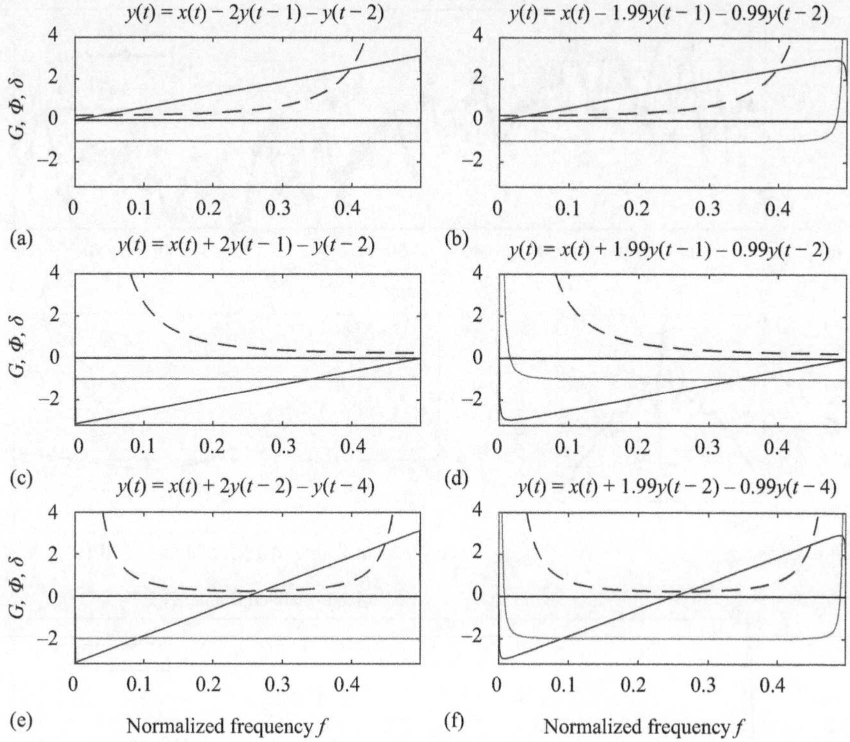

Figure 3.9 Theory III—multiple delays: (a), (b) Lowpass DINGD predictor with GD of −1. Shown are the frequency response function as well as the GD for the lowpass predictor with parameters stated in the text and also provided in the figure titles (b = 1). Again, the left column has theoretical parameters that cause a constant NGD for all frequencies, and the right column has parameters that are more realistic. (The graphs in the left column are identical to the graphs of the cascaded predictor in Figure 3.7(a) and (b), left column); (c), (d) highpass DINGD predictor with GD of −1; (e), (f) bandpass DINGD predictor with GD of −2. The two delays are now twice as large as before, i.e., $\tau_1 = 2$ *and* $\tau_2 = 4$*. For legends, please refer to Figure 3.7*

can be predicted well (Figure 3.10(a)) although it does not constitute an envelope of a high-frequency signal, and that the stability spectrum again indicates instability (Figure 3.11(c)).

- Bandpass predictor with $d_1 = -2, d_2 = 1, \tau_1 = 2, \tau_2 = 4$. The phase, gain, and GD are shown in Figure 3.9(e) and (f). This predictor has a GD of −2. To use this predictor instead of the low- or highpass predictors can have additional advantages in addition to the higher NGD; the behavior of the gain allows for somewhat higher-/lower-frequency components of the signal as in the lowpass/highpass predictors. In addition, compared with the single-delay predictor in Figure 3.6,

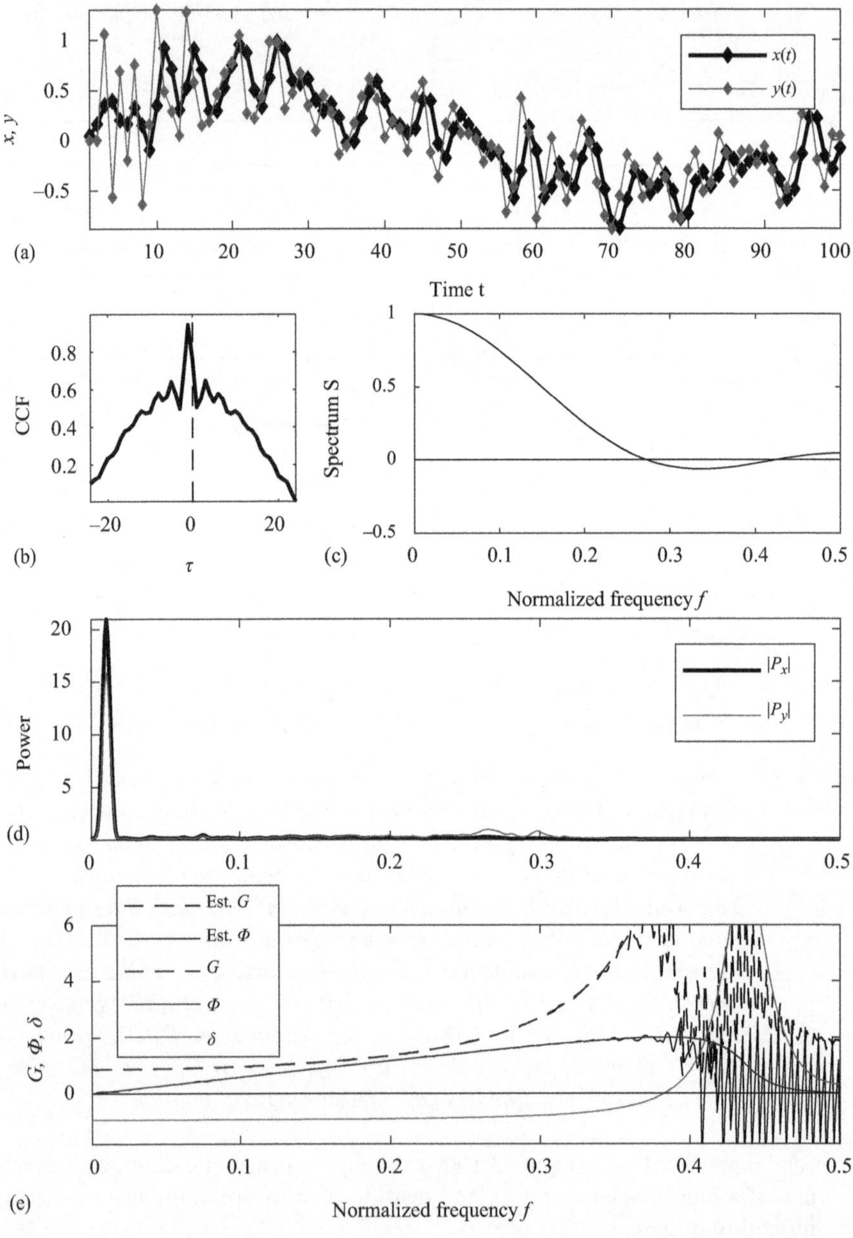

Figure 3.10 Lowpass DINGD predictor and violation of stability criterion: (a) The first 100 data points of a lowpass filtered noise signal x(t) with an added sinusoidal signal with frequency 0.01 and its prediction signal y(t). Due to the very low frequency component of the sinusoidal, relatively long transients of the prediction signal are visible; (b) the CCF has a peak value of CCF(−1) = 0.94. This shows that y(t − 1) ≈ x(t), or, equivalently, y(t) ≈ x(t + 1). Therefore, y(t) is

this two-delay predictor allows for a larger signal bandwidth. This is shown in the numerical example of Figure 3.12. Of all the examples, the signal here appears to be the most complex one. That this signal can be predicted in real time two time points ahead by the simple mechanism of DINGD, (3.31), is remarkable.

3.3 Prediction of manual cursor movement

3.3.1 Description of the algorithm

The DINGD theory will executed by an algorithm that predicts, in real time, the trajectory of smooth voluntary manual movements of a computer input device such as a mouse or sylus on a graphics pad. For achieving good results, a graphics tablet with a stylus is recommended for use rather than a mouse. Touch screens have not been tested but might work as well. Detailed instructions about how to run the MATLAB script on a personal computer or laptop are provided as comments at the beginning of the script. It is provided in Appendix A.

The script first launches a window with a circular pointer that can be controlled by the chosen input device (mouse or stylus). The red marker indicates the signal that is predicted via the DINGD mechanism. Smooth motions of the input device should appear to be predicted by the red marker. Once the experiment is finished, the recorded time series of both the horizontal and vertical screen coordinates of the pointer and marker are analyzed by the script. The results are provided by means of graphs that should be familiar after reading this book chapter, and by numbers in the MATLAB command window. If on average the pointer had been predicted by the marker, the provided GD estimated from the time series should be negative. If the GD is positive,

Figure 3.10 (Continued)

a predictor of $x(t)$; (c) The stability spectrum $S(\omega)$ does not indicate guaranteed stability of this filter as it becomes negative for some frequencies. However, much longer simulations than shown here indicate that the predictor does not "blow up" for this data set. This is plausible, as the stability criterion violation affects frequencies with negligible power in this data set, as can be seen in panel (d); (d) Power spectral density function estimates for $x(t)$ and $y(t)$; (e) The frequency response function as well as corresponding GD. Shown are analytic values as provided in the text as well as gain and phase values estimated from the data, see legend. The large estimation errors of gain and phase for frequencies > 0.35 are due to the lack of input signal power for those frequencies. Signal and prediction parameters: the signal $x(t)$ consists of 1,000 samples of uniformly distributed white noise, low-pass filtered with a seventh-order Butterworth filter with a cutoff frequency of 0.30 with an added sine signal of amplitude 0.5 and frequency 0.01. The parameters of (3.31) are $b = 3.00$, $d_1 = 1.65$, $d_2 = 0.80$, $\tau_1 = 1$, $\tau_2 = 2$

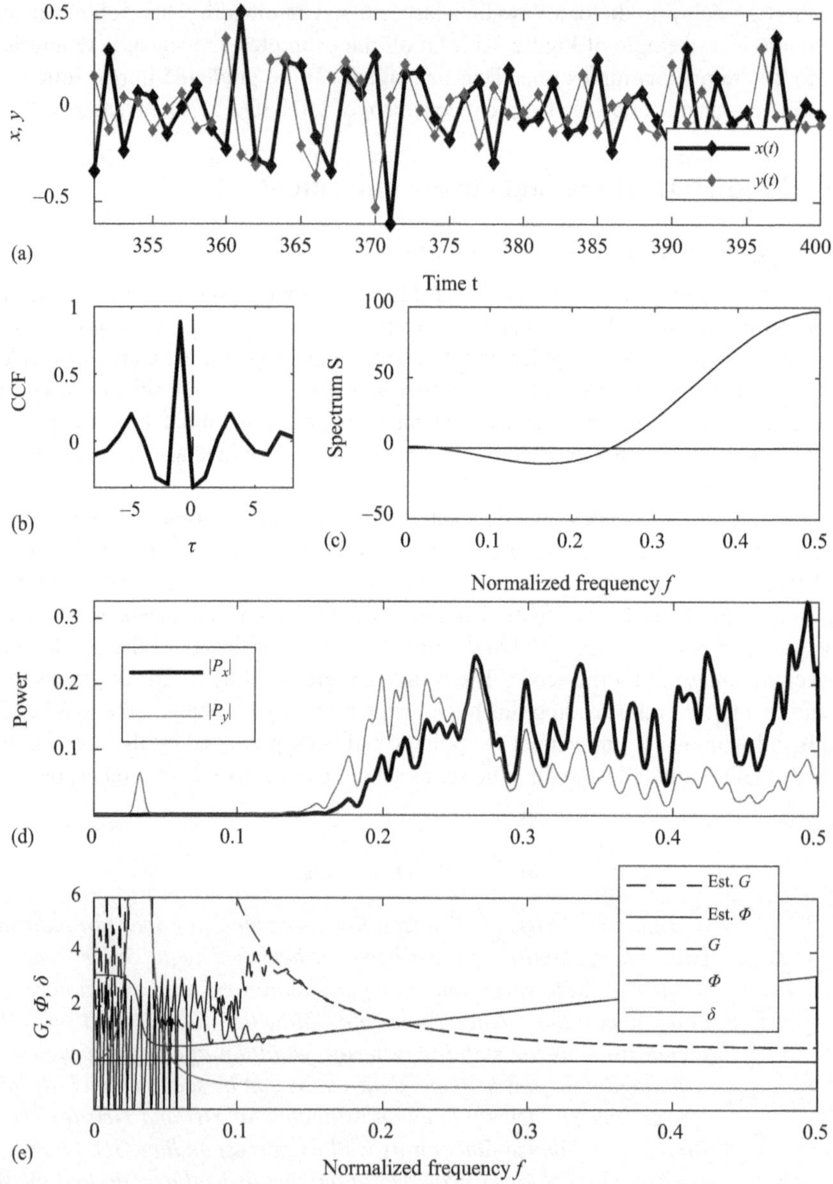

Figure 3.11 Highpass DINGD predictor: (a) A highpass filtered noise signal x(t) and its prediction signal y(t). Out of 1,000 simulated time points, 50 are shown. Similar to Figure 3.6, it is evident that patterns of data points that are not formed by the envelope of an oscillatory signal are predicted well; (b) the CCF has a peak value of $CCF(-1) = 0.89$. This shows that $y(t-1) \approx x(t)$, or, equivalently, $y(t) \approx x(t+1)$. Therefore, y(t) is a predictor of x(t); (c) the stability spectrum $S(\omega)$ does not indicate guaranteed stability of this filter, similarly to the

the marker on average lags the pointer. In this case, the experiment can be repeated with smoother movements. It stands to reason that unpredictable trajectories would cause a positive GD, as described before. After the data analysis, all variables are saved to a file with a time stamp to keep a record of the results that could be used for further analysis. The script allows for several parameter settings as indicated by comments within the script. For example, in order not to bias the subject in its movements, the marker that predicts the pointer can be made invisible.

The predictor used is the simplest possible discrete-time DINGD predictor with only one delay, as given by (3.19), namely

$$y(t) = bx(t) - cy(t - \tau). \tag{3.36}$$

The parameters are $b = 1$, $c = 0.8$, and $\tau = 70$. Therefore, the model is simply

$$y(t) = x(t) - 0.8y(t - 70). \tag{3.37}$$

These parameters have been obtained by trial and error. In addition, the script contains a cutoff to reduce overshooting of the predicted trajectory, but NGDs also should be achievable without that threshold. As one can see, the chosen model could easily be improved; for example, one could introduce additional parameters according to (3.16). This is left as an exercise to the reader.

The model specification of the script consists of the two lines `b=1; c=0.8; tau=70;` for the definition of the parameters, and `y=b*x-c*ytau;` for the model update step, (3.37). The rest of the script defines the initial coordinates, manages the recording of cursor position whenever the cursor is being moved, and provides post-processing of the resulting cursor and prediction trajectories, after a key has been pressed to indicate the end of the experiment.

An example output of the script is shown in Figure 3.13. This script, as it is, might not have much usefulness beyond a demonstration that real-time prediction of natural signals per DINGD is possible. However, one can think of the applications of prediction via DINGD in various fields, such as acceleration of man–machine interface applications including augmented communication [34] or the compensation of delay times in time-delayed control problems [35].

Figure 3.11 (Continued)

previous figure; (d) power spectral density function estimates for $x(t)$ and $y(t)$; (e) the frequency response function as well as corresponding GD. Shown are analytic values as provided in the text as well as gain and phase values estimated from the data, see legend. The large estimation errors of gain and phase for frequencies < 0.15 are due to the lack of input signal power for those frequencies. Signal and prediction parameters: the signal $x(t)$ consists of 1,000 samples of uniformly distributed white noise, low-pass filtered with a seventh-order Butterworth filter with a cutoff frequency of 0.20. The parameters of (3.31) are $b = -2.00$, $c_1 = -1.92$, $c_2 = 0.96$, $\tau_1 = 1$, $\tau_2 = 2$

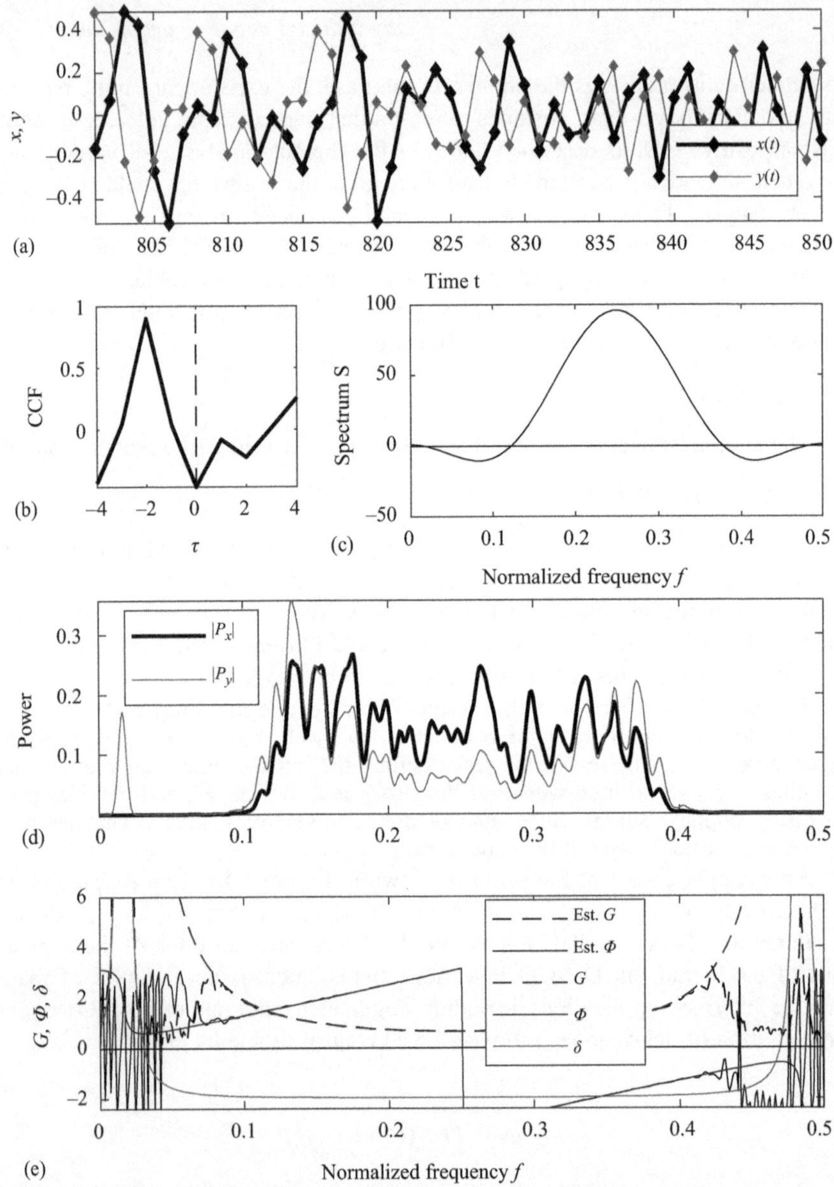

Figure 3.12 Band-limited noise III—bandpass DINGD predictor with GD of −2: (a) A bandpass filtered noise signal x(t) and its prediction signal y(t). Out of 1,000 simulated time points, 50 are shown; (b) the CCF has a peak value of CCF(−2) = 0.90. This shows that y(t − 2) ≈ x(t), or, equivalently, y(t) ≈ x(t + 2). Therefore, y(t) is a predictor of x(t); (c) the stability spectrum S(ω) does not indicate guaranteed stability of this filter, similarly to the previous figure. Again, the predictor is designed such as to be potentially unstable for frequencies outside of the signal power frequencies, as can be verified in panel (d); (d) power spectral density function estimates for x(t) and y(t); (e) the

Figure 3.12 (Continued)

frequency response function as well as corresponding GD. Shown are analytic values as provided in the text as well as gain and phase values estimated from the data, see legend. The large estimation errors of gain and phase for frequencies between 0.1 and 0.4 are due to the lack of input signal power for those frequencies. Remarkably, the GD is constant −2 for all frequencies. Signal and prediction parameters: the signal x(t) consists of uniformly distributed white noise, bandpass filtered with a seventh-order Butterworth filter with cutoff frequencies of 0.12 and 0.38. The parameters of (3.31) are $b = -2.50$, $c_1 = -1.90$, $c_2 = 0.94$, $\tau_1 = 2$, $\tau_2 = 4$

Matlab command window

```
>> CursorPrediction
Press any key when done with the experiment
Sampling time interval = 0.010321 s
Group delay = -0.15482 s
CCF (-0.15482 s) = 0.8715
```

Result figure

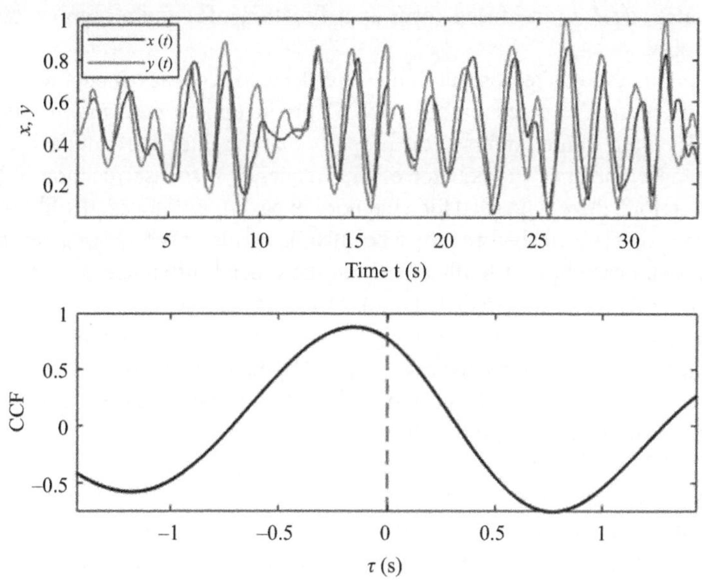

Figure 3.13 Example output of a MATLAB algorithm for the prediction of voluntary manual cursor movements

3.4 Discussion

3.4.1 Prediction by NGD versus conventional prediction

The DINGD signal prediction concept differs from most other predictors or forecasting models, which do not rely on past *predicted* but past *input* signal values. In other words, most conventional time series predictors can be written in the form

$$\text{predicted } x(t+\tau) = f(x(t), \quad x(t-1), \quad x(t-2), \ldots), \tag{3.38}$$

in which $f(.)$ is a specific model of the time series to be predicted. These predictors rely on prior observations rather than prior predictions, and usually contain coefficients resulting from a fit to a learning dataset, such as in the nearest-neighbor-state methods [36,37], or which are continuously being updated, such as in adaptive filtering [38]. (An exception would be systems that use anticipatory synchronization for prediction [33].) In contrast, prediction via DINGD cannot be expressed in this explicit form, as it does not rely on past input signal values $x(t-1), x(t-2), \ldots$ for prediction, but only on previously predicted output values $y(t-1), y(t-2), \ldots$, along with the present input value $x(t)$. For example, discrete-time DINGD predictors can be written as

$$\text{predicted } x(t+\tau) = y(t) = g(x(t), \quad y(t-1), \quad y(t-2), \ldots), \tag{3.39}$$

where $g(.)$ is a function depending on the spectral content of the time series to be predicted. This expression is fundamentally different from the analogous expression for conventional prediction, (3.38). It is apparently not possible to express the predicted value $x(t+\tau)$ in closed form as a function of a finite number of past values of x only.

The previously predicted output values are delayed feedback inputs to the predictor. In this sense, the discrete-time DINGD predictor is a pure *infinite impulse response* (IIR) filter. "Pure" refers to the property that the filter does not have a finite impulse response part, i.e., the counter of the frequency response function is just a constant. Therefore, there is no need for a memory of past *signal* values. IIR filters have been shown to possess NGD before [30]. The DINGD predictor is a *dynamic predictor*; it is still a special case of an IIR filter [39] but the crucial difference is in its usage. Normally, in an IIR filter the predicted $x(t+1)$ would be given by $y(t+1)$, not $y(t)$. This dynamic origin of prediction makes the DINGD predictor so counterintuitive. However, the prediction horizon and accuracy of DINGD predictors seem to be more limited in comparison to model-based approaches. This is the price one has to pay for the privilege not to have to fit a model to data.

The DINGD predictor does not require a learning dataset. Rather, it is a true real-time predictor once the coefficients have been fixed and the dynamics have settled into a steady state. As the nonstationary chirp signal examples have indicated, it is worth looking into the real-time advantages of this prediction scheme compared to adaptive prediction algorithms. Furthermore, it has been demonstrated here that the DINGD predictor is able to predict complex nonsmooth signals, too, adding insights beyond some of the contemporary understanding of

NGD. It would be worth investigating how NGD predictors fit into general theories of prediction [40].

3.4.2 Causality

Although being counterintuitive, the phenomenon of NGD does not violate causality [1,41]. Its performance depends on the choice of parameters and the data, and for improperly chosen parameters and data that do not fulfill the specified stationarity and spectral criteria it might not be predictive or cause instabilities [14,22,23]. Nevertheless, some anticipatory systems can defy the inference of the direction of information flow from data [42–45]. This holds true for NGD systems, too [46].

All NGD filters presented here are causal because in the time domain they only refer to present and past but not to future states. For spatially extended systems, things are more involved because of the time a signal travels, which involves group and phase velocities. It has been shown that there are media that actually require an NGD for some frequencies in order to be, overall, causal media [47]. Furthermore, in media with strong gain for some frequencies, the GD would be negative for zero frequency [47]. And, in any medium, the group velocity is abnormal for a frequency at which the absorption has an absolute maximum [47]. Analogous behaviors can be observed in the examples presented here, in which the filter gain turned out to be smallest for frequencies with NGD. Note that the theoretical GD in several examples presented here alternates between negative and positive values. This is a result of the causality restrictions on the filter's frequency response function. As the phase delay is not bound by causality constraints, it can stay negative for all frequencies [24]. Along these lines, Ref. [14] provides an analysis of how NGD in nonspatially extended systems is limited by causality. In summary, somewhat paradoxically one might say that the predictive properties of DINGD filters arise from their causality.

3.4.3 Anticipatory coupling

Some systems can be predicted in real time by coupling identical system copies to them, if the coupling term contains a time-delayed state of the coupled system. This *anticipatory coupling* [2,42,48–50] specifically refers to the design that the input signal value $x(t)$ at the present time t is compared or "coupled" to time delayed output values of the predicting system, $y(t-\tau)$. Anticipatory coupling includes terms like

$$\Delta^{\tau}(t) = x(t) - y(t-\tau) \tag{3.40}$$

and causes the systems to synchronize [51,52] in a way that the driven system leads the driving system in time, which is called *anticipatory synchronization* [41–43]. In contrast to prediction by DINGD, where the prediction horizon usually is smaller than the used feedback delay, in anticipatory synchronization the prediction horizon and the coupling delay are identical, even for stochastic systems [53,54].

Both the continuous-time as well as the discrete-time DINGD predictor implicitly use the concept of anticipatory coupling, too, although it might be obscured somewhat

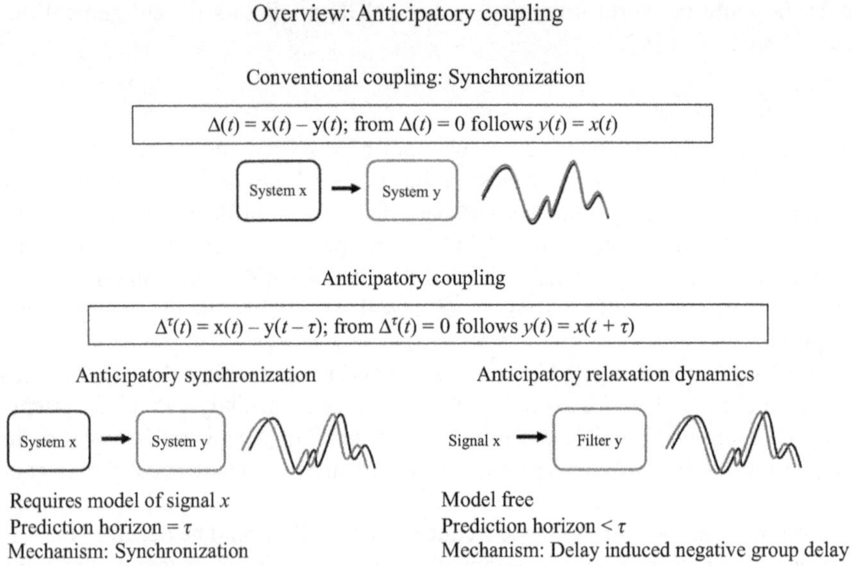

Figure 3.14 An overview about anticipatory coupling schemes their comparison with conventional coupling

and does not stand out as clearly as in anticipatory synchronization. This has been explained in more detail for *anticipatory relaxation dynamics* (ARD) [2]. ARD is caused by delay-induced NGD, too [3]. Whereas in ARD anticipatory coupling is more evident than in DINGD prediction, because the coupling term (3.37) is used directly, in DINGD prediction in general $x(t)$ and $y(t-\tau)$ can have different weights. Figure 3.14 provides an overview over these different cases. Interestingly, in the case of anticipatory synchronization, more effective anticipatory coupling terms than (3.37) have been found [55–60]. It might be worthwhile to use those insights to develop more effective anticipatory coupling terms for prediction by NGD, too.

3.4.4 Biology and engineering

[. . .] we wonder whether some neural systems make use of negative delay circuits to synchronize activity [61]

One of the most important properties of real world neuronal as well as biochemical signaling is time-delays. Rather than being a limitation of such signaling, this is one of the most interesting elements in the computational repertoire of such networks. Integration and differentiation of the time-course of signals now becomes possible, as does short-term storage of information [62]

Indeed, delays should be considered as [. . .] additional media through which evolution, or skilled engineers, can achieve particular dynamical effects [63]

Mechanisms for prediction and control based on anticipatory coupling are already used in physics and engineering [64–91], as well as in the neurosciences [92–102]. However, these mechanisms usually only work with signals generated by known dynamical systems, which have to be known beforehand or estimated from the signal. On the other hand, the DINGD based prediction schemes presented here just require their parameters to be adjusted to the center frequency and bandwidth of the signal. Due to this parsimony, in general prediction via DINGD will probably not outperform these model-based or even off-line prediction schemes, in terms of the prediction horizon and accuracy.

However, the prediction horizons obtained by DINGD might already be sufficient to explain typical time scales of behavioral observations [103–105]. This has been worked out in detail in the analysis of a human motor control experiment [4]. Therefore, even if prediction horizons are relatively short and predictions can be distorted, it still is conceivable that neuronal systems could have implemented prediction by NGD by utilizing synaptic or conduction delays, which are known to affect the dynamics of neuronal information processing [106–108]. In particular, it could be worthwhile to search for signatures of NGD in, for example, the distribution of peak prediction variability in cognitive motor control experiments [109,110]. In general, in prediction via DINGD (or pure IIR filters), the translation process from a sensory input to the internal coding of information needs to happen only once, at the present time point, but not for signals restored from memory. This translation happens, for example, in the retina while generating neuronal spikes, and probably nothing resembling the original sensory signals are stored anywhere in the brain. If the brain predicted by using conventional prediction mechanisms, it would need those original signals, though. Further, the main ingredient for the DINGD predictor are time delays, which are abundant in the central nervous system including the cerebellum. Not only that the cerebellum plays a crucial role for fine motor control [111], but its peculiar neuronal architecture and function [98,112] might provide all the ingredients for biological implementations of multiple DINGD predictors as one of its functions. Finally, prediction by NGD might augment the related concept of neuronal anticipatory synchronization [23,99,100,113], which recently has been used to explain paradoxical observations in brain dynamics [114].

On the other hand, "For a function as fundamental as prediction, the nervous system uses as many tricks as possible, and the phenomena revealing this function may show up in many guises" [115]. In any case, if we can learn more about how biology solves predictive control problems [116–124], these insights could provide a contribution to the growing utilization of anticipatory coupling mechanisms toward science and engineering applications.

3.5 Conclusion

A very simple real-time predictor based on delayed feedback-induced NGD has been described. This DINGD predictor predicts future signal values by using present input and past output, or already predicted, signal values. It thus differs from most other predictors or forecasting models, which do not take into account past *predicted* but past *input* signal values.

Since the DINGD predictor does not require a specific model of the signal to be predicted, it is one of the most parsimonious descriptions of a real-time predictor conceivable. Therefore, the question arises if not nature already has implemented this type of prediction by means of anticipatory coupling. As discussed, there are first hints that this might indeed be the case. The simplicity of the DINGD predictor and the fact that it accomplishes real-time prediction without the need for model fitting or adaptation invites its use in engineering applications, machine learning, and artificial intelligence, too, despite its generally limited prediction horizon and accuracy if compared with model-based or adaptive prediction schemes.

Acknowledgment

I would like to thank Nigel Stepp and Blaise Ravelo for discussions.

Appendix A

A.1 MATLAB script for real-time cursor prediction

This script can be downloaded from https://www.mathworks.com/matlabcentral/fileexchange/64168-cursorprediction--, and there the posted DISCLAIMER applies.

```
% Cursor movement prediction by negative group delay
% Version 12. 10. 2017

% Instructions

% - Start Matlab
% - Start this script with F5 or 'run'
% - Put pointer into the appearing window and click left
%   mouse button or press stylus on the graphic tablet
% - Keep left mouse button / stylus pressed during
%   cursor movement
% - Move cursor slowly and smoothly into random
%   directions, describing circles, figure eights,
%   back and forth movements, etc.
% - When done, release mouse button / stylus
%   and go into Matlab command window
% - Press return and an analysis of the time series
%   will appear (the x and y coordinates are
%   concatenated into a single time series)
% - The data is written to dated mat file to keep a record

% The script was developed on Matlab R2017a and might
% not work properly on older versions.
```

```
function CursorPrediction()

format short
clf reset
close all

% Parameters to be set by user

showprediction=1;
% 1 for showing the predicted pointer, 0 for not showing

threshold=0.01;
% > 0 to prevent too large deviations due to instability,
% 0 if no correction wanted

% Filter coefficients
% b=1; c=0.9; tau=70; % should work in most cases
b=1; c=0.8; tau=70; % less overshooting of predictor

% End of parameter section

figure(1)
plot([0, 1], [0, 1], '.') % cursor plane

i_delay=0*(1:tau);
j_delay=i_delay;
ij_DINGD=[0, 0];
ij=[0, 0];

% Interactive part

tic

set(gcf, 'windowbuttondownfcn', @CP_startit)
set(gcf, 'windowbuttonupfcn', @CP_stopit, 'userdata', [])
set(gcf, 'Renderer', 'painters')
set(gcf, 'Menubar', 'none')

disp('Press any key when done with the experiment')
pause()

dT=toc; dT=dT/length(ij);
disp(['Sampling time interval = ' num2str(dT) ' s'])
% dT usually overestimated by waiting for input

ij_DINGD=ij_DINGD(2:end, :);

save(['Data_', datestr(now, 'dd-mmm-yyyy-HH-MM')])

% Analysis of ij (x, input) and ij_DINGD (y, predictor output)
```

```
% Combine i and j into one time series for analysis
x=[ij(:, 1); ij(:, 2)];
y=[ij_DINGD(:, 1); ij_DINGD(:, 2)];

taxis=transpose((1:length(x))*dT);
N=length(taxis);

% Beginning of figure output
figure(1)
subplot(2, 1, 1)
plot(taxis, x, 'k-')
hold on
plot(taxis, y, 'r-')
axis tight
legend('x(t)', 'y(t)', 'Location', 'northwest')
xlabel('time t [s]')
ylabel('x, y')

% CCF, modified from crosscorr()
taull=round(2*tau);
x1=x-mean(x);
y1=y-mean(y);
nfft=2^(nextpow2(N)+1);
F=fft([x1 y1], nfft);
acf1=ifft(F(:, 1).*conj(F(:, 1)));
acf2=ifft(F(:, 2).*conj(F(:, 2)));
ccf=ifft(F(:, 1).*conj(F(:, 2)));
ccf=ccf([(taull+1:-1:1) (nfft:-1:(nfft-taull+1))]);
ccf=real(ccf)/(sqrt(acf1(1))*sqrt(acf2(1)));
lags=(-taull:taull)';
%
[maxccf, lagind]=max(abs(ccf));
lag=lags(lagind);
tau_c=-lag*dT;
disp(['Group delay = ', num2str(-tau_c) ' s'])
disp(['CCF (', num2str(-tau_c) ' s) = ' num2str(maxccf, 4)])

subplot(2, 1, 2)
plot(lags*dT, ccf, 'k', 'Linewidth', 1.5)
hold on
plot([0, 0], [min(ccf), 1], 'k-')
axis tight
xlabel('\tau [s]')
ylabel('CCF')
hold off

% Nested functions to avoid defining global variables

   function CP_startit(~, ~)
        set(gcf, 'userdata', [] )
        set(gcf, 'Pointer', 'circle')
        set(gcf, 'windowbuttonmotionfcn', @CP_followit)
```

```
    function CP_followit(~, ~)

        CurPnt=get(gca, 'CurrentPoint');
        newpoint=CurPnt(2, 1:2);

        oldpoints=get(gcf, 'userdata');
        pts=[oldpoints; newpoint];

        set(gcf, 'userdata', pts)

        % i, j are screen coordinates,
        % x, y are signals(input and output)

        % DINGD model 1 for horizontal coordinate i
        x=newpoint(1)-.5; % centralization to zero mean
        ytau=i_delay(1);
        y=b*x-c*ytau;

        if threshold>0
            distance=abs(x-y);
            if distance>threshold
                y=y/(1+distance-threshold);
            end
        end

        i_delay=[i_delay(2:end), y];
        i_DINGD=y+.5;

        % DINGD model 2 for vertical coordinate j
        x=newpoint(2)-.5;
        ytau=j_delay(1);
        y=b*x-c*ytau;

        if threshold>0
        distance=abs(x-y);
        if distance>threshold
            y=y/(1+distance-threshold);
        end
    end

    j_delay=[j_delay(2:end), y];
    j_DINGD=y+.5;

    if showprediction
        plot(i_DINGD, j_DINGD, 'r.', 'MarkerSize', 10)
        xlim([0, 1]); ylim([0, 1])
    end

    ij_DINGD=[ij_DINGD; [i_DINGD, j_DINGD]];

  end
end
```

```
  function CP_stopit(~, ~)
      set(gcf, 'Pointer', 'arrow')
      set(gcf, 'windowbuttonmotionfcn', [])
      ij=get(gcf, 'userdata');
      close
   end

end
```

References

[1] M. W. Mitchell and R. Y. Chiao: "Negative group delay and fronts in a causal system: An experiment with very low frequency bandpass amplifiers," *Phys Lett A*, vol. 230, pp. 133–138, 1997.

[2] H. U. Voss: "Signal prediction by anticipatory relaxation dynamics," *Phys Rev E*, vol. 93, pp. 030201(R), 2016.

[3] H. U. Voss: "The leaky integrator with recurrent inhibition as a predictor," *Neural Comput*, vol. 28, pp. 1498–1502, 2016.

[4] H. U. Voss and N. Stepp: "A negative group delay model for feedback-delayed manual tracking performance," *J Comput Neurosci*, vol. 41, pp. 295–304, 2016.

[5] R. D. Driver: "Ordinary and Delay Differential Equations" (Springer-Verlag, New York, 1977).

[6] J. K. Hale: "Theory of Functional Differential Equations" (Springer-Verlag, New York,1977, 2d edn).

[7] I. Győri and G. E. Ladas: "Oscillation Theory of Delay Differential Equations: With Applications" (Oxford University Press, Oxford, 1991).

[8] K. Gopalsamy: "Stability and Oscillations in Delay Differential Equations of Population Dynamics" (Kluwer Academic Publishers, Dordrecht, 1992).

[9] J. Wiener and J. K. Hale: "Ordinary and Delay Differential Equations" (Longman Scientific & Technical; Wiley, Essex, 1992).

[10] T. Erneux: "Applied Delay Differential Equations" (Springer, New York, 2009).

[11] R. P. Agarwal: "Nonoscillation Theory of Functional Differential Equations with Applications" (Springer, New York, 2012).

[12] T. Nakanishi, K. Sugiyama, and M. Kitano: "Demonstration of negative group delays in a simple electronic circuit," *Am J Phys*, vol. 70, pp. 1117–1121, 2002.

[13] M. Kitano, T. Nakanishi, and K. Sugiyama: "Negative group delay and superluminal propagation: An electronic circuit approach," *IEEE J Sel Top Quant*, vol. 9, pp. 43–51, 2003.

[14] M. Kandic, and G. E. Bridges: "Limits of negative group delay phenomenon in linear causal media," *Prog Electromagn Res*, vol. 134, pp. 227–246, 2013.

[15] A. Namajunas, K. Pyragas, and A. Tamasevicius: "An electronic analog of the Mackey-Glass system," *Phys Lett A*, vol. 201, pp. 42–46, 1995.

[16] H. U. Voss: "Real-time anticipation of chaotic states of an electronic circuit," *Int J Bifurcat Chaos*, vol. 12, pp. 1619–1625, 2002.
[17] R. B. Stein: "A theoretical analysis of neuronal variability," *Biophysical Journal*, vol. 5, pp. 173–194, 1965.
[18] S. J. Schiff: "Neural Control Engineering: The Emerging Intersection between Control Theory and Neuroscience" (MIT Press, Cambridge, MA, 2011).
[19] R. E. Plant: "A FitzHugh differential-difference equation modeling recurrent neural feedback," *SIAM J Appl Math*, vol. 40, pp. 150–162, 1981.
[20] E. B. Lee: "Approximation of linear input/output delay differential systems", in L. Markus, K. D. Elworthy, W. N. Everitt, and E. B. Lee (Eds.): *Differential equations, dynamical systems, and control science: A festschrift in honor of Lawrence Markus – Lecture Notes in Pure and Applied Mathematics* (M. Dekker, 1994), pp. 659–682.
[21] T. Munakata, S. Iwama, and M. Kimizuka: "Linear stochastic system with delay: Energy balance and entropy production," *Phys Rev E*, vol. 79, pp. 031104, 2009.
[22] S. Bernard, J. Belair, and M. C. Mackey: "Sufficient conditions for stability of linear differential equations with distributed delay," *Discrete Cont Dyn-B*, vol. 1, pp. 233–256, 2001.
[23] O. Calvo, D. R. Chialvo, V. M. Eguiluz, C. Mirasso, and R. Toral: "Anticipated synchronization: A metaphorical linear view," *Chaos*, vol. 14, pp. 7–13, 2004.
[24] H. U. Voss: "A universal negative group delay filter for the prediction of band-limited signals," https://arxiv.org/abs/1706.07326, pp. 1–11, 2017.
[25] T. Soderstrom and P. Stoica: "On the stability of dynamic-models obtained by least-squares identification," *IEEE T Automat Contr*, vol. 26, pp. 575–577, 1981.
[26] G. A. Williamson: "Adaptive IIR Filters", in V. K. Madisetti, and D. B. Williams (Eds.): *Digital Signal Processing Handbook* (CRC Press LLC, Boca Raton, FL, 1999), pp. Chapter 23.
[27] E. A. Robinson: "Statistical communication and detection; with special reference to digital data processing of radar and seismic signals" (Hafner Pub. Co. 1967).
[28] F. J. Brophy and A. C. Salazar: "Two design techniques for digital phase networks," *Bell Syst Tech J*, vol. 54, pp. 767–781, 1975.
[29] A. T. Chottera and G. A. Jullien: "A linear-programming approach to recursive digital-filter design with linear-phase," *IEEE T Circuits Syst*, vol. 29, pp. 139–149, 1982.
[30] B. Ravelo: "Elementary NGD IIR/FIR systems," *Int J Signal Proc Systems*, vol. 2, pp. 132–138, 2014.
[31] N. Stepp and M. T. Turvey: "Anticipation in manual tracking with multiple delays," *J Exp Psychol Hum Percept Perform*, vol. 43, pp. 914–925, 2017.
[32] C. H. Li and S. Y. Yang: "Synchronization in delayed Cohen-Grossberg neural networks with bounded external inputs," *IMA J Appl Math*, vol. 74, pp. 178–200, 2009.

[33] H. U. Voss: "Synchronization of reconstructed dynamical systems," *Chaos*, vol. 13, pp. 327–334, 2003.

[34] D. M. Brandman, S. S. Cash, and L. R. Hochberg: "Review: Human intracortical recording and neural decoding for brain computer interfaces," *IEEE Trans Neur Syst Rehab Eng*, vol. 99, pp. 1, 2017.

[35] H. U. Voss: "The zero lag time in manual tracking performance," Neuroscience Meeting, pp. Abstract Control Number 1483, Presentation Number 1176.1401, 2017.

[36] M. Casdagli: "Nonlinear prediction of chaotic time-series," *Physica D*, vol. 35, pp. 335–356, 1989.

[37] P. J. Brockwell and R. A. Davis: "Introduction to Time Series and Forecasting" (Springer, New York, 2002, 2nd edn).

[38] H. U. Voss, J. Timmer, and J. Kurths: "Nonlinear dynamical system identification from uncertain and indirect measurements," *Int J Bifurcat Chaos*, vol. 14, pp. 1905–1933, 2004.

[39] J. S. Barlow: "The Cerebellum and Adaptive Control" (Cambridge University Press, New York, 2002).

[40] J. P. Crutchfield: "Between order and chaos," *Nature Physics*, vol. 8, pp. 17–24, 2012.

[41] H. U. Voss: "A backward time shift filter for nonlinear delayed-feedback systems," *Phys Lett A*, vol. 279, pp. 207–214, 2001.

[42] H. U. Voss: "Anticipating chaotic synchronization," *Phys Rev E*, vol. 61, pp. 5115–5119, 2000.

[43] S. D. Pethel, N. J. Corron, Q. R. Underwood, and K. Myneni: "Information flow in chaos synchronization: Fundamental tradeoffs in precision, delay, and anticipation," *Phys Rev Lett*, vol. 90, pp. 254101, 2003.

[44] J. N. Blakely, M. W. Pruitt, and N. J. Corron: "Time shifts and correlations in synchronized chaos," *Chaos*, vol. 18, pp. 013117, 2008.

[45] J. Schumacher, T. Wunderle, P. Fries, F. Jakel, and G. Pipa: "A statistical framework to infer delay and direction of information flow from measurements of complex systems," *Neural Comput*, vol. 27, pp. 1555–1608, 2015.

[46] H. U. Voss: "Negative group delay of causal filters and its impact on effective connectivity analysis" *Conference abstract Emerging Tools for Acquisition and Interpretation of Whole-Brain Functional Data at Janelia*, 2017.

[47] E. L. Bolda, R. Y. Chiao, and J. C. Garrison: "Two theorems for the group velocity in dispersive media," *Phys Rev A*, vol. 48, pp. 3890–3894, 1993.

[48] H. U. Voss: "Fast response by synchronization," in K.-H. Hoffmann (Ed.): *"2nd caesarium – Coupling of Biological and Electronic Systems"* (Springer, Bonn, 2002), pp. 119–126.

[49] K. Pyragas, and T. Pyragiene: "Coupling design for a long-term anticipating synchronization of chaos," *Phys Rev E*, vol. 78, pp. 046217, 2008.

[50] C. Murguia, R. H. B. Fey, and H. Nijmeijer: "Network synchronization of time-delayed coupled nonlinear systems using predictor-based diffusive dynamic couplings," *Chaos*, vol. 25, pp. 023108, 2015.

[51] L. M. Pecora and T. L. Carroll: "Synchronization in chaotic systems," *Phys Rev Lett*, vol. 64, pp. 821–824, 1990.

[52] A. Pikovsky, M. Rosenblum, and J. Kurths: "Synchronization: A Universal Concept in Nonlinear Sciences" (Cambridge University Press, Cambridge, 2001).

[53] M. Ciszak, R. Toral, and C. Mirasso: "Coupling and feedback effects in excitable systems: Anticipated synchronization," *Mod Phys Lett B*, vol. 18, pp. 1135–1155, 2004.

[54] A. A. Budini and M. O. Caceres: "Adiabatic small noise fluctuations around anticipated synchronization: A perspective from scalar master-slave dynamics," *Physica A*, vol. 387, pp. 4483–4496, 2008.

[55] M. Y. Chen and J. Kurths: "Synchronization of time-delayed systems," *Phys Rev E*, vol. 76, pp. 036212, 2007.

[56] D. Ghosh, P. Saha, and A. R. Chowdhury: "On synchronization of a forced delay dynamical system via the Galerkin approximation," *Commun Nonlinear Sci*, vol. 12, pp. 928–941, 2007.

[57] G. Ambika and R. E. Amritkar: "Anticipatory synchronization with variable time delay and reset," *Phys Rev E*, vol. 79, pp. 056206, 2009.

[58] T. Pyragiene and K. Pyragas: "Anticipating of chaotic states via anti-phase diagonal coupling," *Lith J Phys*, vol. 50, pp. 167–172, 2010.

[59] T. W. Huang, D. Gao, C. D. Li, and M. Q. Xiao: "Anticipating synchronization through optimal feedback control," *J Global Optim*, vol. 52, pp. 281–290, 2012.

[60] A. Baraik, H. Singh, and P. Parmananda: "Experimental evidence of enhancement in the anticipation time by cascading," *Phys Lett A*, vol. 378, pp. 1356–1360, 2014.

[61] H. R. Dajani and J. C. H. Lam: "Prediction of pulsatile physiological signals using a negative group delay circuit," *Proceedings of the 1st WSEAS International Conference on Biomedical Electronics and Biomedical Informatics*, pp. 91–96, 2008.

[62] J. M. Bower and D. Beeman: "The Book of Genesis - Exploring Realistic Neural Models with the GEneral NEural SImulation System. Internet Edition." 2003.

[63] P. Baldi and A. F. Atiya: "How delays affect neural dynamics and learning," *IEEE Trans Neural Networks*, vol. 5, pp. 612–621, 1994.

[64] K. Pyragas: "Synchronization of coupled time-delay systems: Analytical estimations," *Phys Rev E*, vol. 58, pp. 3067–3071, 1998.

[65] W. Just, H. Benner, and E. Schöll: "Control of chaos by time-delayed feedback: A survey of theoretical and experimental aspects," *Adv Solid State Phys*, vol. 43, pp. 589–603, 2003.

[66] S. Tang and J. M. Liu: "Experimental verification of anticipated and retarded synchronization in chaotic semiconductor lasers," *Phys Rev Lett*, vol. 90, pp. 194101, 2003.

[67] A. Gonzalez, M. Vazquez-Prada, J. B. Gomez, and A. F. Pacheco: "Using synchronization to improve the forecasting of large relaxations in a cellular-automaton model," *Europhysics Letters*, vol. 68, pp. 611–617, 2004.

[68] M. Kostur, P. Hänggi, P. Talkner, and J. L. Mateos: "Anticipated synchronization in coupled inertial ratchets with time-delayed feedback: A numerical study," *Phys Rev E*, vol. 72, pp. 036210, 2005.

[69] C. M. Gonzalez, M. C. Torrent, and J. Garcia-Ojalvo: "Controlling the leader-laggard dynamics in delay-synchronized lasers," *Chaos*, vol. 17, pp. 033122, 2007.

[70] H. Huijberts, H. Nijmeijer, and T. Oguchi: "Anticipating synchronization of chaotic Lur'e systems," *Chaos*, vol. 17, pp. 013117, 2007.

[71] A. B. Cohen, B. Ravoori, T. E. Murphy, and R. Roy: "Using synchronization for prediction of high-dimensional chaotic dynamics," *Phys Rev Lett*, vol. 1012008.

[72] D. Ghosh, A. R. Chowdhury, and P. Saha: "On the various kinds of synchronization in delayed Dufng-Van der Pol system," *Commun Nonlinear Sci*, vol. 13, pp. 790–803, 2008.

[73] A. Wagemakers, J. M. Buldu, and M. A. F. Sanjuan: "Experimental demonstration of bidirectional chaotic communication by means of isochronal synchronization," *Europhys Lett*, vol. 81, pp. 40005, 2008.

[74] T. M. Hoang: "Transition among synchronous schemes in coupled nonidentical multiple time delay systems," *Chaos Solitons & Fractals*, vol. 42, pp. 938–950, 2009.

[75] B. Schmitzer, W. Kinzel, and I. Kanter: "Pulses of chaos synchronization in coupled map chains with delayed transmission," *Phys Rev E*, vol. 80, pp. 047203, 2009.

[76] S. Y. Xu and Y. Yang: "Predicting dynamic behavior via anticipating synchronization in coupled pendulum-like systems," *J Phys A*, vol. 42, pp. 335207, 2009.

[77] B. Nery and R. Ventura: "Online event segmentation in active perception using adaptive strong anticipation," *Biol Insp Cogn Architectures*, vol. 221, pp. 86–91, 2010.

[78] N. Stepp and M. T. Turvey: "On strong anticipation," *Cogn Syst Res*, vol. 11, pp. 148–164, 2010.

[79] F. Chiarello, L. Ursini, and M. Santagiustina: "Securing wireless infrared communications through optical chaos," *IEEE Photonic Tech L*, vol. 23, pp. 564–566, 2011.

[80] B. Nery and R. Ventura: "A dynamical systems approach to online event segmentation in cognitive robotics," *PALADYN J Behavioral Robotics*, vol. 2, pp. 18–24, 2011.

[81] S. Y. Xu, Y. Tang, H. D. Sun, Z. G. Zhou, and Y. Yang: "Characterizing the anticipating chaotic synchronization of RCL-shunted Josephson junctions," *Int J Nonlin Mech*, vol. 47, pp. 1124–1131, 2012.

[82] A. Alvarez-Aguirre, N. van de Wouw, T. Oguchi, and H. Nijmeijer: "Predictor-based remote tracking control of a mobile robot," *IEEE T Contr Syst T*, vol. 22, pp. 2087–2102, 2014.

[83] F. Havlak and M. Campbell: "Discrete and continuous, probabilistic anticipation for autonomous robots in urban environments," *IEEE T Robot*, vol. 30, pp. 461–474, 2014.

[84] T. J. McKetterick and L. Giuggioli: "Exact dynamics of stochastic linear delayed systems: Application to spatiotemporal coordination of comoving agents," *Phys Rev E*, vol. 90, pp. 042135, 2014.
[85] A. Ray and A. R. Chowdhury: "Robust synchronization of delay differential systems," *Int J Mod Phys C*, vol. 25, pp. 1350100, 2014.
[86] R. Yeniceri, S. Kilinc, and M. E. Yalcin: "Attack on a chaos-based random number generator using anticipating synchronization," *Int J Bifurcat Chaos*, vol. 25, pp. 1550021, 2015.
[87] H. Eberle, S. Nasuto, and Y. Hayashi: "Anticipating synchronisation for robot control," in N. F. Lepora, A. Mura, Mangan, M., P. F. M. J. Verschure, M. Desmulliez, and T. J. Prescott (Eds.): *Biomimetic and Biohybrid Systems, Living Machines 2016* (Springer International Publishing Switzerland, 2016), pp. 424–428.
[88] R. Jaimes-Reategui, V. P. Vera-Avila, R. Sevilla-Escoboza, G. Huerta-Cuellar, C. E. Castaneda-Hernandez, R. Chiu-Zarate, and A. N. Pisarchik: "Synchronization of unidirectionally delay-coupled chaotic oscillators with memory," *Eur Phys J-Spec Top*, vol. 225, pp. 2707–2715, 2016.
[89] S. J. Nasuto and Y. Hayashi: "Synapses in digital medium: Computational investigations of neural basis of anticipation," *Anticipation across Disciplines*, vol. 29, pp. 187–201, 2016.
[90] N. Ciobanu, S. Rusu, and V. Tronciu: "Anticipated synchronization of semiconductor lasers with passive dispersive reflectors," *Rom Rep Phys*, vol. 692017.
[91] T. Pyragiene and K. Pyragas: "Anticipatory synchronization via low-dimensional filters," *Phys Lett A*, vol. 381, pp. 1893–1898, 2017.
[92] R. C. Miall, D. J. Weir, D. M. Wolpert, and J. F. Stein: "Is the cerebellum a Smith predictor?," *J Motor Behav*, vol. 25, pp. 203–216, 1993.
[93] D. A. Engstrom, J. A. S. Kelso, and T. Holroyd: "Reaction-anticipation transitions in human perception-action patterns," *Hum Movement Sci*, vol. 15, pp. 809–832, 1996.
[94] B. Mehta and S. Schaal: "Forward models in visuomotor control," *J Neurophysiol*, vol. 88, pp. 942–953, 2002.
[95] M. Ozer, M. Uzuntarla, and S. N. Agaoglu: "Effect of the sub-threshold periodic current forcing on the regularity and the synchronization of neuronal spiking activity," *Phys Lett A*, vol. 360, pp. 135–140, 2006.
[96] A. B. Cohen, B. Ravoori, T. E. Murphy, and R. Roy: "Using Synchronization for Prediction of High-Dimensional Chaotic Dynamics," *Phys Rev Lett*, vol. 101, pp. 154102, 2008.
[97] N. Stepp and M. T. Turvey: "Anticipating synchronization as an alternative to the internal model," *Behav Brain Sci*, vol. 31, pp. 216–217, 2008.
[98] B. Pollok, J. Gross, D. Kamp, and A. Schnitzler: "Evidence for anticipatory motor control within a cerebello-diencephalic-parietal network," *J Cog Neurosci*, vol. 20, pp. 828–840, 2008.
[99] T. Pyragiene and K. Pyragas: "Anticipating spike synchronization in nonidentical chaotic neurons," *Nonlinear Dynamics*, vol. 74, pp. 297–306, 2013.

[100] J. M. Sausedo-Solorio and A. N. Pisarchik: "Synchronization of map-based neurons with memory and synaptic delay," *Phys Lett A*, vol. 378, pp. 2108–2112, 2014.

[101] F. Montani, O. A. Rosso, F. S. Matias, S. L. Bressler, and C. R. Mirasso: "A symbolic information approach to determine anticipated and delayed synchronization in neuronal circuit models," *Philos T R Soc A*, vol. 373, pp. 20150110, 2015.

[102] D. de Santos-Sierra, A. Sanchez-Jimenez, M. A. Garcia-Vellisca, A. Navas, and J. A. Villacorta-Atienza: "Effects of spike anticipation on the spiking dynamics of neural networks," *Frontiers in Computational Neuroscience*, vol. 9, pp. 144, 2015.

[103] R. Nijhawan and S. Wu: "Compensating time delays with neural predictions: are predictions sensory or motor?," *Philos T R Soc A*, vol. 367, pp. 1063–1078, 2009.

[104] M. Bar: *Predictions in the Brain: Using Our Past to Generate a Future* (Oxford University Press, New York, 2011).

[105] H. U. Voss and N. Stepp: "Signatures of negative group delay dynamics in neuroscience data," *Oral presentation, Neuroscience* 2016, *abstract number 10422*016.

[106] S. M. Crook, G. B. Ermentrout, and M. C. Vanier: "The role of axonal delay in the synchronization of networks of coupled cortical oscillators," *J Comput Neurosci*, vol. 4, pp. 161–172, 1997.

[107] L. P. Shayer and S. A. Campbell: "Stability, bifurcation, and multistability in a system of two coupled neurons with multiple time delays," *SIAM J Appl Math*, vol. 61, pp. 673–700, 2000.

[108] A. Washburn, R. W. Kallen, C. A. Coey, K. Shockley, and M. J. Richardson: "Harmony From Chaos? Perceptual-Motor Delays Enhance Behavioral Anticipation in Social Interaction," *J Exp Psychol Human*, vol. 41, pp. 1166–1177, 2015.

[109] J. G. Milton: "The delayed and noisy nervous system: implications for neural control," *J Neural Eng*, vol. 8, pp. 065005, 2011.

[110] M. Rohde, L. C. J. van Dam, and M. O. Ernst: "Predictability is necessary for closed-loop visual feedback delay adaptation," *J Vision*, vol. 14, pp. 4, 2014.

[111] R. C. Miall and G. Z. Reckess: "The cerebellum and the timing of coordinated eye and hand tracking," *Brain Cogn*, vol. 48, pp. 212–226, 2002.

[112] V. Braitenberg: "Is the cerebellar cortex a biological clock in the millisecond range?," *Prog Brain Res*, vol. 25, pp. 334–346, 1967.

[113] M. Ciszak, F. Marino, R. Toral, and S. Balle: "Dynamical mechanism of anticipating synchronization in excitable systems," *Phys Rev Lett*, vol. 93, pp. 114102, 2004.

[114] F. S. Matias, L. L. Gollo, P. V. Carelli, S. L. Bressler, M. Copelli, and C. R. Mirasso: "Modeling positive Granger causality and negative phase lag between cortical areas," *Neuroimage*, vol. 99, pp. 411–418, 2014.

[115] R. Nijhawan: "Visual prediction: Psychophysics and neurophysiology of compensation for time delays," *Behav Brain Sci*, vol. 31, pp. 179–198; discussion 198–239, 2008.

[116] M. Kawato: "Internal models for motor control and trajectory planning," *Curr Opin Neurobiol*, vol. 9, pp. 718–727, 1999.

[117] D. M. Wolpert and Z. Ghahramani: "Computational principles of movement neuroscience," *Nat Neurosci*, vol. 3, pp. 1212–1217, 2000.

[118] J. R. Flanagan, P. Vetter, R. S. Johansson, and D. M. Wolpert: "Prediction precedes control in motor learning," *Curr Biol*, vol. 13, pp. 146–150, 2003.

[119] H. U. Voss, B. D. McCandliss, J. Ghajar, and M. Suh: "A quantitative synchronization model for smooth pursuit target tracking," *Biol Cybern*, vol. 96, pp. 309–322, 2007.

[120] K. L. Downing: "Predictive models in the brain," *Connect Sci*, vol. 21, pp. 39–74, 2009.

[121] S. A. Kane, A. H. Fulton, and L. J. Rosenthal: "When hawks attack: animal-borne video studies of goshawk pursuit and prey-evasion strategies," *J Exp Biol*, vol. 218, pp. 212–222, 2015.

[122] M. Mischiati, H. T. Lin, P. Herold, E. Imler, R. Olberg, and A. Leonardo: "Internal models direct dragonfly interception steering," *Nature*, vol. 517, pp. 333–338, 2015.

[123] Y. Hayashi, J. Blake, and S. J. Nasuto: "Anticipatory engineering: Anticipation in sensory-motor systems of human," *Anticipation across Disciplines*, vol. 29, pp. 275–282, 2016.

[124] S. A. Oprisan and D. I. Austin: "A generalized phase resetting method for phase-locked modes prediction," *PLoS ONE*, vol. 12, pp. e0174304, 2017.

Chapter 4

Microwave transversal and recursive filter-based negative group delay circuits and non-Foster elements

Chung-Tse Michael Wu[1]

Abstract

This chapter presents some latest developments in microwave transversal and recursive filter-based negative group delay (NGD) circuits as well as their applications in realizing non-Foster elements. By using transversal and recursive-filter approaches, the desired amount of NGD can be synthesized with enhanced bandwidth. The fully distributed topology of the proposed NGD circuits makes it feasible to scale up to higher frequency ranges such as millimeter waves. In addition, it is demonstrated that the transversal-filter-based NGD circuit can realize non-Foster elements with unconditional stability, enabling a new approach to realizing stable non-Foster elements.

Keywords

Anomalous dispersion, dispersion engineering, negative group delay (NGD), non-Foster elements, recursive filters, transversal filters.

4.1 Introduction

The phenomenon of negative group delay (NGD) has recently gained much attention in the microwave community and can be utilized in many practical applications. For instance, NGD circuits can be incorporated to enhance the efficiency of the feed-forward amplifier, realize broadband phase shifters, and shorten delay lines [1–3]. This results from the fact that excessive positive group delays introduced by circuit components and electrical connections in electronic systems can be compensated.

[1]Department of Electrical and Computer Engineering, Rutgers University, USA

Studies of NGD phenomenon can be traced back to the 1970s, when analytical investigation and theoretical analysis of the Gaussian light propagation in the NGD medium was first proposed [4]. In the 1980s, the first experimentation of NGD effect was made in the optical domain [5]. Thereafter in the 1990s, theoretical and experimental investigations in the field of electronic and microwave engineering confirmed that certain passive circuit topologies are susceptible to generate NGD phenomenon [6,7]. For low-frequency electronic circuits, the NGD phenomenon has exhibited its feasibility using simple lowpass and bandpass amplifiers [8–10]. On the other hand, in the microwave regime, because NGD occurs at the frequency range where the maximum absorption or attenuation occurs [11], bandstop structures are used to realize NGD circuitries. Based on either series or shunt *RLC* resonators, various kinds of active and passive NGD circuits have been proposed and demonstrated [12–16].

In spite of successful demonstration of NGD, the lumped element approach may not be very suitable to operate at higher frequency ranges that go beyond their self-resonances. In an effort to create a fully distributed topology of NGD circuits, recently, transmission line-based NGD circuits have been proposed and demonstrated to realize the prescribed amount of NGD [17–20]. In this chapter, recent developments of distributed-topology NGD circuits using microwave transversal and recursive filters are discussed. It is shown that these distributed topology-based methods can synthesize the desired NGD to be utilized in electronic circuits. Furthermore, non-Foster elements realized by transversal filter-based NGD circuits will be introduced, enabling a new perspective of synthesizing stable negative inductance and capacitance in the microwave regime.

4.2 Transversal filter-based NGD circuits

The first kind of distributed NGD circuits is based on microwave transversal filters, which can be viewed as a microwave equivalent of finite impulse response filter widely used in the digital domain. Transmission lines in these structures are treated as delay elements to provide necessary time delay among the stages. By properly choosing the weighting/coupling coefficients along with delay elements, desired amount of NGD can be synthesized at the filters' output port.

Figure 4.1 illustrates a general diagram of a continuous time transversal filter. It consists of delay lines (τ), weighting coefficients (b_N), and summing operators. The relation of input $x(t)$ and output $y(t)$ in time domain of a transversal filter with N weighting elements or N sections can be expressed as [21]:

$$y(t) = \sum_{n=1}^{N} b_n \cdot x\left(t - \sum_{m=1}^{n} (\tau_{im} + \tau_{om})\right) \tag{4.1}$$

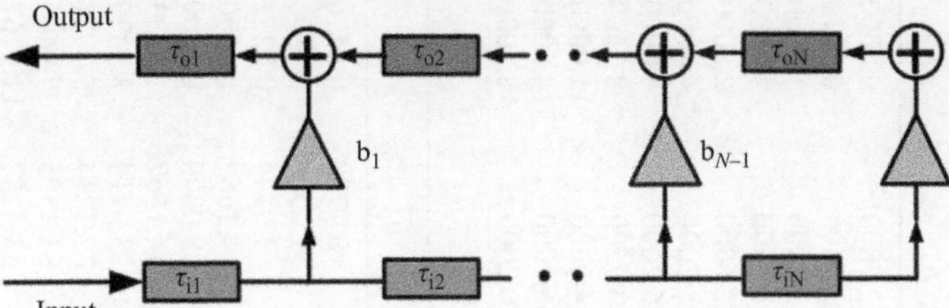

Figure 4.1 Schematic of a transversal filter [17]

Its transfer function in the frequency domain can then be obtained by taking the Fourier transform. Furthermore, if we let $\tau_{im} = \tau_{om} = \tau_0$, it follows that the transfer function becomes

$$H(\omega) = \sum_{n=1}^{N} b_n \cdot e^{-j2n\omega\tau_0} \tag{4.2}$$

By definition, the group delay of (4.2) at $\omega = \omega_0$ can be written as follows:

$$\tau_g = -\frac{\partial \angle H(\omega)}{\partial \omega}|_{\omega=\omega_0} \tag{4.3}$$

From (4.2), it can be observed that the phase response of the transfer function can be manipulated by altering the tap coefficients (b_N). It follows that for an N-stage transversal filter with maximally flat NGD responses, the coefficients b_n can be solved by the following equations with $\Omega_0 = \omega_0\tau_0 = \pi/2$:

$$|H_d(\Omega = \Omega_0)| = \left| \sum_{n=1}^{N} b_n \cdot e^{-j2n\Omega_0} \right| = 1 \tag{4.4}$$

$$-\frac{\partial \angle H_d(\Omega)}{\partial \Omega}|_{\Omega=\Omega_0} = \tau_{g0} \tag{4.5}$$

$$-\frac{\partial^{2n-3} \angle H_d(\Omega)}{\partial \Omega^{2n-3}}|_{\Omega=\Omega_0} = 0, n = 3, 4, \ldots, (N-1) \tag{4.6}$$

Equation (4.6) implies that the higher-order derivatives of group delay are set to zero in order to obtain a maximally flat response. The results calculated from (4.4)–(4.6) are listed in Table 4.1, which gives the exact coefficients of maximally flat NGD using transversal-filter topology for $N=2$, 3, 4, 5, and 6 sections, with various amount of NGD (τ_{g0}). It is also noticed that the amount τ_{g0} is in terms of the sampling period T. In this case, the sampling period T is equal to τ_0, and therefore the NGD will become a multiple of τ_0.

Table 4.1 *Coefficients for maximally flat NGD response*

	N = 2		N = 3			N = 4			
τ_{g0}	b_1	b_2	b_1	b_2	b_3	b_1	b_2	b_3	b_4
−1.0	2.5000	1.5000	3.5000	3.5000	1.0000	4.2000	5.4000	2.7000	0.5000
−1.5	2.7500	1.7500	4.0625	4.3750	1.3125	5.0469	7.0656	3.7406	0.7219
−2.0	3.0000	2.0000	4.6667	5.3333	1.6667	6.0000	9.0000	5.0000	1.0000
−3.0	3.5000	2.5000	6.0000	7.5000	2.5000	8.2500	13.7500	8.2500	1.7500
−4.0	4.0000	3.0000	7.5000	10.0000	3.5000	11.0000	19.8000	12.6000	2.8000
−5.0	4.5000	3.5000	9.1667	12.8333	4.6667	14.3000	27.3000	18.2000	4.2000
−6.0	5.0000	4.0000	11.0000	16.0000	6.0000	18.2000	36.4000	25.2000	6.0000
−7.0	5.5000	4.5000	13.0000	19.5000	7.5000	22.7500	47.2500	33.7500	8.2500
−8.0	6.0000	5.0000	15.1667	23.3333	9.1667	28.0000	60.0000	44.0000	11.0000
−9.0	6.5000	5.5000	17.5000	27.5000	11.0000	34.0000	74.8000	56.1000	14.3000
−10.0	7.0000	6.0000	20.0000	32.0000	13.0000	40.8000	91.8000	70.2000	18.2000

	N = 5					N = 6					
τ_{g0}	b_1	b_2	b_3	b_4	b_5	b_1	b_2	b_3	b_4	b_5	b_6
−1.0	4.7143	7.0714	4.7143	1.5714	0.2143	5.1071	8.5119	6.8095	3.0952	0.7738	0.0476
−1.5	5.8039	9.5594	6.7922	2.3719	0.3352	6.4014	11.7920	10.1074	4.8340	1.2568	0.1396
−2.0	7.0714	12.5714	9.4286	3.4286	0.5000	7.9444	15.8889	14.4444	7.2222	1.9444	0.2222
−3.0	10.2143	20.4286	16.7143	6.5000	1.0000	11.9167	27.0833	27.0833	14.5833	4.1667	0.5000
−4.0	14.3000	31.2000	27.3000	11.2000	1.8000	17.3333	43.3333	46.6667	26.6667	8.0000	1.0000
−5.0	19.5000	45.5000	42.0000	18.0000	3.0000	24.5556	66.1111	75.5556	45.3333	14.1667	1.8333
−6.0	26.0000	64.0000	61.7143	27.4286	4.7143	34.0000	97.1429	116.5714	72.8571	23.5714	3.1429
−7.0	34.0000	87.4286	87.4286	40.0714	7.0714	46.1429	138.4286	173.0357	111.9643	37.3214	5.1071
−8.0	43.7143	116.5714	120.2143	56.5714	10.2143	61.5238	192.2619	248.8095	165.8730	83.1395	7.9444
−9.0	55.3714	152.2714	161.2286	77.6286	14.3000	80.7500	261.2500	348.3333	238.3333	83.4167	11.9167
−10.0	69.2143	195.4286	211.7143	104.0000	19.5000	104.5000	348.3333	476.6667	333.6667	119.1667	17.3333

In fact, the closed-form solution can be obtained analytically using the same approach for bandpass linear phase filters provided in [22,23]. Let $t' = \tanh(j\Omega)$, and we can write

$$e^{-j2\Omega} = \frac{1-t'}{1+t'} \tag{4.7}$$

Then (4.2) can be expressed in terms of t as

$$H_d = e^{-j2\Omega} \times \sum_{m=0}^{M=N-1} a_m \cdot e^{-j2m\Omega} = e^{-j2\Omega} \times \sum_{m=0}^{M=N-1} a_m \cdot \left(\frac{1-t'}{1+t'}\right)^m$$
$$= e^{-j2\Omega} \times \frac{\sum_{m=0}^{M=N-1} a_m (1+t')^{M-m}(1-t')^m}{(1+t')^M} = e^{-j2\Omega} \times H(t') \tag{4.8}$$

It can be proved that the denominator of $H(t')$, i.e., $(1+t')^M$, has a constant delay of M at all frequencies [24]. Furthermore, it has been shown that if the numerator of $H(t')$ has a maximally flat group delay about $t' = \infty$, i.e., $\Omega = \pi/2$, it must be of the form of a symmetrical Jacobi polynomial, $Q_R{}^{(\alpha)}(t')$, which has a delay of α [22]. The explicit form of $Q_R{}^{(\alpha)}(t')$ can be expressed as

$$Q_R{}^{(\alpha)}(t) = \sum_{r=0}^{R} \frac{\binom{R+\alpha}{r}\binom{\alpha-1-r}{R-r}(1+t)^r(1-t)^{R-r}}{2\binom{2R-1}{R}} \tag{4.9}$$

Thus, the total delay τ_g of H_d depicted in (4.8) will be $M-\alpha+2$ or $N-\alpha+1$. Furthermore, by comparing (4.9) and the numerator of $H(t)$ in (4.8), we can obtain the coefficients:

$$a_{N-m-1} = \frac{\binom{N+\alpha-1}{m}\binom{\alpha-1-m}{N-m-1}}{2\binom{2N-3}{N-1}} = b_{N-m} \tag{4.10}$$

where b_{N-m} refers to the coefficients shown in Figure 4.1 and Table 4.1. To give a quantitative example, if we have a transversal filter with $N=4$ and we let the index $\alpha = 12$, then the total delay will be $4-12+1=-7$. The coefficients, according to (4.10), are $(a_0, a_1, a_2, a_3) = (b_1, b_2, b_3, b_4) = (455, 945, 675, 165)/20 = (22.75, 47.25, 33.75, 8.25)$, which results in an identical solution obtained by (4.9) and listed in Table 4.1 for $N=4$ and $\tau_{g0} = -7$.

As an illustrative example, assuming the sampling period $T = \tau_0 = 0.25$ *ns* and τ_{g0} is chosen to be -4, we can obtain the corresponding coefficients for different section numbers from Table 4.1. Figure 4.2 plots the frequency responses of the transversal-filter-based NGD circuits with different numbers of section. Since we

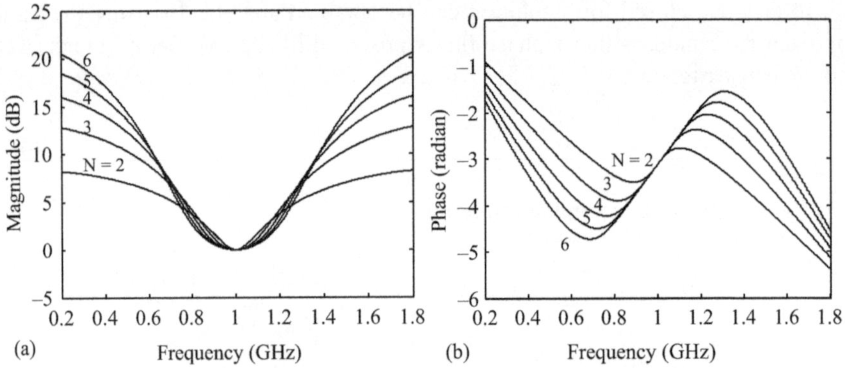

Figure 4.2 (a) Amplitude and (b) phase responses of various sections for transversal-filter-based NGD circuits [17]

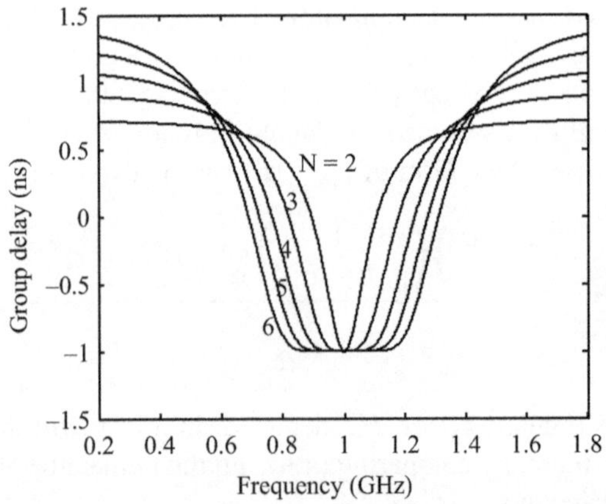

Figure 4.3 Group delay of various numbers of sections for transversal filters [17]

choose the sampling period to be 0.25 ns, at 1 GHz, the electrical length of each section becomes $\pi/2$, which is equal to Ω_0 that we set to solve (4.4)–(4.6). Therefore, at 1 GHz, the magnitudes of the transfer functions are all unity or 0 dB as shown in Figure 4.2(a). The phases maintain a constant and positive slope around the center frequency 1 GHz; one can observe that from Figure 4.2(b), as the section number increases, the bandwidth of this constant positive slope of phase also increases.

The corresponding group delays for various numbers of sections are plotted in Figure 4.3. NGDs with maximally flat responses can be clearly observed from the figure. As a consequence of the phase responses shown in Figure 4.2(b), the bandwidth of the flat NGD responses can be augmented when more sections are used in the transversal filter. Moreover, the amount of NGD is 1 ns, which results from the sampling period of 0.25 ns and τ_{g0} of -4 that we choose for this case. It is worth mentioning that the bandwidth of NGD can theoretically be kept

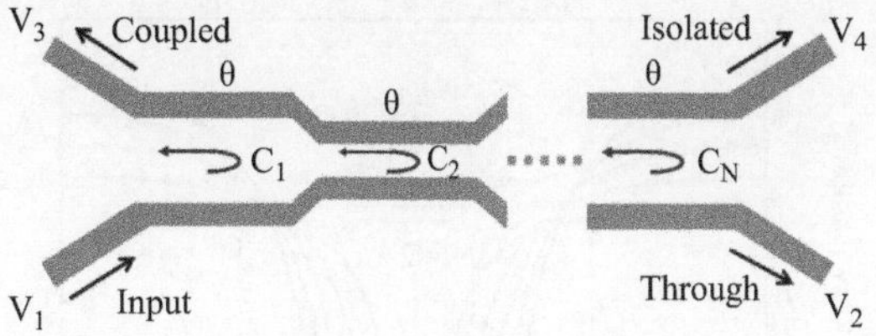

Figure 4.4 An N-section directional coupler [17]

increasing by adding more sections to the transversal filter. In addition, it is noted that here we have made an effort in making the group delay maximally flat or equivalently generating a linear phase response. On the other hand, the amplitude response is uniquely related to the phase response through Hilbert transforms in the case of single input–output path networks, i.e., minimum-phase networks [25]. Therefore, in the present case, the amplitude is determined once the condition of maximally flat group delay is imposed. Nevertheless, one can always slightly loosen the maximally flat group delay condition to adjust the amplitude response so that it can satisfy the desired requirement.

4.2.1 Passive realization

A simple realization of a passive transversal filter in microwave regimes is to use a multisection directional coupler as shown in Figure 4.4. Assuming that the coupling is weak ($C_n \leq -10$ dB) and each section is quarter-wavelength long ($\theta = \pi/2$) at the center frequency, the total voltage transfer function of the coupled port of an N-section multisection directional coupler can be expressed as [26]:

$$\frac{V_3}{V_1} = \sum_{n=1}^{N} (jC_n \sin\theta) e^{-j(2n-1)\theta} \tag{4.11}$$

where C_n is the coupling factor for nth section of the directional coupler, and θ is the electrical length of coupled lines. It is noted that (4.11) has the similar mathematical form as (4.2) if we map θ to Ω. Although the coefficients of (4.11) have a $j\sin\theta$ dependency, it will not affect the mutual ratios among each coefficients since the $j\sin\theta$ dependency will be cancelled out. The main difference between (4.11) and (4.2) is that since (4.11) starts from $e^{-j\theta}$ instead of $e^{-j2\theta}$, it will have one less pole compared to (4.2), which will result in an NGD increased by one sampling period. Using (4.11), Figure 4.5 plots the group delays in terms of the sampling periods versus the electrical length θ for various sections of directional couplers, in which the coupling factors are the same as the coefficients used previously, i.e., those for $\tau_{g0} = -4$. From Figure 4.5, we can find that the NGD increases by one sampling period to -5 as mentioned above, and we also observe that as more sections of directional coupler are used, the bandwidth of NGD will increase accordingly.

To verify the NGD circuits based on transversal-filter concepts, we fabricated three different microstrip line-based multisection directional couplers for $N = 2$, 3,

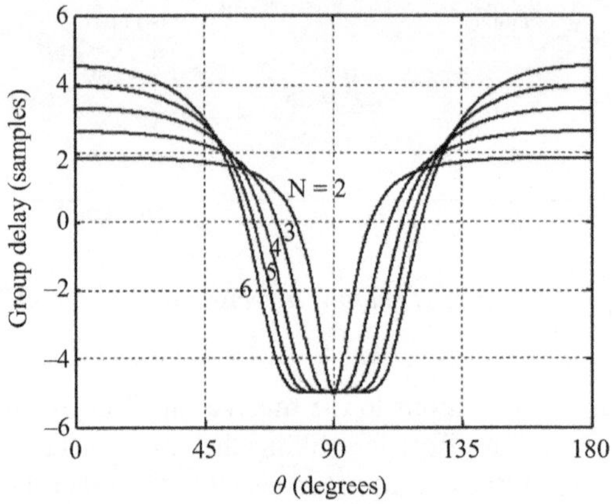

Figure 4.5 Group delays of various sections of direction coupler calculated from (4.11) [17]

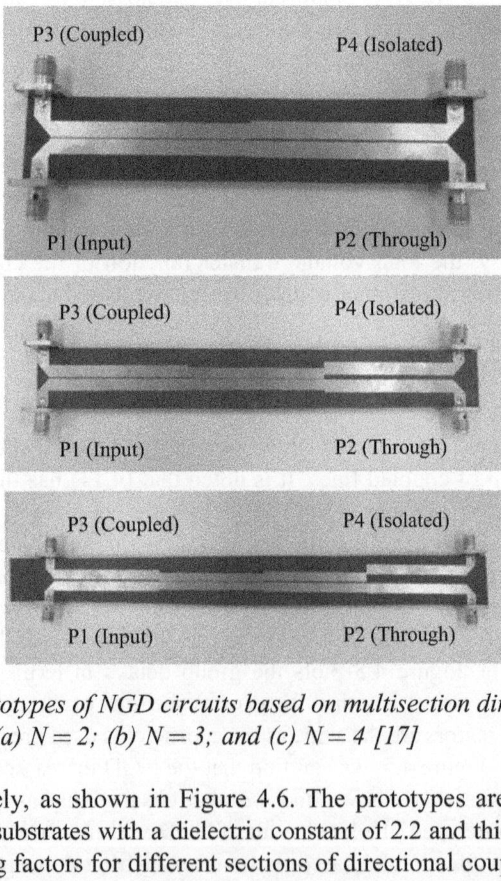

Figure 4.6 Prototypes of NGD circuits based on multisection directional couplers for (a) $N=2$; (b) $N=3$; and (c) $N=4$ [17]

and 4, respectively, as shown in Figure 4.6. The prototypes are fabricated using RT/duroid 5880 substrates with a dielectric constant of 2.2 and thickness of 62 mils.

The coupling factors for different sections of directional couplers ($N=2$, 3, 4) are based on the coefficients listed in Table 4.1 for $\tau_{g0} = -4$, as shown in (4.12),

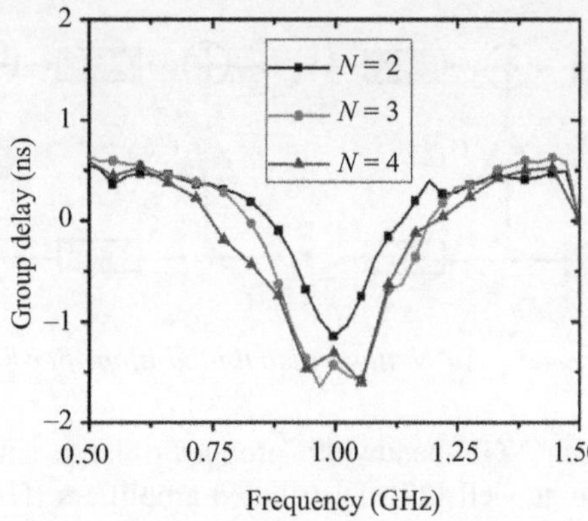

Figure 4.7 Measured group delay for NGD circuits based on multisection directional couplers [17]

in which the maximum coefficients are normalized to −10 dB, which is 0.3162 in linear scale.

$$\mathbb{C}_{N=2} = [0.3162, 0.2372] \tag{4.12a}$$

$$\mathbb{C}_{N=3} = [0.2372, 0.3162, 0.1107] \tag{4.12b}$$

$$\mathbb{C}_{N=4} = [0.1757, 0.3162, 0.2012, 0.0447] \tag{4.12c}$$

It is noted that the coefficients in (4.12) are calculated from the ideal transversal filters and may be used as a starting point to find out the actual dimensions of the directional couplers shown in Figure 4.6. To elaborate further, although (4.11) is a good approximation for a multisection directional coupler with a low coupling ratio, it might not accurately reflect the phase response due to the tiny secondary coupling from the output line to input line. This can be an issue especially when the coupling is strong, as the phase response is more sensitive than the amplitude response. Therefore, in practice, it is necessary to do a post-layout optimization of the structure with a full-wave simulation using the ideal coefficients given in Table 4.1 as starting numbers.

The center frequency of the directional couplers is designed to be 1 GHz, which means the electrical length θ is $\pi/2$ at 1 GHz. Figure 4.7 plots the measured group delay of the fabricated prototypes using printed circuit board (PCB) technology with various number of stages for directional couplers. Overall, with more sections used in directional couplers, the bandwidth of NGD increases. Moreover, as mentioned above, a flat response of NGD can be achieved by using the proper coupling coefficients for each section of the directional coupler.

4.2.2 Active realization

Although a multisection directional coupler can provide a very simple way to realize the form of a transversal filter, it suffers from tremendous loss in NGD

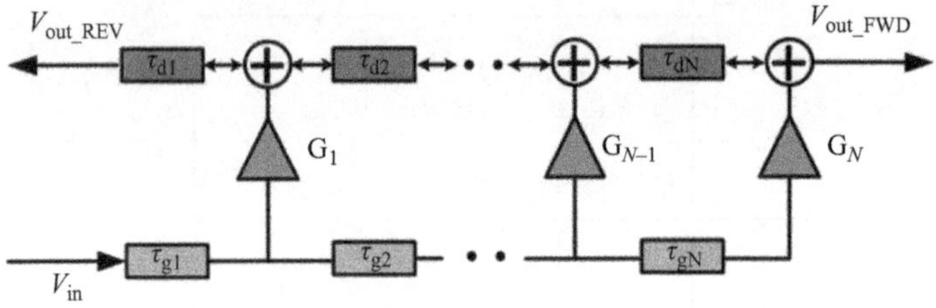

Figure 4.8 An N-stage distributed amplifier [17]

region, especially when NGD-bandwidth product is large since the out-of-band gain needs to be large as well [27]. Distributed amplifiers (DAs) stand out as an excellent choice to provide amplification for the NGD circuits, being essentially one type of transversal-filer in active fashion. Figure 4.8 illustrates the diagram of an N-stage DA, in which G_k is the gain coefficient of the kth tap amplifier, τ_{gk} and τ_{dk} indicate the time delay of the kth section of the gate line and drain line, respectively. Furthermore, if we choose that $\tau_{gi} = \tau_{di} = \tau_0$, then the transfer functions of the forward and reverse port with respect to the input can be expressed as

$$H_{\mathrm{FWD}}(\omega) = \frac{V_{\mathrm{out\_FWD}}}{V_{\mathrm{in}}} = \sum_{k=1}^{N} G_k e^{-j\omega(N\tau_0)} \tag{4.13a}$$

$$H_{\mathrm{REV}}(\omega) = \frac{V_{\mathrm{out\_REV}}}{V_{\mathrm{in}}} = \sum_{k=1}^{N} G_k e^{-j\omega(2k\tau_0)} \tag{4.13b}$$

One can immediately identify that (4.13b) has the same mathematical form as (4.2) and can further observe that if we use the coefficients that we have derived in Table 4.1 for the gain coefficients G_k in (4.13), we will have NGD response at the DA's reverse port, while maintaining a flat gain at its forward port as G_k's merely sum up constructively as indicated in (4.13a).

It is noted that the coefficients G_k are in fact related and proportional to the transconductance (g_m) of an active device. Therefore, the idea can be quickly examined using the circuit simulator in advanced design system (ADS) for the ideal situation. Figure 4.9 shows the schematics in ADS for a two-stage DA, in which the unit-cell contains ideal transmission lines with the characteristic impedance of 50 Ω and phase of 90°, as well as an ideal voltage-control-current-source (VCCS) with a transconductance of g_m (S). Furthermore, if we choose the amounts of g_m for the two-stage to be large enough as (0.4 S, 0.3 S), obtained from Table 4.1 ($N=2$, $\tau_{g0} = -4$) with a factor of one-tenth, then we will obtain the responses shown in Figure 4.10. Figure 4.10(a) plots the forward (S_{41}) and reverse gain (S_{31}) of the DA; as expected, the reverse gain reaches minimum at the center frequency of 1 GHz during NGD region as seen in Figure 4.10(b), whereas the forward gain maintains a flat response. It can be seen that if the transconductance of active devices is large enough, even though the reverse gain reaches the minimum, it will still have gain. In fact, the NGD-bandwidth product is proportional to the maximum of out-of-band

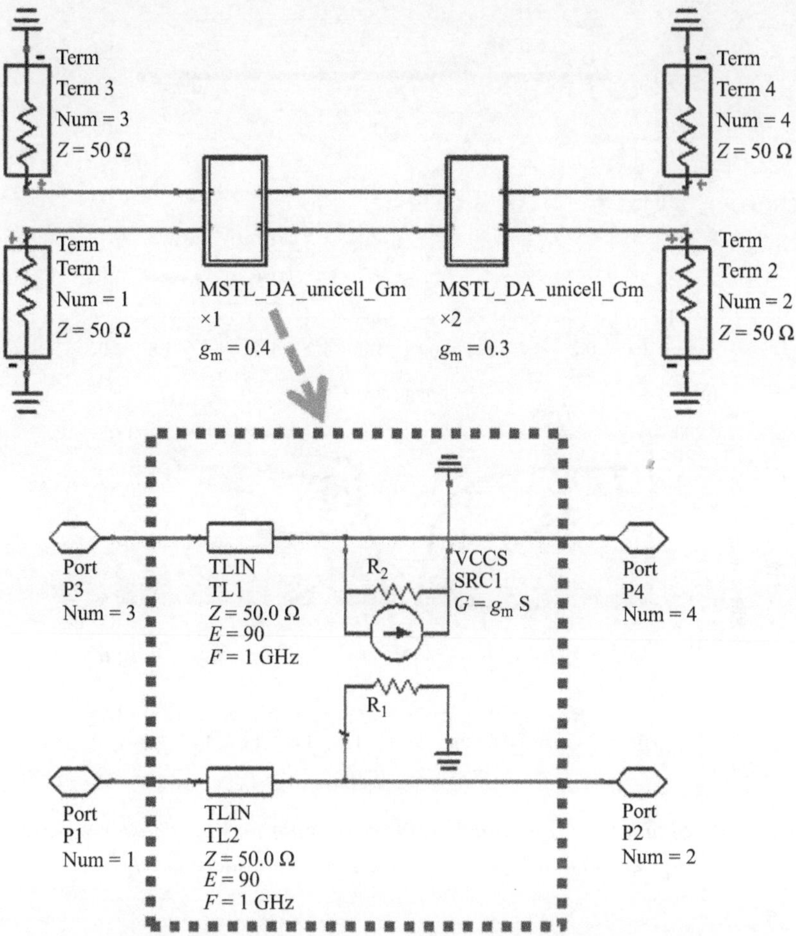

Figure 4.9 Schematics of an ideal two-stage DA [17]

gain [20]; it follows that if NGD-bandwidth product is small, then it does not require so much gain from the amplifiers.

As a proof-of-concept, we fabricated a DA using commercial off-the-shelf discrete transistors NEC NE3210S01. As shown in Figure 4.11, the DA contains two transistors with individual gate biases to form a two-stage NGD circuit. The electrical length between each stage is designed to be quarter-wavelength at the center frequency of 1 GHz. The maximum transconductance of NE3210S01, according to the data sheet, can be found to be around 0.075 S, in which the bias condition is $V_g = -0.3$ V and $V_d = 2$ V; therefore, we use this number as the maximum in our gain coefficients calculation. Figure 4.12 shows the measured *S*-parameters of the fabricated two-stage DA. As expected, the forward gain shows a flat response of around 10 dB gain, whereas the reverse gain has a minimum of −7.8 dB at the center frequency 1 GHz where NGD occurs. In addition, the return

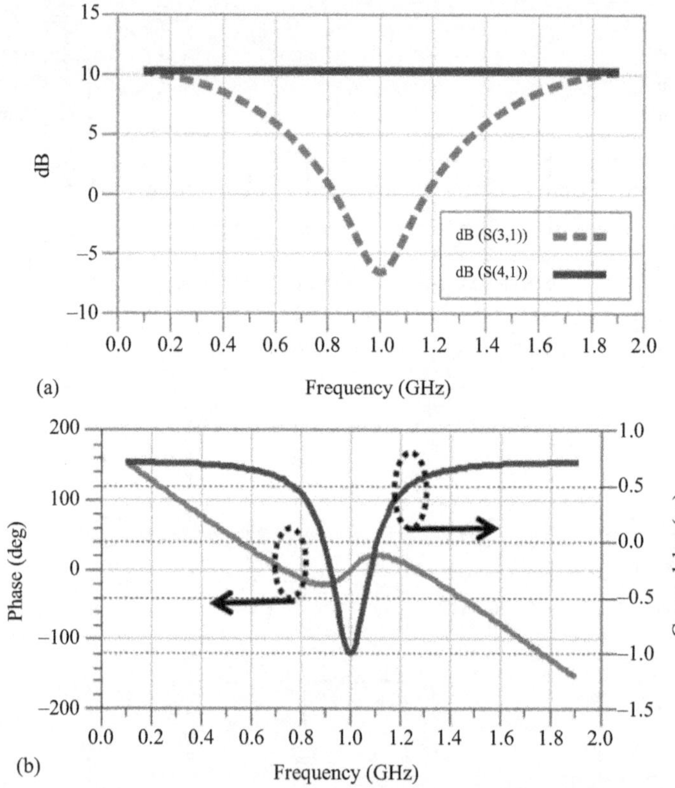

Figure 4.10 Simulated (a) forward and reverse gain and (b) phase response and group delay at the reverse port of an ideal DA [17]

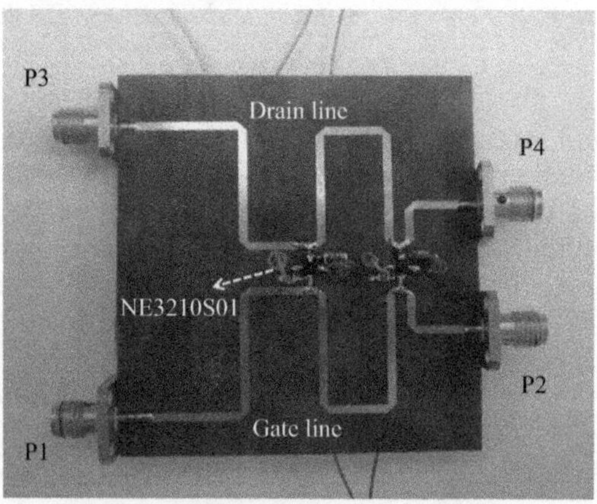

Figure 4.11 Prototype of a two-stage NGD circuits based on DAs [17]

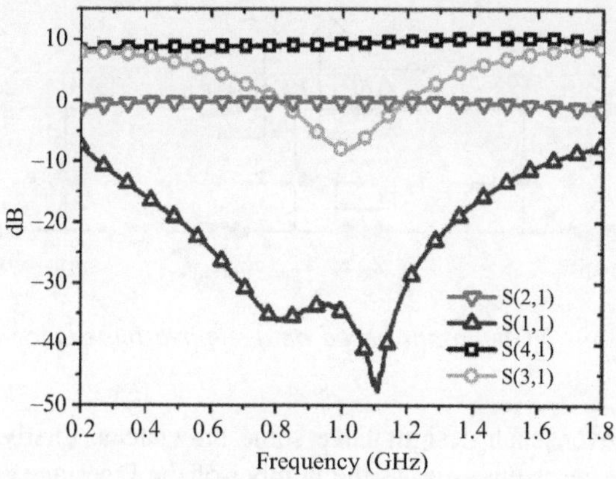

Figure 4.12 Measured S-parameters of the fabricated two-stage negative group delay distributed amplifier (NGDDA) [17]

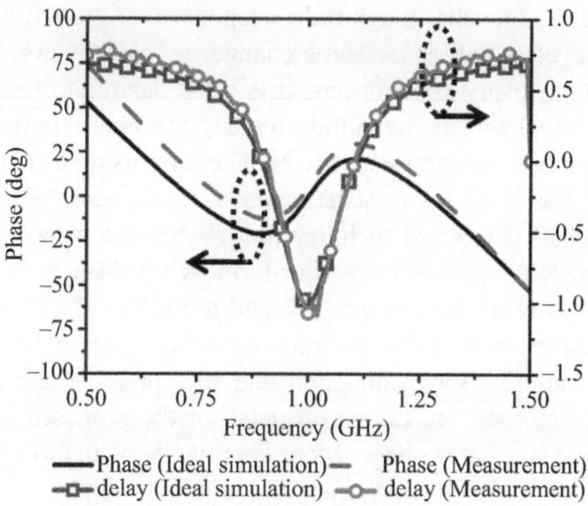

Figure 4.13 Comparison between the ideal circuit simulation and measured results of the two-stage NGDDA [17]

loss is >30 dB and the insertion loss is around 0.2 dB in the frequency of interest. The phase and group delay of the DA are plotted in Figure 4.13, indicating a great agreement between the ideal circuit simulation and measurement. The maximum group delay of the two-stage DA is −1 ns at 1 GHz and the NGD region of which is 0.9–1.1 GHz.

4.2.3 Voltage-insensitive stepped-impedance DA-based NGD circuit

In the aforementioned active transversal filter-based NGD scheme, DAs are applied to achieve the required NGD by varying the transconductance ratio of transistors in

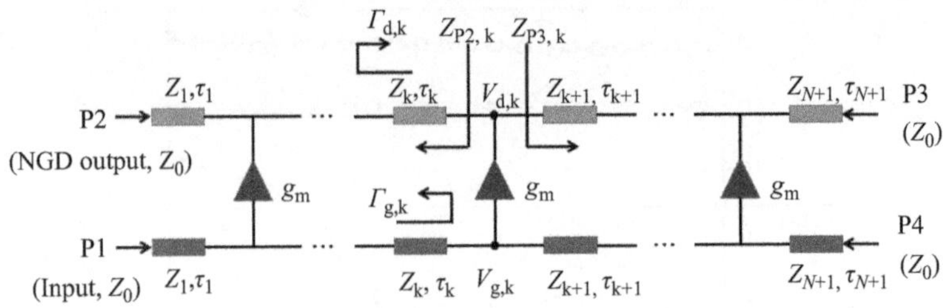

Figure 4.14 NGD circuit based on a stepped-impedance DA [28]

each stage. However, such designs have some drawbacks. Firstly, the number of required bias voltage is the same as the number of the DA stage. Therefore, when the stage number increases, excessive numbers of gate voltages are needed and may be difficult to implement for the overall circuit design. Secondly, it is noted that the transistor transconductance is very sensitive to the gate voltage, in which some fluctuation in the bias voltage may bring degradation to the NGD characteristics.

In addition, although the magnitude responses are configurable through adjusting the bias level, it will cause some change in group delays. This might not be advantageous, as in many applications, it is often desired to weaken the correlation between the NGD and its magnitude response to facilitate the adjustment.

These issues can be alleviated with the NGD circuit using a stepped-impedance DA topology [28]. Based on a two-stage stepped-impedance DA, only one single voltage for the gate bias is selected such that the two transistors have the same transconductance. Meanwhile, quarter-wave transformers (QWTs) with different characteristic impedances are applied to the gate and drain line of DA to modulate the voltage gain coefficients. In so doing, the ratio of voltage gain in different stages can meet the NGD conditions. Both simulation and measurement show that once the transistor channel is pinched on, the group delay of the proposed circuit is almost constant with respect to the gate voltage. More importantly, with this configuration, the group delay and gain/loss characteristics are no longer relevant.

Figure 4.14 shows the NGD circuit using an N-stage stepped-impedance DA. For the kth stage, the characteristic impedances of gate and drain transmission line are both equal to Z_k, with τ_k being the associated time delay. The NGD phenomenon can be observed between P2 (reversed port) and P1 (input port). Assuming $\tau_k = \tau_0$, in order to synthesize the desired NGD, the transfer function should satisfy

$$\text{minimize } \max\left\{\left|\sum_{k=1}^{N} G_k e^{-j\omega 2k\tau_0} - e^{jb\omega}\right|\right\} \tag{4.14}$$

where G_k is the voltage gain coefficient for the kth transistor, $-b$ is the desired value of group delay, and τ_0 indicates the time delay of the transmission line sections [18]. It has already been proved that by achieving a specific G_k ratio, the NGD response can be realized. However, different from [17,18], this structure utilizes stepped-impedance quarter-wave lines to implement the specific G_k ratio. It is noted at the center frequency, each line segment all have the identical phase delay, i.e., $\omega_0\tau_0 = \pi/2$. Therefore, around the center frequency, based on the transmission line theory, G_k can be expressed as:

$$G_k = g_m \frac{Z_{p2,k}Z_{p3,k}}{Z_{p2,k} + Z_{p3,k}} \prod_{i=1}^{k} \left[\frac{1+\Gamma_{g,i}}{1-\Gamma_{g,i}}\right] \prod_{j=1}^{k} \left[\frac{1+\Gamma_{d,i}}{1-\Gamma_{d,i}}\right] \quad (4.15)$$

where g_m is the transconductance, $Z_{P2,k}$ and $Z_{P3.k}$ are output impedances seen to Port 2 and Port 3, $\Gamma_{g,k}$ and $\Gamma_{d,k}$ are reflection coefficients for the kth transistor, and they can be calculated as:

$$\Gamma_{g,k} = \frac{Z_{p3,k} - Z_k}{Z_{p3,k} + Z_k} \quad (4.16a)$$

$$\Gamma_{d,k} = \frac{Z_{p2,k-1} - Z_k}{Z_{p2,k-1} + Z_k} \quad (4.16b)$$

From (4.15) and (4.16), it can be observed that if we let g_m be a constant, G_k can be solely controlled by the characteristic impedances of gate and drain lines. In addition, $Z_1 = Z_{N+1}$ is enforced, so the number of unknown characteristic impedance is N. The ratio of G_k can be applied to obtain the corresponding $N-1$ equations and the last equation can be achieved through the absolute value of G_k. In so doing, for an N-stage DA, there are N unknown characteristic impedances and N corresponding equations, and therefore the characteristic impedances can be solved. Since the PCB process is fairly stable, this impedance ratio can be very accurate. As the ratio of G_k now is no longer determined by the transconductance g_m, the NGD characteristic becomes insensitive to the gate voltage. It is worth mentioning that the previous case in Figure 4.8 can be viewed as a special case when all characteristic impedances of lines are 50 Ω, where the ratio of the gain coefficients is achieved solely by transconductances, as can be seen from (4.15).

Another feature we can observe is that the transconductance g_m of transistor is now only related to the gain response of the overall circuit. NGD circuits based on this method can therefore have a steady group delay with a configurable gain. Figure 4.14 shows the simulated results of an NGD circuit based on a two-stage DA using ideal VCCS transistors. In this design, we have $Z_1 = Z_3 = 40$ Ω, $Z_2 = 33.7$ Ω, with G_1/G_2 equal to 44/39. It can be seen that the circuit exhibits a group delay of -1 ns at most when the slope of phase is positive. Additionally, the gain between the reversed output port and input port increases when the transconductance increases, as shown in Figure 4.15(a). The group delay response remains completely the same as shown in Figure 4.15(b) under different transconductance conditions, which is in agreement with the previous theoretical analysis. Figure 4.15(c) plots the NGD responses for $N = 2$, 3, and 4, where the corresponding Z_k and G_k are listed in Table 4.2. It can be seen that when the number of order increases, the bandwidth increases as well. Figure 4.15(d) shows the NGD and magnitude characteristics when Z_2 varies, from which it can be seen that when the amount of NGD increases, the insertion loss also increases.

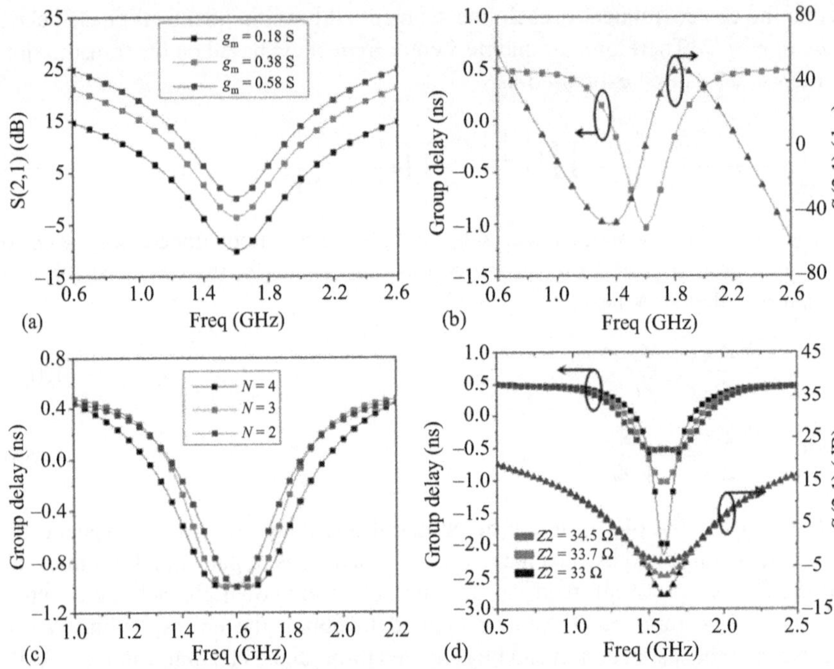

Figure 4.15 Simulated results of an ideal stepped-impedance DA-based NGD circuit: (a) magnitude versus transconductance g_m ($N = 2$); (b) phase response and NGD remains unchanged for different g_m ($N = 2$); (c) NGD for stepped-impedance DA with $N = 2$, 3, 4; and (d) NGD and gain performance when Z_2 varies ($N = 2$) [28]

Table 4.2 Z_K and G_K when $N = 2$, 3, and 4 [28]

	Z_1	Z_2	Z_3	Z_4	Z_5
$N=2$	40	33.7	40	N/A	N/A
$N=3$	50.4	53.1	64	50.4	N/A
$N=4$	49.5	41.5	64.8	60.95	49.5
	G_1	G_2	G_3	G_4	
$N=2$	7.1	6.3	N/A	N/A	
$N=3$	5.4	12.4	7.8	N/A	
$N=4$	4.6	8.1	11.1	7.2	

To validate the proposed concept, we fabricate the entire NGD circuit using RT/Duroid 5880 with a dielectric constant of 2.2 and thickness of 0.508 mm as shown in Figure 4.16. The lengths of all line segments are all 31 mm, and the widths are 1.5 mm for the input/output port line segments, and 3.1 mm for the middle-stage

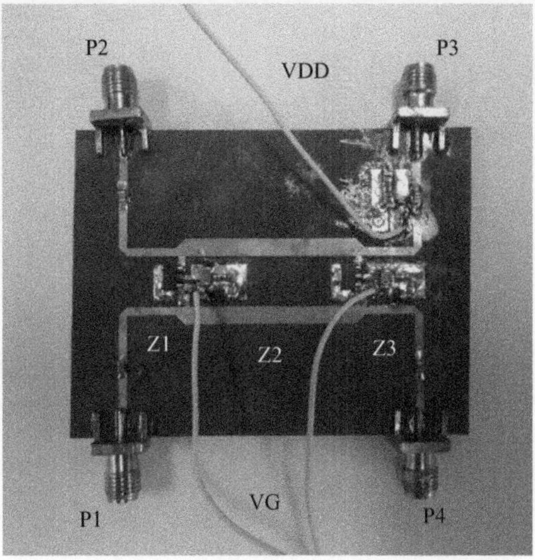

Figure 4.16 Prototype of the stepped-impedance DA-based NGD circuit [28]

lines. For the proof-of-concept, the NGD circuit is based on a two-stage DA structure, in which we use Avago's enhancement mode HEMT, ATF50189, as the active device for the DA to achieve a relatively higher transconductance.

In order to ensure the stability, a 3.9 nH shunt-inductor is placed at the gate to make the transistor unconditionally stable throughout the entire band (0–20 GHz). Due to parasitic effects in these transistors, the characteristic impedance of the transmission line is optimized according to the simulation model provided from the datasheet. Both of the transistors operate at the same biasing point. The drain voltage is 4.5 V and the single gate bias varies from 0.3 to 0.6 V for the validation.

The simulated results carried out by the ADS and the experimental results are plotted in Figure 4.17. Figure 4.17(a) shows the simulated and measured frequency response for the proposed NGD circuit. The gate voltage in this case is set to 0.5 V for the both transistors. It can be seen that the simulation and measurement show good consistency with each other. The maximum NGD is −1 ns and the circuit can exhibit the NGD characteristic in 1.6–1.73 GHz. The magnitude where the maximum NGD occurs is −5.3 dB. This result proves that the proposed structure can indeed realize the desired NGD by modulating the impedance value of the transmission line of DA under the same bias condition. Although this prototype does not provide gain, it can be seen from Figure 4.15(a) that if the devices with higher g_m are utilized, positive gain can be realized.

Figure 4.17(b) plots the measured NGD at three different gate voltages. The bias voltage varies from 0.4 to 0.6 V with 0.1 V step. It can be seen from the figure that the three NGD curves have almost exactly the same shape under different bias values. Especially at 0.5 and 0.6 V, the two curves completely overlap each other. On the other hand, the structure in the previous work [18] shows a 2 ns

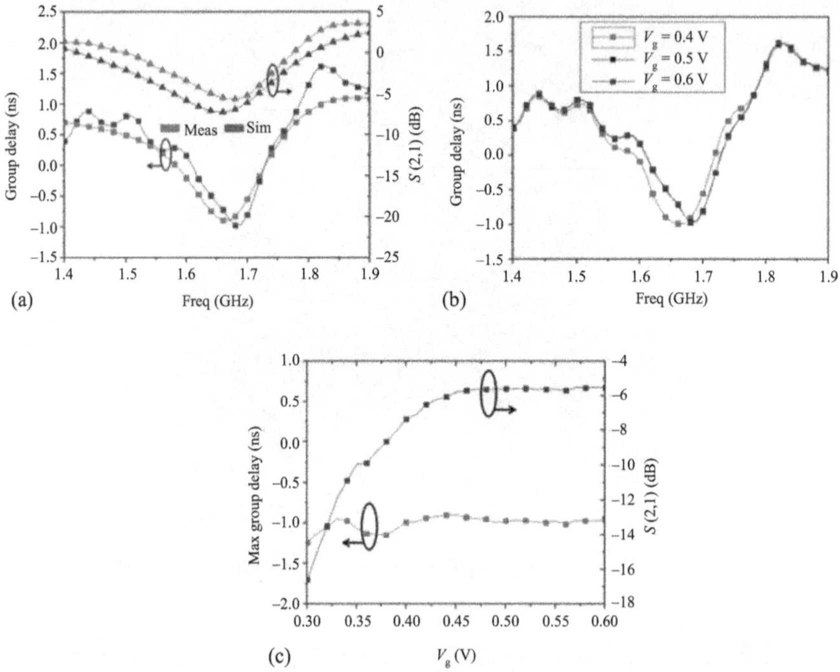

Figure 4.17 Measured results of the proposed stepped-impedance DA-based NGD circuit: (a) group delay and dB (S(2,1)) when $V_g = 0.5$ V; (b) group delay versus different bias condition; and (c) maximum NGD and minimum dB (S(2,1)) under different bias conditions [28]

variation of NGD under a gate voltage change of 0.03 V. It is therefore apparent that the NGD response of the proposed stepped-impedance DA structure is insensitive to the bias voltage and to the transistor transconductance.

Figure 4.17(c) shows the measured maximum NGD and gain/loss performance of the proposed circuit versus gate voltages. The transistor channel is pinched on at $V_g = 0.3$ V and the maximum testing gate bias voltage is 0.6 V. It can be seen that within this range, the maximum NGD change is <0.2 ns. It is also noted that when the transconductances of the transistors become larger, the proposed NGD circuit exhibits a 10 dB dynamic tuning range in the magnitude. This proves that the stepped-impedance design is able to decorrelate the relationship between NGD and the magnitude response, demonstrating a good agreement with the theory.

4.3 Recursive filter-based NGD circuits

In addition to transversal filters, microwave recursive filters can also be used to generate NGD of predetermined values in the desired frequency band by using only distributed components [29]. To illustrate, multistage QWTs are used to generate

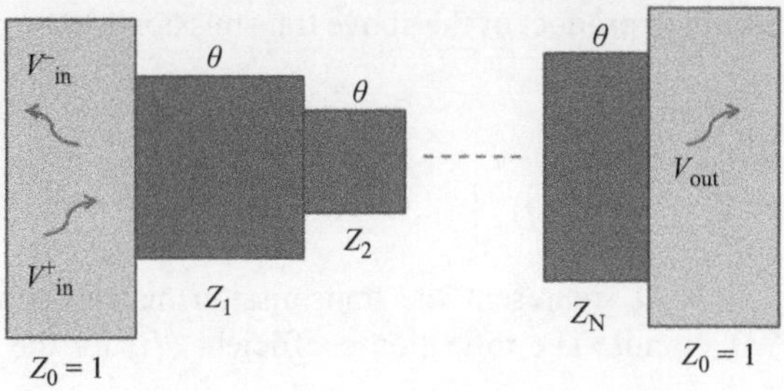

Figure 4.18 An N-stage QWT [29]

NGD of predetermined values as shown in Figure 4.18. Essentially, it behaves as a recursive filter since the reflected waves will bounce back and forth among the impedance interfaces and form an infinite impulse response (IIR). Following this concept, we will show a comprehensive methodology to synthesize the desired NGD by treating the multistage QWT as an IIR filter. In addition, it is worth mentioning that there are no lumped components present, which makes this design easy to fabricate and to be scaled up to higher frequency.

A microstrip-line-based transmission line structure is used to realize QWTs. As depicted in the Figure 4.18, the QWTs have the characteristic impedance of Z_1, Z_2, Z_3, …, Z_N with an identical electrical length of θ. The impedances of all QWTs have been normalized to the termination impedance of 50 Ω. The design flow is as follows: First, we need to obtain the characteristic impedance of each of the microstrip transmission line QWT that can generate the desired group delay. Once we obtain the characteristic impedance, the width and length of the transmission line QWT can be calculated according to the synthesis procedure [26]. In order to solve for the desired characteristic impedances, we first relate them to the transmission matrix or ABCD matrix parameters. The relation between the characteristic impedance of the transmission line and the transmission matrix parameters for the ith stage of QWT is as follows:

$$\begin{bmatrix} \cos\theta & jZ_i\sin\theta \\ \left(\dfrac{j}{Z_i}\right)\sin\theta & \cos\theta \end{bmatrix} = \begin{bmatrix} A_i & jB_i \\ jC_i & D_i \end{bmatrix} \tag{4.17}$$

Furthermore, when cascading N stages of QWT we will obtain

$$\prod_{i=1}^{N}\begin{bmatrix} \cos\theta & jZ_i\sin\theta \\ \left(\dfrac{j}{Z_i}\right)\sin\theta & \cos\theta \end{bmatrix} = \prod_{i=1}^{N}\begin{bmatrix} A_i & jB_i \\ jC_i & D_i \end{bmatrix} \tag{4.18}$$

where N is the number of stages in the circuit, Z_i corresponds to the impedance of the ith QWT, and θ is the electrical length of each stage of the transmission line which is equal to $\pi/2$ at the center frequency ($\theta = \beta l$, where $l = \lambda/4$ and $\beta = 2\pi/\lambda$).

We can then represent the product of the above transmission matrix parameter for N stage as follows:

$$\prod_{i=1}^{N}\begin{bmatrix} A_i & jB_i \\ jC_i & D_i \end{bmatrix} = \begin{bmatrix} A_n & jB_n \\ jC_n & D_n \end{bmatrix} \tag{4.19}$$

where A_n, B_n, C_n, and D_n represent the transmission matrix parameters for the entire N-stage QWT circuit. The reflection coefficient (Γ) for the N-stage circuit can be written as:

$$\Gamma = \frac{V_{\text{in}}^-}{V_{\text{in}}^+} = \frac{(A_n - D_n) + j(B_n - C_n)}{(A_n + D_n) + j(B_n + C_n)} \tag{4.20}$$

The phase of the reflection coefficient can be expressed as

$$\angle\Gamma = \tan^{-1}\frac{(B_n - C_n)}{(A_n - D_n)} - \tan^{-1}\frac{(B_n + C_n)}{(A_n + D_n)} \tag{4.21}$$

In order to obtain get the prescribed maximally flat group delay, we need to obtain the characteristic impedance of each of the transmission line [Z_1, Z_2, Z_3... Z_N] by solving the following equations similar to (4.4)–(4.6):

$$\left|\Gamma\left(\theta = \frac{\pi}{2}\right)\right| = \Gamma_p \tag{4.22}$$

$$-\frac{\partial\angle\Gamma(\theta)}{\partial\theta}\bigg|_{\theta=\frac{\pi}{2}} = \tau_{gp} \tag{4.23}$$

$$-\frac{\partial^{2n-3}\angle\Gamma(\theta)}{\partial\theta}\bigg|_{\theta=\frac{\pi}{2}} = 0, \text{where } n = 3, 4, \ldots, (N-1) \tag{4.24}$$

Here, Γ_p is the prescribed magnitude, and τ_{gp} is the prescribed NGD with a unit of the sampling period T. In order to have the maximally flat response, we set all the higher-order derivatives shown in (4.24) to zero. To give a quantitative example, let us take the two-stage QWTs case ($N = 2$) and assume τ_{gp} to be -4 and Γ_p to be 0.04 at the center frequency 1 GHz. Since the sampling period T of a 90° ($\pi/2$) delay line at 1 GHz is 0.25 ns, we can generate group delay of $-4 \times 0.25 = -1$ ns. Similarly, by changing τ_{gp} to -8, we can obtain group delay of -2 ns and so on. It is noted that we terminate our device with standard 50 Ω load, and the characteristic impedance of the transmission lines obtained here are normalized to the load impedance. Table 4.3 shows the different values of normalized characteristic impedances for various group delays by setting Γ_p equal to 0.04 derived from (4.22)–(4.24). After obtaining the characteristic impedances, we can easily calculate the dimension of QWT.

Figure 4.19 shows the simulated results of three different NGD at 1 GHz for a two-stage QWT using the tabulated coefficients. In addition, it is worth mentioning that one can also design a three-stage transformer to generate prescribed NGD as shown in Figure 4.20 with the coefficients obtained from Table 4.3. For a given NGD, the bandwidth can be enhanced when more stages are cascaded.

Table 4.3 Normalized characteristic impedance for two and three stages ($\Gamma_p = 0.04$) [29]

τ_{gp}(T)	$N=2$		$N=3$		
	Z_1	Z_2	Z_1	Z_2	Z_3
−1	1.0833	1.0408	1.10527	1.0834	1.02024
−2	1.1052	1.0618	1.15044	1.15051	1.0409
−3	1.1275	1.0833	1.20535	1.23801	1.06904
−4	1.1503	1.1052	1.27094	1.34942	1.1051
−5	1.1735	1.1274	1.34819	1.489	1.14954
−6	1.1971	1.1501	1.43799	1.66171	1.20276
−7	1.2211	1.1732	1.54106	1.87306	1.26507
−8	1.2455	1.1966	1.65783	2.12887	1.33657
−9	1.2703	1.2205	1.78835	2.43493	1.41715
−10	1.2956	1.2448	1.9323	2.79674	1.50646

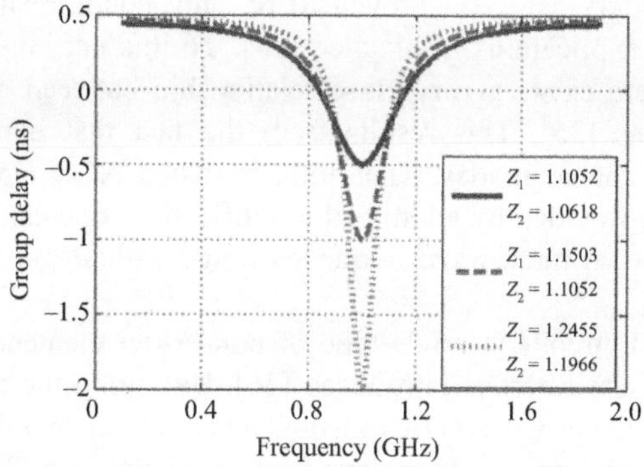

Figure 4.19 Simulated results for prescribed group delay of −0.5, −1, and −2 ns for a two-stage QWT (N = 2) [29]

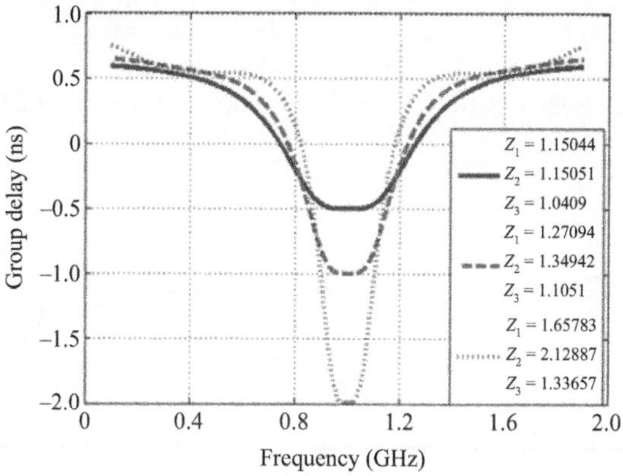

Figure 4.20 Simulated results for prescribed group delay of −0.5, −1, and −2 ns for a three-stage (N = 3) QWT [29]

4.4 Non-Foster element using active transversal-filter-based NGD circuit

Non-Foster elements represent negative capacitors, inductors, or a combination of both. Unlike conventional reactive elements that obey Foster theorem, the reactance of non-Foster elements decreases with respect to the frequency in the operating bandwidth [30]. Compared with traditional inductors (L) or capacitors (C), non-Foster elements can be used to overcome several restrictions. For instance, the Bode–Fano limitation can be broken, making it possible to achieve a very broadband matching. Such matching method has been utilized in electronically small ultra-wide-band antennas [31]. In addition, by periodically loading series negative inductors and shunt negative capacitors, a low-dispersion transmission line operating in the fast wave region can be realized [32]. Furthermore, negative capacitors can be employed in an electrostatic discharge protection unit to lower the parasitic capacitance and extend the speed of a broadband amplifier [33].

However, conventional realization of non-Foster elements using impedance converters or inverters has been proven to be only conditionally stable at best, which limits their applications in practice [34]. To this end, it has been recently introduced that there exists a very close relationship between NGD devices and non-Foster elements [35]. This results from the fact that non-Foster elements also exhibit NGD characteristics. Therefore, by using NGD circuits with proper loss compensation provided by additional amplification, one can realize unilateral non-Foster elements at microwave frequency range without running into stability issues.

Utilizing NGD circuits, a novel type of non-Foster element using DA-based microwave transversal-filters is introduced [36]. By tuning the transconductances of different stages, the proposed DA-based NGD circuit can exhibit unilateral non-Foster characteristics between its reverse port and input port. The simulated and measured *S*-parameters show that the proposed structure can emulate an ideal non-Foster T-network. More importantly, as the proposed structure is based on a DA, good input and output impedance matching, as well as unconditional stability can be achieved, making it feasible to be integrated in practical circuitries and scaled up to higher frequency.

A two-port non-Foster T-network element composed of a shunt negative capacitor and two series negative inductors is shown in Figure 4.21. Assuming that

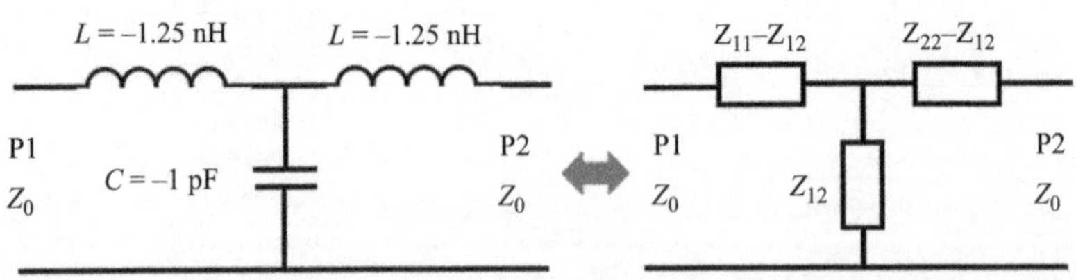

Figure 4.21 The non-Foster T-network and its equivalent Z-parameter [36]

the two inductors are identical, the relationship between the component values and the Z-parameters can be obtained as:

$$L = \frac{\text{imag}(Z_{11} - Z_{12})}{2\pi f} \tag{4.25}$$

$$C = \frac{1}{2\pi f(\text{imag}(Z_{12}))} \tag{4.26}$$

where f is the working frequency. The Z-parameters can be derived from the S-parameter as follows [26]:

$$Z_{11} = Z_0 \frac{(1 + S_{11})(1 + S_{22}) + S_{21}S_{12}}{(1 - S_{11})(1 - S_{22}) - S_{21}S_{12}} \tag{4.27}$$

$$Z_{12} = Z_0 \frac{2S_{12}}{(1 - S_{11})(1 - S_{22}) - S_{21}S_{12}} \tag{4.28}$$

As can be seen from (4.25)–(4.28), the S-parameters of the two-port network are directly related to the negative C and L. Accordingly, in order to realize the non-Foster T-network, the following three requirements need to be satisfied:

1. The absolute values for magnitude and phase of S_{21} should both match with those of the ideal non-Foster network at the desired frequency. It is noted because of the active topology, the network is nonreciprocal, and thus the condition that S_{12} is equal to S_{21} is imposed in the calculation to extract the equivalent non-Foster values.
2. The input and output matching, i.e., S_{11} and S_{22}, of the proposed circuit should be well matched with the ideal non-Foster network, in which it is often desired both ports are matched to 50 Ω.
3. The emulating circuit should exhibit NGD characteristics in the operating band. Furthermore, since the group delay is determined by the derivative of phase of S_{21} with respect to the frequency [8,9], this implies that the slope of the phase response for the NGD circuit should be the same as the ideal non-Foster network. Based on these requirements, one is able to realize a non-reciprocal non-Foster T-network in its operating bandwidth.

The proposed circuit based on a DA is shown in Figure 4.22. The non-Foster characteristics is realized through P1 to P2. The characteristic impedances (Z_0) of

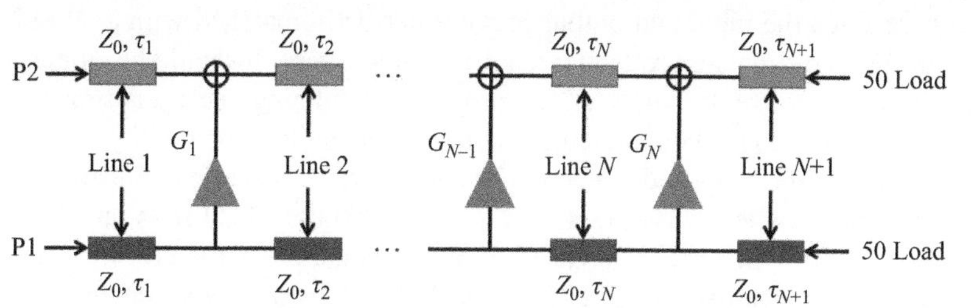

Figure 4.22 Schematic of the proposed non-Foster network using a DA-based active transversal filter [36]

all the transmission lines are set to be 50 Ω. Thanks to the transmission line topology of DA, the impedance matching can be achieved in both two ports, thereby satisfying the aforementioned second requirement. The transfer function between P1 and P2 can be expressed as:

$$H(\omega) = \sum_{k=1}^{N} G_k e^{-j\omega \sum_{i=1}^{k} 2\tau_i} \tag{4.29}$$

where G_k is the gain coefficient of the kth tap amplifier, τ_i indicates the time delay of the kth section of transmission line. From (4.29), one can first obtain a specific absolute phase value of S_{21} at the desired center frequency to match the desired ideal non-Foster network by adjusting the values of τ_i, while the value of G_k can be determined later to match the desired amplitude response.

In addition, for the third requirement, the following formulation should be satisfied to realize a group delay of $-b$:

$$\text{minimize max}\left\{\left|\sum_{k=1}^{N} G_k e^{-j\omega \sum_{i=1}^{k} 2\tau_i} - e^{jb\omega}\right|\right\} \tag{4.30}$$

As τ_i has been decided above, in order to satisfy (4.30), the convex optimization technique is applied to solve for the coefficient set $G = \{G_1, G_2, \ldots, G_N\}$. In so doing, one can obtain the ratio among the gain coefficients for a specific NGD value to match the ideal non-Foster network. By adjusting all G_k proportionally, the magnitude of S_{21} can be set to the desired value as the desired non-Foster element, while maintaining a constant NGD, and the requirement 1 and 3 can be satisfied.

The simulated results based on the proposed structure are shown in Figure 4.23 to validate the aforementioned theoretical analysis. The target values of C and L are −1 pF and −1.25 nH, respectively. In this simulation, a two-stage DA is used to achieve the desired non-Foster characteristics, in which the gain coefficients are $G_1 = 2$, $G_2 = 1.025$. The phase of Line 1 and 3 is 77° and that of Line 2 is 90° at the center frequency of 1.4 GHz. It can be seen from Figure 4.23 that within the operating bandwidth (1.3–1.5 GHz), the phase, magnitude, and group delay characteristics all show a good agreement between the ideal non-Foster T-network element and the proposed emulating circuit. The reflection coefficients are not given here since the input and output port are perfectly matched with a 50 Ω load, and therefore the simulated reflection coefficients are zero. It should be emphasized that, similar to conventional DAs, by properly stabilizing the transistor at each stage, the proposed non-Foster element exhibits unconditional stability.

To validate the proposed concept, the entire circuit is fabricated using Rogers RO5880 with a dielectric constant of 2.2 and thickness of 20 mils, as shown in Figure 4.24. The NGD circuit is based on a two-stage DA structure, in which we use Avago's enhancement mode HEMT, ATF50189, as the active device for high transconductance. In order to ensure the stability, a shunt-inductor is placed at the gate to ensure that the transistor is unconditionally stable throughout the entire band (0–20 GHz). As there are some intrinsic parasitic effects in transistors, the dimensions

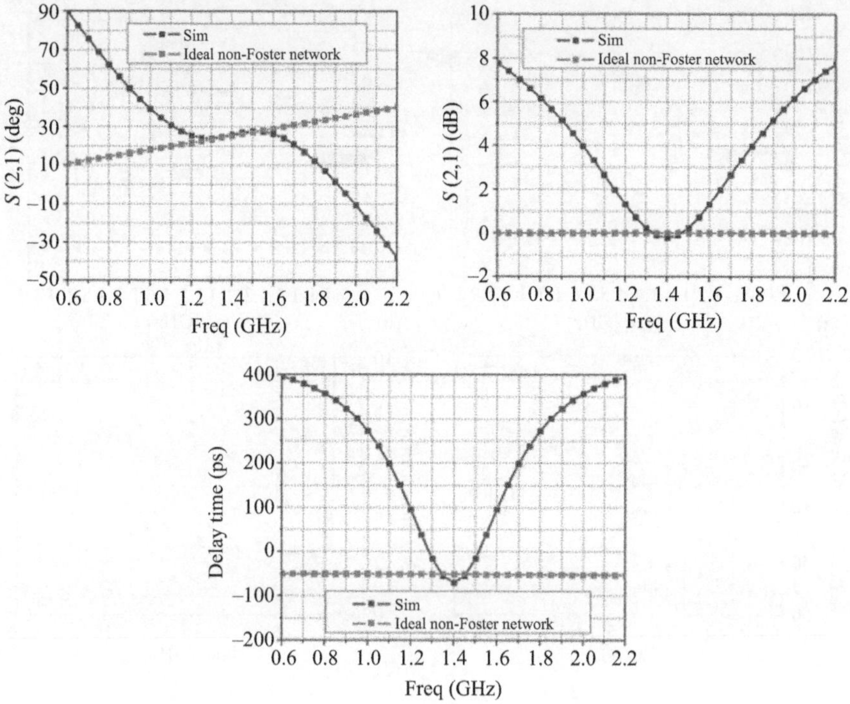

Figure 4.23 Simulated results comparing the proposed non-Foster element and an ideal T-network with a shunt C = −1 pF and two series L = −1.25 nH [36]

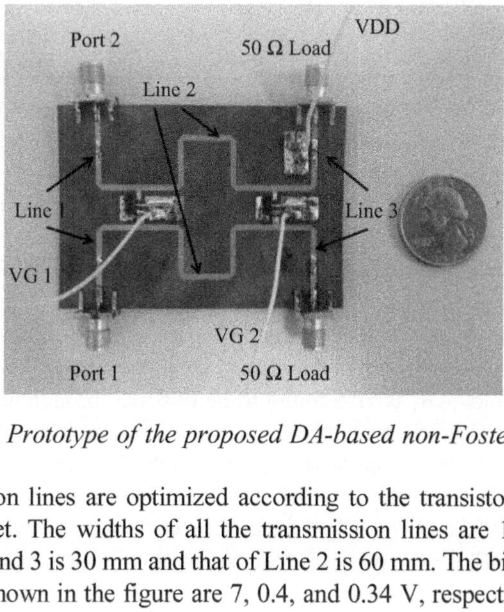

Figure 4.24 Prototype of the proposed DA-based non-Foster circuit [36]

of the transmission lines are optimized according to the transistor model from the vendor's datasheet. The widths of all the transmission lines are 1.54 mm, and the length of Line 1 and 3 is 30 mm and that of Line 2 is 60 mm. The bias voltages VDD, VG1, and VG2 shown in the figure are 7, 0.4, and 0.34 V, respectively.

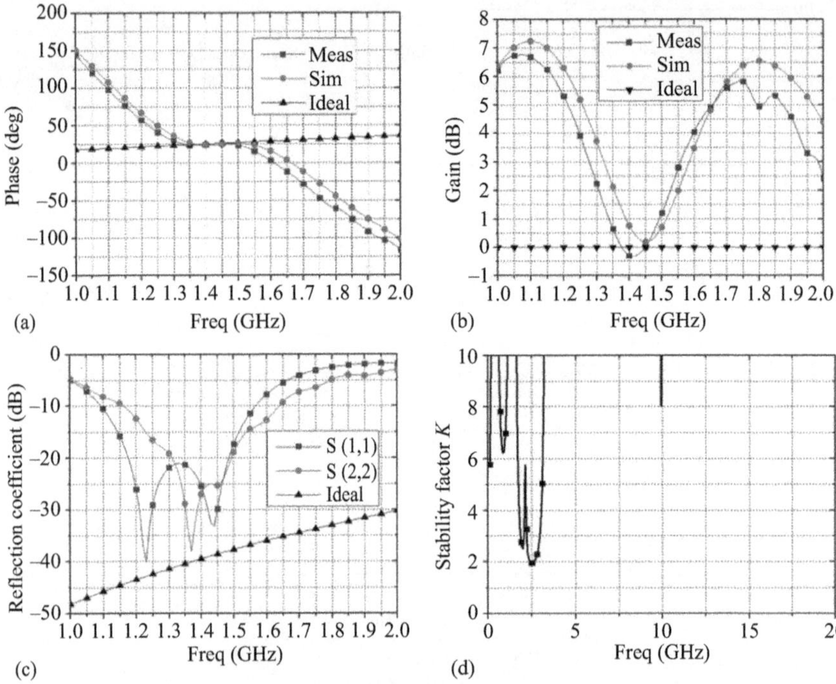

Figure 4.25 Experimental results of the proposed non-Foster circuit compared with an ideal non-Foster T-network shown in Figure 4.21: (a) the phase response; (b) the magnitude response; (c) the input/output reflection coefficient; and (d) stability factor of the proposed non-Foster circuit [36]

The experimental and simulated results are shown in Figure 4.25, in which the operating frequency for the proposed non-Foster element is around 1.35–1.45 GHz. Figure 4.25(a) and (b) shows that the phase of S_{21} is 25° and the magnitude is around 0 dB with less than ± 0.7 dB variation. From the figure, it can also be seen that the slopes of the phase are identical, indicating the circuit has the same NGD as the ideal non-Foster element. The measured reflection coefficients of the proposed circuit are shown in Figure 4.25(c). It can be seen that the input and output reflection coefficients are <-20 dB, showing an excellent input and output matching. The stability factor K is plotted in Figure 4.25(d), indicating the circuit is unconditionally stable throughout the entire frequency band, along with the fact that the auxiliary factor B_1 is also found to be positive throughout the band [37].

Based on (4.25)–(4.26), the equivalent components of the non-Foster T-network are extracted using the measured S-parameter as plotted in Figure 4.26. It is observed that in the operating frequency, the difference between the series inductor and the ideal one (-1.25 nH) is <0.1 nH. And the difference between the shunt capacitor and the ideal one (-1 pF) is <0.1 pF, indicating a very small fluctuation in the reactance

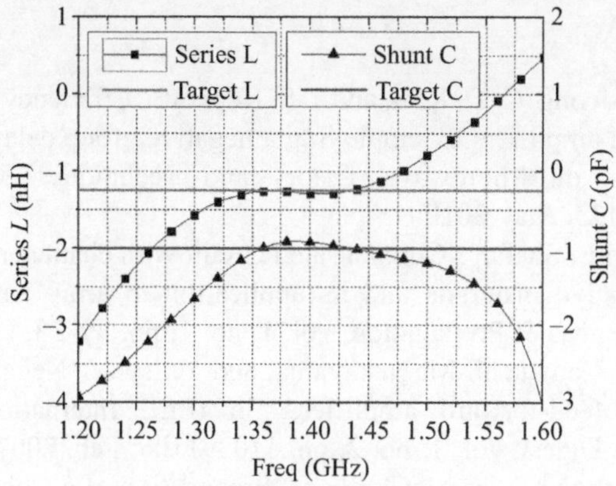

Figure 4.26 Extracted equivalent shunt capacitor and series inductor [36]

Table 4.4 Comparison with other existing non-Foster techniques [36]

	Negative impedance converters (NICs) and inverters (NII)	Lumped NGD circuit-based non-Foster elements	Proposed transversal-filter-based non-Foster elements
Operating frequency	Mostly in HF, VHF, and UHF bands	<3 GHz (limited by self-resonance of lumped elements)	Distributed topology can be scaled up to microwave and millimeter-wave range
Stability	Conditionally stable	Unconditionally stable	Unconditionally stable
Bandwidth	Few hundred MHz at Frequency < 1GHz	Few hundred MHz	Few hundred MHz to few GHz if more stage cascaded

for the proposed circuit. Table 4.4 compares the proposed transversal-filter-based non-Foster element with other realizations. It can be seen that the proposed NGD circuit provides unconditional stability and can be potentially scaled up to a higher frequency range such as millimeter-waves.

4.5 Conclusion

In this chapter, some recent developments of NGD circuits based on microwave transversal and recursive filters are presented and discussed. The fully distributed nature of these NGD circuits makes them feasible to further operate at high frequency ranges, such as millimeter-wave or even terahertz regimes. It is also shown that the transversal-filter-based NGD circuit can be utilized to form non-Foster elements with unconditional stability, which is expected to have a great impact on realizing high frequency, stable negative inductors and capacitors.

References

[1] H. Choi, Y. Jeong, C. D. Kim, and J. S. Kenney, "Efficiency enhancement of feedforward amplifiers by employing a negative group delay circuit," IEEE Transactions on Microwave Theory and Techniques, vol. 58, no. 5, pp. 1116–1125, May 2010.
[2] S. S. Oh and L. Shafai, "Compensated circuit with characteristics of lossless double negative materials and its application to array antennas," Microwaves, Antennas & Propagation, vol. 1, no. 1, pp. 29–38, Feb. 2007.
[3] H. Noto, K. Yamauchi, M. Nakayama, and Y. Isota, "Negative group delay circuit for feed-forward amplifier," in IEEE International Microwave Symposium Digest, vol. 1, no. 2, pp. 1103–1106, Feb. 2007.
[4] G. B. Garrett and D. E. McGumber, "Propagation of a Gaussian light pulse through an anomalous dispersion medium," Physical Review A, vol. 1, no. 2, pp. 305–313, Feb. 1970.
[5] S. Chu and S. Wong, "Linear pulse propagation in an absorbing medium," Physical Review Letters, vol. 48, no. 11, pp. 738–741, Mar. 1982.
[6] S. Lucyszyn, I. D. Robertson, and A. H. Aghvami, "Negative group delay synthesiser," Electronics Letters, vol. 29, no. 9, pp. 798–800, Apr. 1993.
[7] S. Lucyszyn and I. D. Robertson, "Analog reflection topology building blocks for adaptive microwave signal processing applications," IEEE Transactions on Microwave Theory and Techniques, vol. MTT-43, no. 3, pp. 601–611, Mar. 1995.
[8] M. W. Mitchell and R. Y. Chiao, "Negative group delay and 'fronts' in a causal system: An experiment with very low frequency bandpass amplifiers," Physics Letters A, vol. 230, no. 3–4, pp.133–138, Jun. 1997.
[9] M. W. Mitchell and R. Y. Chiao, "Causality and negative group delays in a simple bandpass amplifier," American Journal of Physics, vol. 66, no. 1, pp. 14–19, Jan. 1998.
[10] M. Kitano, T. Nakanishi, and K. Sugiyama, "Negative group delay and superluminal propagation: An electronic circuit approach," IEEE Journal of Selected Topics in Quantum Electronics, vol. 9, no. 1, pp. 43–51, Jan. 2003.
[11] E. L. Bolda and R.Y. Chiao, "Two theorems for the group velocity in dispersive media," Physical Review A, vol. 48, no. 5. pp. 3890–3894, Nov. 1993.
[12] B. Ravelo, A. Perennec, and M. Le Roy, "Synthesis of broadband negative group delay active circuits," in IEEE International Microwave Symposium Digest, pp. 2177–2180, Jun. 2007.
[13] H. Choi, Y. Jeong, J. Lim, S. Y. Eom, and Y. B. Jung, "A novel design for a dual-band negative group delay circuit," IEEE Microwave Wireless Component Letter, vol. 21, no. 1, pp. 19–21, Jan. 2011.
[14] S. Lucyszyn. I. D. Robertson, and A. H. Aghvami, "Negative group delay synthesiser," Electronics Letters, vol. 29, no. 9, pp. 798–800, Apr. 1993.

[15] M. Kandic and G. E. Bridges, "Asymptotic limits of negative group delay in active resonator-based distributed circuits," IEEE Transactions on Circuits and Systems I: Regular Papers, vol. 58, no. 8, pp. 1727–1735, Aug. 2011
[16] O. Siddiqui, M. Mojahedi, and G. Eleftheriades, "Periodically loaded transmission line with effective negative refractive index and negative group velocity," IEEE Transactions on Microwave Theory and Techniques, vol. 52, no. 5, pp. 1449–1454, May 2004.
[17] C.-T. M. Wu and T. Itoh, "Maximally flat negative group delay circuit: A microwave transversal filter approach," IEEE Transactions on Microwave Theory and Techniques, vol. 62, no. 6, pp. 1330–1342, Jun. 2014.
[18] C.-T. M. Wu, S. Gharavi, B. Daneshrad, and T. Itoh, "A dual-purpose reconfigurable negative group delay circuit based on distributed amplifiers," IEEE Microwave and Wireless Components Letters, vol. 23, no. 11, pp. 593–595, Nov. 2013.
[19] C.-T. M. Wu, S. Gharavi, and T. Itoh, "Negative group delay circuit based on a multisection asymmetrical directional coupler," Asia-Pacific Microwave Conference 2013, Seoul, Korea, Nov. 2013.
[20] M. Kandic, "Asymptotic limits of negative group delay phenomenon in linear causal media," Doctoral dissertation, Department of Electrical & Computer Engineering, University of Manitoba, Sep. 2011.
[21] A. Borjak, P. P. Monteiro, J. J. O'Reilly, and I. Darwazeh, "High speed generalized distributed-amplifier-based transversal-filter topology for optical communication systems," IEEE Transactions on Microwave Theory and Techniques, vol. 45, no. 8, pp.1453–1457, Aug. 1997.
[22] J. D. Rhodes, "The design and synthesis of a class of bandpass linear phase filters," IEEE Transactions on Microwave Theory and Techniques, vol. 17, no. 4, pp. 189–204, Apr. 1969.
[23] T. A. Abele, "Transmission line filters approximating a constrained delay in a maximally flat sense," IEEE Transactions on Circuit Theory, CT-14, pp. 298–306, Sep. 1967.
[24] S. O. Scanlan and J. D. Rhodes, "Microwave networks with constant delay," IEEE Transactions on Circuit Theory, vol. CT-14, no. 3, pp. 290–297, Sep. 1967.
[25] A. V. Oppenheim, R. W. Schafer, and J. R. Buck, Discrete-Time Signal Processing 2nd Ed. Upper Saddle River, NJ: Prentice-Hall, 1999.
[26] D. M. Pozar, Microwave Engineering, 3rd Ed., Toronto: John Wiley & Sons, 2005.
[27] M. Kandic and G. E. Bridges, "Limits of negative group delay phenomenon in linear causal media," Progress in Electromagnetics Research, vol. 134, pp. 227–246, Nov. 2013.
[28] T. Zhang, R. Xu, and C.-T. M. Wu, "Voltage-insensitive negative group delay circuit based on a stepped-impedance distributed amplifier." IEEE Microwave and Wireless Components Letters, vol. 27, no. 11, pp. 1007–1009, Nov. 2017.
[29] M. A. Ali and C.-T. M. Wu, "Negative group delay circuit based on microwave recursive filters," International Journal of Microwave Science and Technology, vol. 2015, Article ID 646029, pp. 1–6, Aug. 2015.

[30] R. M. Foster, "A reactance theorem," Bell System Technical Journal, vol. 3, no. 2, pp. 259–267, Apr. 1924.
[31] M.-C. Tang, T. Shi, and R. W. Ziolkowski, "Electrically small, broadside radiating Huygens source antenna augmented with internal non-Foster elements to increase its bandwidth," IEEE Antennas and Propagation Society, vol. 99, no. 99, pp. 1–1, Aug. 2016.
[32] A. Poggio and P. Mayes, "A unique broadband dipole antenna design utilizing conjugate reactance loading," in Antennas and Propagation Society International Symposium, vol. 7, pp. 250–258, Dec. 1969.
[33] S. Galal and B. Razavi, "40-Gb/s amplifier and ESD protection circuit in 0.18- m CMOS technology," IEEE Journal of Solid-State Circuits, vol. 39, no. 12, pp. 2389–2396, Dec. 2004.
[34] J. G. Linvil, "Transistor negative-impedance converters." Proceedings of the IRE, vol. 40, no. 6, pp. 725–729, Jun. 1953.
[35] H. Mirzaei and G. V. Eleftheriades, "Realizing non-Foster reactive elements using negative group delay networks," IEEE Transactions on Microwave Theory and Techniques, vol. 61, no. 12, pp. 4322–4332, Dec. 2013.
[36] T. Zhang, R. Xu, and C.-T. M. Wu, "Unconditionally stable non-Foster element using active transversal-filter-based negative group delay circuit." IEEE Microwave and Wireless Components Letters, vol. 27, no. 10, pp. 921–923, Oct. 2017.
[37] G. Gonzalez, *Microwave transistor amplifiers: analysis and design*. Prentice-Hall, Inc., Upper Saddle River, NJ, 1996.

Chapter 5
Hybrid coupler-based NGD circuit

Blaise Ravelo[1]

Abstract

A theory of negative group delay (NGD) topology based on hybrid coupler is investigated. It acts as a passive microwave circuit topology. The NGD cell under study is constituted by a hybrid coupler with one of the output and isolated access ways connected in feedback. The feedback circuit is constituted by a lossy transmission line (TL). Two configurations of cells with interconnected isolated-coupled and isolated-direct accesses are investigated. The equivalent *S*-parameter model of the proposed structure is established. It is derived mathematically that the fully distributed hybrid coupler-based topology is susceptible to exhibit NGD phenomenon. The fundamental properties of the innovative NGD topology as the NGD level-bandwidth product are formulated. The synthesis relations enabling to determine the NGD circuit parameters in function of the desired NGD value are extracted. The relation between the NGD level and the insertion loss is elaborated. Parametric numerical analyses in function of the TL attenuation and time delay were performed. The NGD concept feasibility is highlighted by a very good correlation between the theoretical calculation and the numerical simulations. In addition, the NGD phenomena were also occurred with the consideration of microstrip lines in feedback with 90°- and 180°-hybrid couplers. The NGD topology under study is potentially useful for the signal delay correction in the RF system.

Keywords

Negative group delay (NGD), hybrid coupler, circuit theory, characterization method, *S*-parameter modeling.

5.1 Introduction

The time delay is one of the key parameters of communication systems. The time delay affects undeniably the quality of signal processing and the data transmission

[1]ESIGELEC, France

chains. Different methodologies were proposed to control the time delay in the various systems as electronic, RF, and optical devices [1–6]. For example, delay line synthesizers with passive and active circuits were introduced [1,2]. Furthermore, a tentative technique to synthesize a delay insensitive effect was proposed [3]. During the RF circuit design, in addition to the noise issue, the effect of group delay remains a bottleneck in the reception phase and the reduction of the propagation channel influence [4,5]. The same issue needs to be overcome with the optical communication devices [6,7]. A phase-control technique was developed but the reliability is critical due to the system oscillator phase noise. So far, the phase noise remains the limitation factor in such a system. Moreover, to minimize the channel delay, a technique based on the cyclostationarity was presented in [8]. But such a technique is limited to very narrow band application and cannot be adapted to most of nonlinear effects as in active circuit.

Nevertheless, different design and implementation concepts of delay lines were proposed by microwave design engineers. So far, the most expanded delay lines are based on the use of transmission lines (TLs). An example of application with radio frequency identification tag integrating TL is proposed in [9]. The delay line can be found also in the miniaturize circuits as a CMOS-based all-pass approximated time delay presented in [10]. A wider application dedicated to the realization of multiband delay time synthesizer for the frequency translation technique was introduced in [11].

Similar to the influence onto the microwave systems, the time delays are susceptible to play a major role in the field of digital circuit signal integrity engineering [12,13]. Nowadays, the performances as the speed of digital and mixed system are limited by the propagation delay. More practically, the electrical interconnection delays are the main factors to distort the data signal in the printed circuit boards. Considering the electrical interconnect effects, the microelectronic circuit designers integrate and implement regularly the classical repeater elements [14]. However, those classical delay lines and repeaters consists basically in adding further delay in order to reduce the signal delay differences which induce implicitly the synchronization issues.

In counterintuitive, it was investigated recently that the signal recovery with propagation delay reduction can be performed with negative group delay (NGD) circuit [15,16]. The existence of NGD lines inspired from the negative refractive index metamaterials was demonstrated in the 2000s [17,18]. It was introduced in [19] that the NGD phenomenon can be generated through the resonant absorption effect. Therefore, RF/microwave research design and manufacturing engineer were progressively interested to the development of NGD circuits and their potential applications [15–30]. Despite the NGD asymptotic limitation as stated in [20,21], various topologies of NGD passive and active circuits [22,27] were developed recently. However, the NGD active circuit noise factor and power consumption need to be improved in order to be prominently integrable [15,23,24]. In addition, the use of lumped and active components as inductance literally limits the operation frequencies. Therefore, distributed element-based NGD circuits were proposed very recently [25–30]. More precisely, an NGD circuit based on the use of multisection asymmetrical directional coupler is proposed in [28]. With the similar concept, a transmission-type doublet coupler-based NGD circuit was experimented in [29,30]. However, the complexity and theoretical approach of the NGD circuits proposed in [28–30] could be a major factor limiting their exploitations. Therefore, simpler NGD approach is necessary to expand the NGD circuit applications.

For this reason, a complete theory on the innovative NGD topology based on the basic microwave function as hybrid coupler is established in this chapter. Classically, the RF coupling effect is exploited by the microwave circuit designer to increase the performances of the filter, antenna, mixer, and various RF communication systems. The coupler which has been established since the 1980s [31], constitutes one of the most popular RF function. However, the coupling effect will be explored in the present chapter to exhibit the NGD function. To accomplish this, the classical coupling effect as reported in [32–34] combined with the feedback approach will be exploited to realize the NGD topology.

The chapter is organized in three main sections. Section 5.2 is focused on the theoretical concept. The *S*-parameter analytical modeling of the NGD topology will be described. Afterwards, the basic properties of the NGD function will be established. Section 5.3 will be focused on the illustrative examples of validation results. Finally, the conclusion of the chapter is drawn in Section 5.4.

5.2 Theory on the hybrid coupler-based NGD passive structure

The hybrid coupler under consideration consists of a 4-way passive circuit. The innovative NGD passive topology is essentially composed of a hybrid coupler. The overall circuit is configured with one of the output ways interconnected to the isolated access. Similar to the classical systems, for the sake of stability, the feedback chain should be a passive circuit. For the analytical approach, the feedback circuit *S*-parameter is assumed to be defined as:

$$[R] = \begin{bmatrix} r & t \\ t & r \end{bmatrix} \tag{5.1}$$

with r and t are complex variables representing reflection and transmission coefficient, respectively. Consequently, the global structure can be assumed as a two-port circuit.

Figure 5.1 represents the two configurations to be investigated in this chapter. In Figure 5.1(a) (resp. Figure 5.1(b)), we have a configuration constituted by the output coupled (resp. direct) way interconnect with the isolated one.

Thanks to the mathematical manipulation of the matrix system generated from substitution of the input and output wave powers through the two *S*-parameter matrices, the equivalent global two-port *S*-parameter was determined as:

$$[S] = \begin{bmatrix} s_{11} & s_{12} \\ s_{21} & s_{22} \end{bmatrix} \tag{5.2}$$

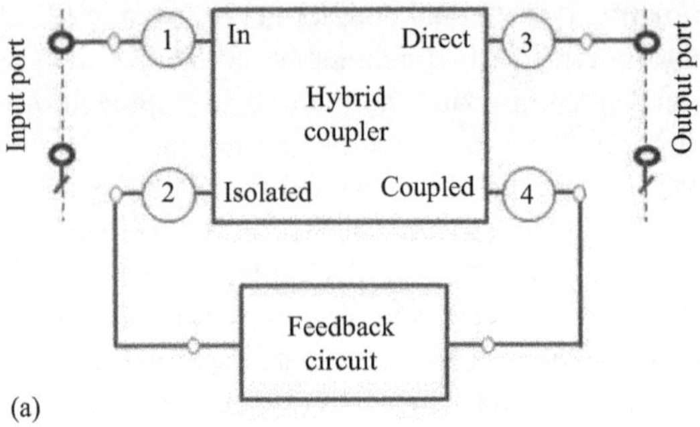

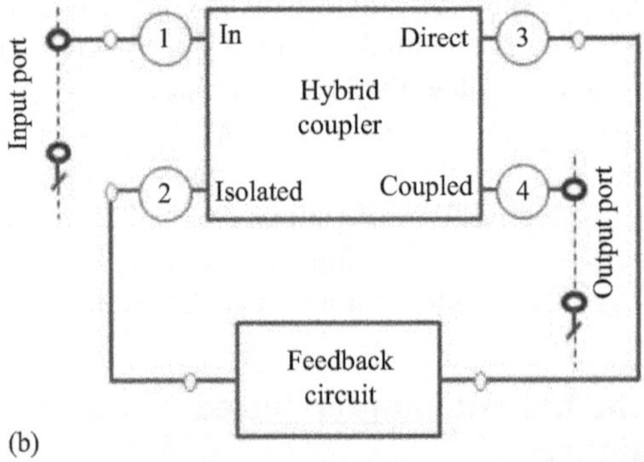

Figure 5.1 Configuration of the hybrid coupler-based topologies under study constituted by interconnected: (a) coupled-isolated and (b) direct-isolated ways

of the NGD topology. For the sake of analytical simplification, the hybrid coupler is supposed as lossless. In function of the coupler ways interconnection, different *S*-parameter models can be obtained.

To characterize these two topologies, the group delay expressions from the *S*-parameter models will be established in the following sections. The particular cases of 90°- and 180°-hybrid couplers will be considered.

5.2.1 Investigation on the topology constituted by isolated and coupled ways interconnected hybrid coupler

As aforementioned, the present theory is built with the *S*-parameter modeling. To do this, the inter-combination of the hybrid coupler and the feedback chain *S*-parameters will be considered. Then, the innovative topology will be characterized from the extraction of transmission and group delay.

5.2.1.1 *S*-parameter modeling of the hybrid coupler with isolated and coupled way interconnected

The ideal 90° hybrid coupler considered in this study is assumed modeled by the following *S*-parameter expression:

$$[S]_{\text{coupler}} = \begin{bmatrix} 0 & a & j \cdot b & 0 \\ a & 0 & 0 & j \cdot b \\ j \cdot b & 0 & 0 & a \\ 0 & j \cdot b & a & 0 \end{bmatrix} \tag{5.3}$$

In this expression, the parameter a represents the coupling factor and:

$$b = \sqrt{1 - a^2}. \tag{5.4}$$

Similarly, with the same parameters, the 180° hybrid coupler will be modeled by the *S*-matrix:

$$[S]_{coupler} = \begin{bmatrix} 0 & a & b & 0 \\ a & 0 & 0 & -b \\ b & 0 & 0 & a \\ 0 & -b & a & 0 \end{bmatrix} \tag{5.5}$$

The equivalent *S*-parameter model proposed in (5.2) can be determined analytically with the inter-combination of the hybrid coupler *S*-parameter defined in (5.3) and (5.5). The analytical calculations were carried out by substituting the wave powers at the hybrid coupler and feedback line access ports. Accordingly, the reflection and transmission elements of the global matrix generalized in (5.2) can be expressed as:

$$s_{11} = s_{22} = \frac{b^2 \cdot r}{a^2 r^2 - (a \cdot t - 1)^2} \tag{5.6}$$

$$s_{12} = s_{21} = \frac{a(a^2 + b^2)(t^2 - r^2) - (2a^2 + b^2)t - a}{(a \cdot t - 1)^2 - a^2 r^{22}} \tag{5.7}$$

It is noteworthy that the same *S*-parameter responses are realized with the feedback topology constituted by 90°- and 180°-hybrid couplers. As expected, the system is perfectly matched when the feedback system is ideally matched:

$$r = 0 \Rightarrow s_{11} = s_{22} = 0 \tag{5.8}$$

Before the circuit characterization, the frequency-dependent responses corresponding to the transmission parameters expressed in (5.7) will be established in the following sections.

5.2.1.2 Transmission parameter magnitude and group delay responses of the hybrid coupler feedback topology with interconnected isolated and coupled way

The considered configuration of the present case of the feedback structure is presented in Figure 5.2.

Similar to the general structure, it acts as a two-port circuit with input source and output loads $R_0 = 50\ \Omega$, which represents the reference impedance. By hypothesis, the feedback chain is considered as a lossy TL presenting characteristic impedance Z_c, attenuation α, and time-delay τ_0. For the sake of simplicity, the feedback TL is supposed ideally matched ($r = 0$). Meanwhile, the S-parameter introduced in (5.2) can be transformed as:

$$[S]_{TL} = \begin{bmatrix} 0 & \alpha \cdot e^{-j\omega\tau_0} \\ \alpha \cdot e^{-j\omega\tau_0} & 0 \end{bmatrix} \tag{5.9}$$

It yields that the coupler-based circuit previous global expression of the S-parameter model will be transformed as follows. Therefore, by taking:

$$a = b = \frac{\sqrt{2}}{2}, \tag{5.10}$$

the frequency-dependent expression of the transmission parameter magnitude will become:

$$S_{21}(\omega) = |S_{21}(j\omega)| = \frac{\sqrt{2(1+\alpha^2)^2 - 6\sqrt{2}\alpha^3\cos(\omega\tau_0) + 5\alpha^2}}{\sqrt{2 - 2\sqrt{2}\alpha\cos(\omega\tau_0) + \alpha^2}} \tag{5.11}$$

where:

$$j = \sqrt{-1} \tag{5.12}$$

is the imaginary complex number and ω is the angular frequency variable. Similarly, the transmission phase and the transmission group delay can be defined as:

$$\varphi(\omega) = \arg[S_{21}(j\omega)] \tag{5.13}$$

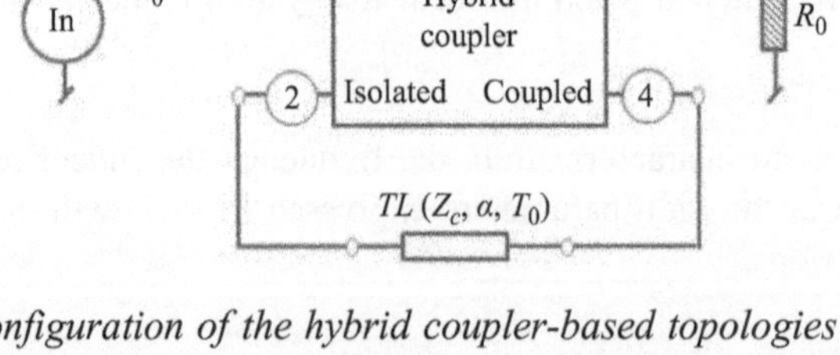

Figure 5.2 Configuration of the hybrid coupler-based topologies under study constituted by interconnected coupled-isolated ways through a TL

$$\tau(\omega) = \frac{-\partial\varphi(\omega)}{\partial\omega} \tag{5.14}$$

More explicitly, the global transmission parameter phase can be expressed as:

$$\varphi_{21}(\omega) = \left\{ \begin{array}{l} \arctan\left[\dfrac{\alpha(3\sin(\omega\tau_0) - \sqrt{2}\alpha\sin(2\omega\tau))}{\sqrt{2} - 3\alpha\cos(\omega\tau_0) + \sqrt{2}\alpha^2\cos(2\omega\tau_0)}\right] \\ -\arctan\left[\dfrac{4\alpha\sin(\omega\tau_0)(\sqrt{2} - \alpha\cos(\omega\tau)}{2 - 2\sqrt{2}\alpha\cos(\omega\tau_0) + \alpha^2\cos(2\omega\tau_0)}\right] \end{array} \right\} \tag{5.15}$$

In the continuation, by using the analytical definition proposed in (5.14), the derived group delay can be written as:

$$\tau(\omega) = \tau'(\omega) - \tau''(\omega) \tag{5.16}$$

with

$$\tau'(\omega) = \frac{\alpha\cdot\tau_0 \left[\begin{array}{l} \alpha^4 - 4\sqrt{2}\alpha^2(\alpha+2)\cos(\omega\tau_0) \\ +4\alpha^2(\cos(2\omega\tau_0)+2)+4 \end{array}\right]}{\begin{array}{l} 2\alpha^3 - 2\sqrt{2}(3\alpha^2+2)\cos(\omega\tau_0) \\ +4\alpha(\cos(2\omega\tau_0)+2) \end{array}} \tag{5.17}$$

$$\tau''(\omega) = \frac{\alpha\cdot\tau_0 \left[\begin{array}{l} 2\alpha^4 - 6\sqrt{2}\alpha^2(\alpha+1)\cos(\omega\tau_0) \\ +\alpha^2(4\cos(2\omega\tau_0)+9)+2 \end{array}\right]}{\begin{array}{l} 4\alpha^3 - 3\sqrt{2}(3\alpha^2+1)\cos(\omega\tau_0) \\ +\alpha(4\cos(2\omega\tau_0)+9) \end{array}}. \tag{5.18}$$

Based on this expression of group delay, we can demonstrate the possibility to generate the NGD phenomenon, which can be traduced by the mathematical condition $\tau(\omega) < 0$. To do this, the behavior of the group delay at the particular frequencies will be analyzed in the next expression.

5.2.1.3 NGD characterization of the topology under study

At very low frequencies ($\omega \approx 0$), the transmission gain magnitude expressed in (5.11) will be transformed as:

$$S_{21}\left(\omega = \frac{2m\pi}{\tau_0}\right) = \frac{|\sqrt{2}\alpha - 1|}{|\alpha - \sqrt{2}|} \tag{5.19}$$

with $m = \{0, 1, 2, \dots\}$ is an integer. The group delay, respectively, expressed in (5.16) will be transformed as:

$$\tau\left(\omega = \frac{2m\pi}{\tau_0}\right) = \frac{\alpha\tau_0}{(\sqrt{2}\alpha - 1)(\sqrt{2} - \alpha)} \tag{5.20}$$

It can be understood from this formula that the cell shown in Figure 5.2 is susceptible to generate NGD phenomenon at very low frequencies. Moreover, this parameter illustrates the influence of the inherent periodicity effect from the distributed element constituting the structure. It can be demonstrated mathematically that the NGD phenomenon can be realized under the condition:

$$a \leq \frac{\sqrt{2}}{2} \tag{5.21}$$

It is noteworthy that this corresponds also to the minimal value of the group delay. The optimal value of the angular frequency is periodically defined by

$$\omega(m) = \frac{2m\pi}{\tau_0} \tag{5.22}$$

Furthermore, at the particular frequencies:

$$\omega(m) = \frac{(8m+1)\pi}{4\tau_0} \tag{5.23}$$

the transmission coefficient magnitude expressed in (5.11) will be transformed as:

$$S_{21}\left(\omega = \frac{\pi}{4\tau_0}\right) = \sqrt{2a^2 - 2a + 1} \tag{5.24}$$

Similarly, the group delay expressed in (5.14) will be transformed as:

$$\tau\left(\omega = \frac{\pi}{4\tau_0}\right) = \frac{a(3a - a^2 - 1)\tau_0}{(a^2 - 2a + 2)(2a^2 - 2a + 1)} \tag{5.25}$$

Moreover, the circuit can present the NGD cutoff frequency if the root of the equation $\tau(\omega) = 0$ exists. After mathematical analysis, it is established that the corresponding NGD cutoff frequencies are written as:

$$\omega_c(m) = \frac{1}{\tau_0}\left[m\pi + \arctan\left(\frac{3a}{\sqrt{2}}\right)\right], \tag{5.26}$$

with $m = \{0, 1, 2, \ldots\}$ is an integer. As the group delay expressed in (5.14) can be negative, it can be stated that the topology under study can be assumed as a typically low-pass and band-pass NGD circuits. It is due to the periodical NGD frequencies defined in (5.22). It means that the baseband NGD bandwidth is equal to:

$$\Delta\omega = \omega_c \tag{5.27}$$

The fundamental characteristic of the hybrid coupler-based NGD topology can be quantified with NGD level-bandwidth product:

$$|\tau_{opt}|\Delta\omega = \frac{a \arctan\left(\frac{3a}{\sqrt{2}}\right)}{|(\sqrt{2}a - 1)(\sqrt{2} - a)|} \tag{5.28}$$

5.2.1.4 Feedback line parameters synthesis method in function of the given insertion loss

A synthesis relation on the feedback circuit attenuation can be determined by inverting (5.19):

$$\alpha = \frac{1 - \sqrt{2}S_{21}}{\sqrt{2} - S_{21}} \tag{5.29}$$

It implies that the optimal group delay formulated in (5.20) can be expressed in function of the optimal insertion loss as:

$$\tau_{opt}(S_{21}) = \frac{(1 - \sqrt{2}S_{21})(S_{21} - \sqrt{2})}{S_{21}}\tau_0 \tag{5.30}$$

Figure 5.3 illustrates the graphical relation corresponding to the NGD optimal value and level-bandwidth product in function of the insertion loss S_{21}.

This abacus summarizes the performance of the NGD topology under investigation. It can be noticed that the NGD level-bandwidth product is better when the insertion loss is decreasing. Furthermore, the group delay is negative when S_{21} is in the range −20 dB to −3 dB. Then, its absolute value $|\tau_{opt}|$ is inversely proportional to S_{21}.

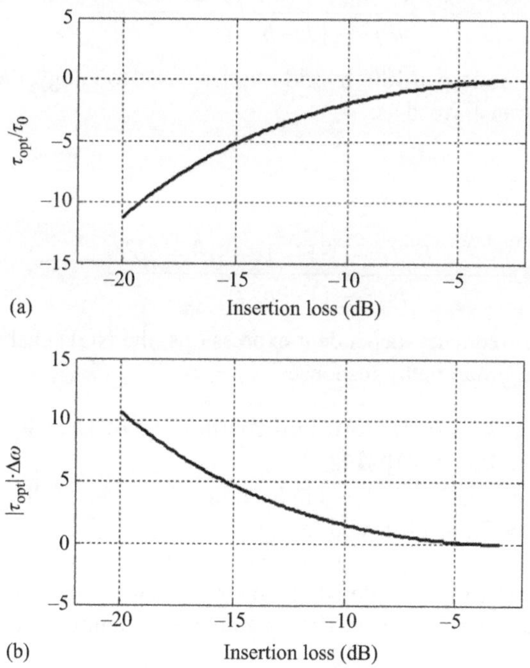

Figure 5.3 Characteristic of optimal NGD (dashed line) and the NGD level-bandwidth product (solid line) versus insertion loss for the topology under consideration

5.2.2 Investigation on the topology constituted by a hybrid coupler with interconnected isolated and direct ways coupler

Similarly, to the previous case, the present analysis will be carried with the *S*-parameter modeling. In this case, the topology under study is assumed to be constituted by a lossless hybrid coupler with interconnected isolated and direct access ways. After the determination of the global structure transmission parameter, the NGD characterization will be performed.

5.2.2.1 *S*-parameter modeling of the interconnected isolated and direct ways topology

In difference with Section 5.2.1.1, in this case, the equivalent model of the present topology is determined with the combination of the wave power through the output direct way of the coupler. The equivalent *S*-parameter model was determined with the substitution of wave powers from matrices (5.1) and (5.3) or (5.5). Consequently, the equivalent reflection and transmission elements for the case if 90° hybrid coupled are expressed as:

$$s_{11} = s_{22} = \frac{a^2 r}{b^2 r^2 - (j + b \cdot t)^2} \tag{5.31}$$

$$s_{12} = s_{21} = \frac{(a^2 + 2b^2)t + jb[1 + (a^2 + b^2)(r^2 - t^2)]}{b^2 r^2 - (j + b \cdot t)^2} \tag{5.32}$$

However, for the case of 180° hybrid coupler-based topology, these *S*-parameter elements will be transformed as:

$$s_{11} = s_{22} = \frac{a^2 r}{b^2 r^2 - (1 + b \cdot t)^2} \tag{5.33}$$

$$s_{12} = s_{21} = \frac{b[(a^2 + b^2)(r^2 - t^2) - 1] - (a^2 + 2b^2)t}{b^2 r^2 - (1 + b \cdot t)^2} \tag{5.34}$$

Based on the frequency-dependent expressions, the NGD characterization will be made from the group delay responses.

5.2.2.2 NGD characterization and properties of the 90° hybrid coupler-based topology

Similar to the configuration explored in Section 5.2.1.2, the feedback circuit can be assumed as a lossy TL as presented in Figure 5.4.

To realize the frequency response analyses, the hybrid coupler *S*-parameter model defined in (5.3) is considered parametrized with a and b defined in (5.10). The frequency-dependent transmission parameter magnitude and phase of the topology shown in Figure 5.4 are defined as:

$$S_{21}(\omega) = \frac{\sqrt{\begin{array}{l}2(1+\alpha^2)^2 - 6\sqrt{2}\alpha^3 \sin(\omega\tau_0) \\ +\alpha^2(5 - 4\cos(2\omega\tau_0))\end{array}}}{\sqrt{2 - 2\sqrt{2}\alpha \sin(\omega\tau_0) + \alpha^2}} \tag{5.35}$$

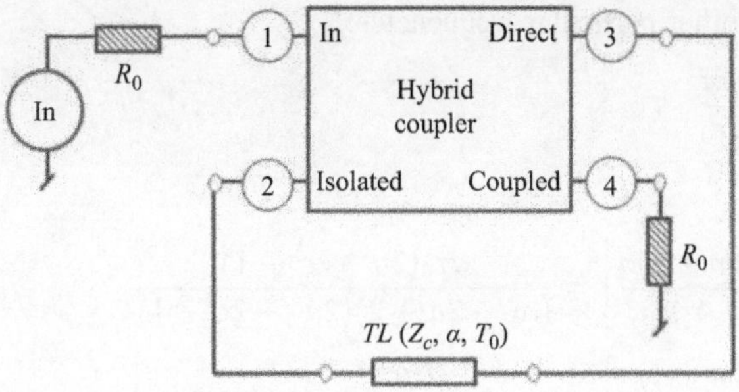

Figure 5.4 Configuration of the hybrid coupler-based topologies under study constituted by interconnected direct-isolated ways through a TL

$$\varphi_{21}(\omega) = \left\{ \begin{array}{l} \arctan\left[\dfrac{\alpha(\sqrt{2}\alpha\sin(2\omega\tau) - 3\cos(\omega\tau_0))}{\begin{array}{c}\sqrt{2}\alpha^2\cos(2\omega\tau_0)+ \\ 3\alpha\sin(\omega\tau_0) - \sqrt{2}\end{array}}\right] \\ -\arctan\left[\dfrac{4\alpha\cos(\omega\tau_0)(\sqrt{2} - \alpha\sin(\omega\tau)}{\begin{array}{c}\alpha^2\cos(2\omega\tau_0)+ \\ 2\sqrt{2}\alpha\cos(\omega\tau_0) - 2\end{array}}\right] \end{array} \right\} \tag{5.36}$$

The corresponding group delay:

$$\tau(\omega) = \tau'(\omega) - \tau''(\omega) \tag{5.37}$$

can be expressed as:

$$\tau'(\omega) = \frac{\dfrac{\alpha\cdot\tau_0}{2}\left\{\begin{array}{l}\alpha^4(\cos^2(2\omega\tau_0) - 2) + 4\sqrt{2}\alpha^3\sin(\omega\tau_0) \\ -4\alpha^2(3 - 2\cos^2(\omega\tau_0)) + 8\sqrt{2}\sin(\omega\tau_0) \\ -4\end{array}\right\}}{\begin{array}{c}\alpha^4 + \sqrt{2}\alpha^2[2\alpha\cos(2\omega\tau_0) - 3\sin(\omega\tau_0)] \\ 2\alpha(\cos(2\omega\tau_0) - 2) + 2\sqrt{2}\sin(\omega\tau_0)\end{array}} \tag{5.38}$$

$$\tau''(\omega) = \frac{\alpha\cdot\tau_0\left\{\begin{array}{l}2(\alpha^4 + 1) - 6\sqrt{2}\alpha(\alpha^2 + 1)\sin(\omega\tau_0) \\ +\alpha^2[9 - 4\cos(\omega\tau_0)]\end{array}\right\}}{\begin{array}{c}-4\alpha^3 - 9\sqrt{2}\alpha^2\sin(\omega\tau_0)+ \\ \alpha[9 - 4\cos(\omega\tau_0)] - 3\sqrt{2}\sin(\omega\tau_0)\end{array}} \tag{5.39}$$

From (5.39), the NGD behavior of the proposed topology was performed in function of the TL parameters based. Accordingly, at very low frequencies, the group delay of the 90° hybrid coupler-based topology is always positive:

$$\tau\left(\omega \approx \frac{m\pi}{\tau_0}\right) = \frac{3\alpha^2\tau_0}{(\alpha^2 + 2)(2\alpha^2 + 1)} \tag{5.40}$$

But at the other particular frequencies:

$$\omega(m) = \frac{m\pi}{4\tau_0}, \tag{5.41}$$

we have:

$$\tau\left[\omega = \frac{(2m+1)\pi}{4\tau_0}\right] = \frac{\alpha\tau_0(3\alpha - \alpha^2 - 1)}{(\alpha^2 - 2\alpha + 2)(2\alpha^2 - 2\alpha + 1)} \tag{5.42}$$

$$\tau\left[\omega = \frac{(2m+1)\pi}{\tau_0}\right] = \frac{\alpha\tau_0}{(\sqrt{2}\alpha - 1)(\sqrt{2} - \alpha)} \tag{5.43}$$

Mathematically, these two expressions can be negative in the certain values of the feedback line attenuation α. It is important to note that with the present topology, the minimal value of the group delay is realized at:

$$\omega_{\text{opt}}(m) = \frac{(4m+1)\pi}{2\tau_0} \tag{5.44}$$

At these particular frequencies, the transmission gain magnitudes are written as:

$$S_{21}\left[\omega = \frac{(2m+1)\pi}{4\tau_0}\right] = \frac{\sqrt{2\alpha^2 - 2\alpha + 1}}{\sqrt{\alpha^2 - 2\alpha + 2}} \tag{5.45}$$

$$S_{21}\left[\omega = \frac{(2m+1)\pi}{\tau_0}\right] = \frac{|\sqrt{2}\alpha - 1|}{|\alpha - \sqrt{2}|} \tag{5.46}$$

Moreover, the cutoff frequencies are given by

$$\begin{cases} \omega_1(m) = \dfrac{1}{\tau_0}\left[m\pi + \arctan\left(\dfrac{3\alpha}{\sqrt{2}}\right)\right] \\ \omega_2(m) = \dfrac{1}{\tau_0}\left[m\pi \pm \arctan\left(\sqrt{2 - 5\alpha^2 + 2\alpha^4}\right)\right] \end{cases}. \tag{5.47}$$

5.2.2.3 NGD characterization and properties of the 180° hybrid coupler-based topology

The frequency-dependent transmission parameter magnitude and phase of the structure shown in Figure 5.3 can be established from (5.35). Accordingly, we obtain the following expressions of the case of 180° hybrid coupler-based topology:

$$S_{21}(\omega) = \frac{\sqrt{\begin{array}{l}2(1+\alpha^2)^2 + 6\sqrt{2}\alpha^3\cos(\omega\tau_0) \\ +\alpha^2(5 + 4\cos(2\omega\tau_0))\end{array}}}{\sqrt{2 + 2\sqrt{2}\alpha\cos(\omega\tau_0) + \alpha^2}} \tag{5.48}$$

$$\varphi_{21}(\omega) = \left\{ \begin{array}{l} \arctan\left[\dfrac{\alpha(\sin(2\omega\tau_0) + 2\sqrt{2}\alpha\sin(\omega\tau_0))}{\begin{array}{c}2 + 2\sqrt{2}\alpha\cos(\omega\tau_0)\\ +\alpha^2\cos(2\omega\tau_0)\end{array}}\right] \\ -\arctan\left[\dfrac{\alpha(\sqrt{2}\alpha\sin(2\omega\tau_0) + 3\sin(\omega\tau_0))}{\begin{array}{c}\sqrt{2} + 3\alpha\cos(\omega\tau_0)\\ +\alpha^2\sqrt{2}\cos(2\omega\tau_0)\end{array}}\right] \end{array} \right\} \tag{5.49}$$

The group delay response can be determined from this transmission phase via (5.13):

$$\tau'(\omega) = \frac{2\alpha\cdot\tau_0\left[\begin{array}{c}\alpha^3 + 4\alpha +\\ \sqrt{2}(3\alpha^2+2)\cos(\omega\tau_0)\end{array}\right]}{\begin{array}{l}\alpha^4 + 8\alpha^2 + 4 + 4\sqrt{2}\alpha(\alpha^2+2)\cos(\omega\tau_0)\\ +4\alpha^2(\cos(2\omega\tau_0)+2)\end{array}} \tag{5.50}$$

$$\tau''(\omega) = \frac{\alpha\cdot\tau_0\left[\begin{array}{l}4\alpha^3 + 3\sqrt{2}(3\alpha^2+1)\cos(\omega\tau_0)\\ +\alpha(9+4\cos(2\omega\tau_0))\end{array}\right]}{\begin{array}{l}2(1+\alpha^4) + 6\sqrt{2}\alpha(1+\alpha^2)\cos(\omega\tau_0)\\ +\alpha^2(9+4\cos(2\omega\tau_0))\end{array}} \tag{5.51}$$

With analog way as the previous cases, the NGD characterization can be drawn from this expression. At very low frequencies, the group delay of the symmetrical 180° directional coupler topology is always positive:

$$\tau\left(\omega \approx \frac{2m\pi}{\tau_0}\right) = \frac{\alpha\tau_0}{(\sqrt{2}\alpha+1)(\sqrt{2}+\alpha)} \tag{5.52}$$

However, at the feedback TL triple quarter- and half- wavelength frequencies, the group delays can be calculated and are expressed as:

$$\tau\left[\omega = \frac{(4m+3)\pi}{\tau_0}\right] = \frac{\alpha\tau_0(3\alpha - \alpha^2 - 1)}{(\alpha^2 - 2\alpha + 2)(2\alpha^2 - 2\alpha + 1)} \tag{5.53}$$

$$\tau\left[\omega = \frac{(2m+1)\pi}{\tau_0}\right] = \frac{\alpha\tau_0}{(\sqrt{2}\alpha - 1)(\sqrt{2} - \alpha)} \tag{5.54}$$

Meanwhile, the circuit is susceptible to generate NGD phenomenon in certain ranges of α. At these particular frequencies, the transmission gain magnitudes are written as:

$$S_{21}\left[\omega = \frac{(4m+3)\pi}{\tau_0}\right] = \frac{\sqrt{2\alpha^2 - 2\alpha + 1}}{\sqrt{\alpha^2 - 2\alpha + 2}} \tag{5.55}$$

$$S_{21}\left[\omega = \frac{(2m+1)\pi}{\tau_0}\right] = \frac{|\sqrt{2}a - 1|}{|a - \sqrt{2}|} \tag{5.56}$$

Moreover, the NGD cutoff frequencies, mathematically defined as the root of the equation $\tau(\omega) = 0$, are given by:

$$\begin{cases} \omega_1(m) = \dfrac{1}{\tau_0}\left[m\pi - \arctan\left(\dfrac{3a\sqrt{2}}{2}\right)\right] \\ \omega_2(m) = \dfrac{1}{\tau_0}\left[m\pi \pm \arctan\left(\sqrt{2 - 5a^2 + 2a^4}\right)\right] \end{cases} \tag{5.57}$$

with $m = \{1, 2, \ldots\}$.

The relevance of this theoretical approach will be checked in Section 5.3 with the application example analyses with parametric variations of the feedback line parameters.

5.3 Validation results with parametric analyses

To verify the feasibility of the NGD topology elaborated previously, illustrative applications will be examined. Doing this, simulations with the electronic circuit simulator tool ADS from Keysight Technologies® will be carried out. Comparisons between the simulated and theoretical calculated frequency responses will be discussed. First, parametric analyses with variation of the feedback line attenuation and time-delay will be explored.

5.3.1 Demonstration of the NGD phenomenon with parametric analyses of the hybrid coupler feedback line

The two configurations of topologies with isolated-coupled and isolated-direct outputs interconnected will be analyzed in this paragraph. In this case, the hybrid coupler components are supposed ideal and lossless. The feedback circuit is modeled by the black box expressed in (5.9). The group delay and insertion loss responses from the simulated structures by varying the feedback line attenuation and time-delay will be examined in the following sections.

The parametric analyses presented in this section were generated from simulations in the ADS schematic environment.

5.3.1.1 Application results with the topology constituted by the 90° hybrid coupler with interconnected isolated and coupled accesses

In this case, a lossless 90° hybrid directional coupler is considered during the simulations. By varying the attenuation *a* from –10 to –6 dB with the fixed time delay $\tau_0 = 0.5$ ns, the first step of parametric simulations was carried out. The simulated group delay responses are displayed in Figure 5.5(a). As predicted in

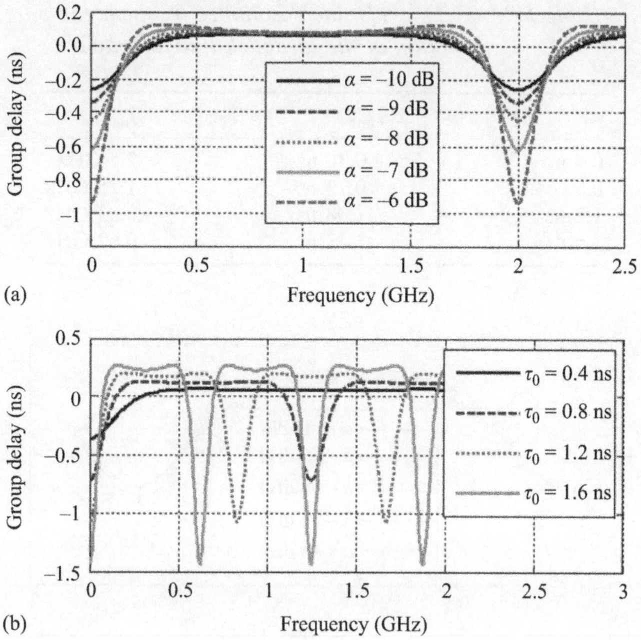

Figure 5.5 Group delay responses versus feedback line: (a) attenuation and (b) time-delay

Table 5.1 NGD level and insertion loss in function of the feedback line attenuation

a_{dB}	τ_{opt}	f_c	$S21_{opt}$
−10 dB	−0.26 ns	180 MHz	−5.96 dB
−9 dB	−0.33 ns	205 MHz	−6.55 dB
−8 dB	−0.45 ns	223 MHz	−7.33 dB
−7 dB	−0.63 ns	242 MHz	−8.39 dB
−6 dB	−0.94 ns	260 MHz	−9.92 dB

theory, it can be noticed that the simulated structure behaves as a low-pass NGD circuit for the different parameters of the feedback line.

As expected and correlated with the analytical calculated results addressed in Table 5.1, a very good correlation between the simulations and theory is realized.

In theory, it can be underlined that the fully passive circuit is susceptible to exhibit the NGD phenomenon from DC to f_c as reported in Table 5.1. The NGD optimal values are varied from –0.26 to –0.94 ns for the attenuation swept from –10 to –6 dB. Furthermore, by varying the feedback line time delay τ_0 from 0.4 to 1.6 ns and maintaining the attenuation fixed $a = -8$ dB, the simulated group delay responses are plotted in Figure 5.5(b).

Table 5.2 *NGD levels and resonance frequency in function of the feedback line time delay*

τ_0	τ_{opt}	f_{opt}
0.4 ns	−0.36 ns	2.5 GHz
0.8 ns	−0.72 ns	1.25 GHz
1.2 ns	−1.08 ns	0.83 GHz
1.6 ns	−1.43 ns	0.62 GHz

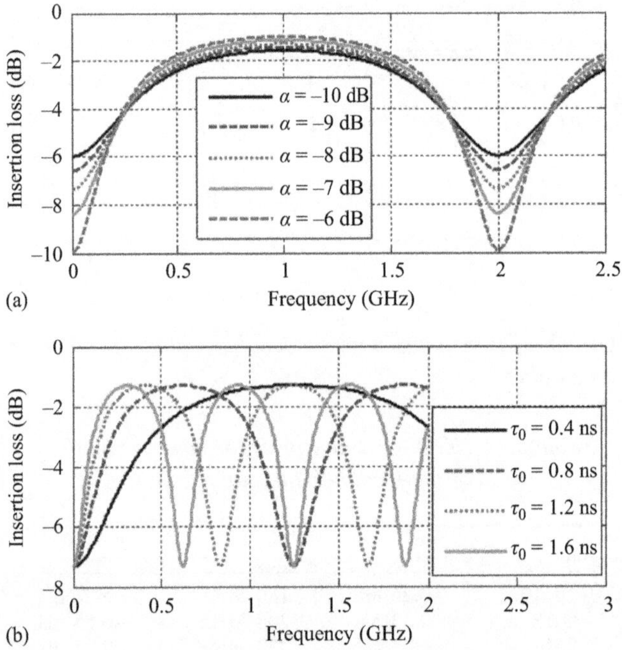

Figure 5.6 Insertion losses versus (a) attenuation and (b) time-delay

Again, a good agreement between the theoretical results drawn in Table 5.2 and the simulations was realized.

It can be noticed that the group delay is negative in base band frequency up to the NGD cutoff frequencies f_c and varied from –0.36 to –1.43 ns when the feedback time delay is swept from 0.4 to 1.6 ns.

The simulated circuit insertion losses are plotted in Figures 5.6. As predicted in Figure 5.3 established earlier in Section 5.2.4, these insertion losses are inversely proportional to the feedback line attenuation and time delay. The insertion loss optimal value varies from –5.9 dB to about –10 dB when varying the feedback line attenuation. Then, it is constant and equal to –7.3 dB and independent to the time-delay variation.

5.3.1.2 Application results with the NGD topology composed of the interconnected isolated and direct accesses 180° hybrid coupler

In this case, the 180° hybrid asymmetrical coupler is implemented in the simulated circuit. The isolated and direct accesses were interconnected. The feedback line attenuation and time delay were swept as in the previous case to perform the parametric analyses.

Once again, as expected in the theory, it can be underlined that the simulated circuit is susceptible to exhibit the NGD phenomenon as illustrated in Figure 5.7(a) and (b). As highlighted with the calculated results addressed in Table 5.3, the NGD

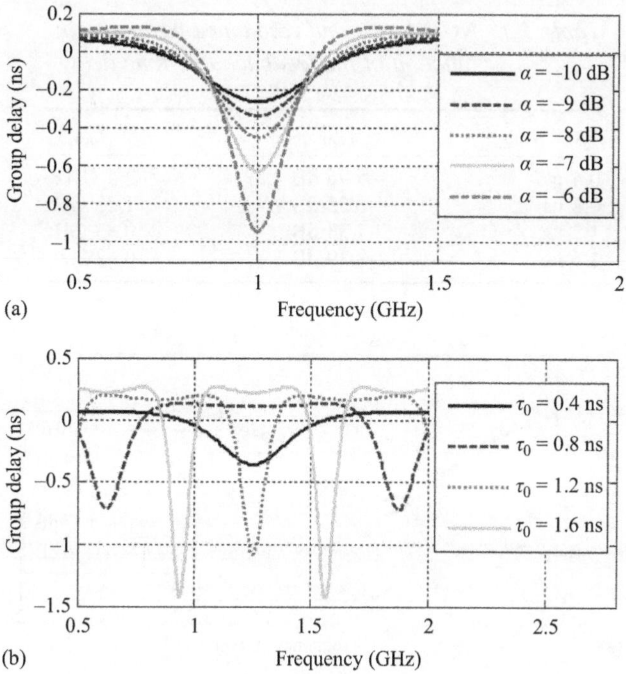

Figure 5.7 Group delay responses versus (a) attenuation and (b) time delay

Table 5.3 NGD levels and insertion loss in function of the feedback line attenuation

a_{dB}	τ_{opt}	$S_{21_{opt}}$
−10 dB	−0.26 ns	−5.96 dB
−9 dB	−0.33 ns	−6.55 dB
−8 dB	−0.45 ns	−7.33 dB
−7 dB	−0.63 ns	−8.39 dB
−6 dB	−0.94 ns	−9.92 dB

optimal values are varied from –0.26 to –0.9 ns when the attenuation swept from –10 to –6 dB.

As shown in Figure 5.7(b), the group delay optimal values are varied from –0.36 to –1.34 ns for the time delay swept from 0.4 to 1.6 ns. In addition, the theoretical calculation of the optimal group delay and the corresponding NGD center frequency presented in Table 5.4 are in good correlation with the simulations.

In the present configuration, different values of feedback line attenuation and time-delay have been considered. Accordingly, the obtained insertion losses are plotted in Figure 5.8.

As predicted earlier with the theoretical results shown in Figure 5.3 of Section 5.2.4, it can be seen that these insertion losses are inversely proportional to

Table 5.4 NGD levels and resonance frequency in function of the feedback line time delay

τ_0	τ_{opt}	f_{opt}
0.4 ns	−5.96 dB	2.5 GHz
0.8 ns	−6.55 dB	1.25 GHz
1.2 ns	−7.33 dB	0.83 GHz
1.6 ns	−8.39 dB	0.62 Hz

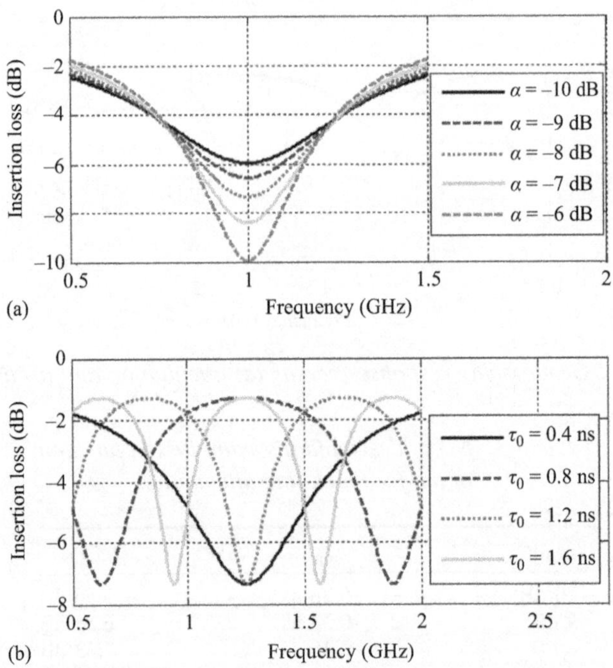

Figure 5.8 Insertion losses versus (a) attenuation and (b) time delay

the feedback line attenuation. Moreover, as predicted in expression (5.24), the insertion loss is independent of the feedback line attenuation time delay.

5.3.1.3 Discussion on the responses of the NGD topology consisted of a hybrid coupler in feedback with a microstrip line

In this case, the feedback circuit is constituted by a microstrip line printed on FR4-epoxy substrate. The relative permittivity and thickness are, respectively, $\varepsilon_r = 4.4$ and $h = 1.6$ mm. The metallization is supposed presenting conductivity $\sigma = 0.05$ MS with thickness of about 35 μm. The microstrip lines present a physical width $w = 3$ mm and lengths $d = \{15 \text{ mm}, 20 \text{ cm}, 25 \text{ cm}\}$. As can be seen from Figure 5.9, it corresponds to the average time delay τ_0, respectively, approximately of about {0.93 ns, 1.25 ns, 1.55 ns}.

5.3.1.4 Application results with the NGD topology constituted by a 90° hybrid coupler with interconnected isolated and coupled accesses

Figure 5.10 represents the simulated circuit schematic configuration. The feedback line is presented by the microstrip line with physical width $w = 3$ mm or characteristic impedance Z_c of about 50 Ω. The time delay is defined with the physical lengths $d = \{15 \text{ cm}, 20 \text{ cm}, 25 \text{ cm}\}$. As defined earlier in Figure 5.1(a) of Section 5.2, the overall circuit main accesses are represented by the in- and

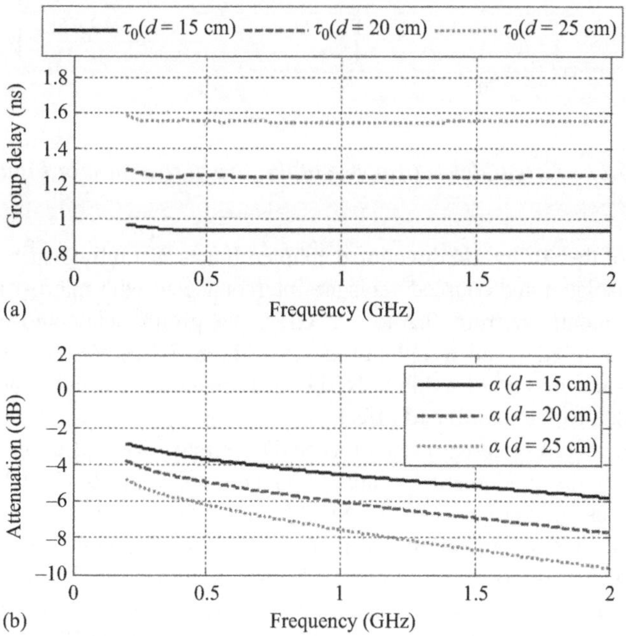

Figure 5.9 Frequency-dependent attenuation and time delay of the considered microstrip line

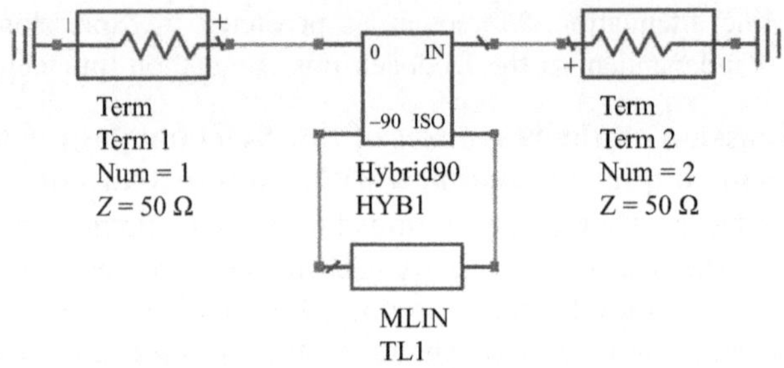

Figure 5.10 Schematic of the simulated circuit composed of 90° hybrid coupler with isolated and coupled ways interconnected with a microstrip line

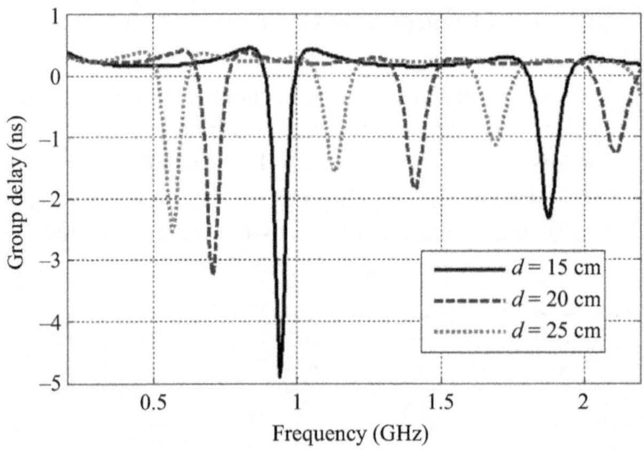

Figure 5.11 Group delay response of the structure shown in Figure 5.10

direct-ways of the 90° hybrid directional coupler. It acts as a fully distributed passive circuit. The isolated and coupled ways are interconnected with microstrip line.

After simulations from 0.2 to 2.2 GHz, the group delay and *S*-parameter responses monitored, respectively, in Figures 5.11 and 5.12 were obtained. It can be pointed out that the circuit exhibits NGD phenomena in the multiband frequency principally due to the distributed effect.

The considered circuit presents the NGD optimal levels of about –2.4, –3.3, and –4.8 ns at the fundamental center frequencies of about 0.57, 0.71, and 0.94 GHz. With these numerical tests, the NGD bandwidths are approximately of about 0.12 GHz. The average insertion loss is of about –7.5, –9, and –11 dB in the NGD frequency band. It is worth noting that in difference with the theoretical prediction, the NGD levels change with the frequency principally due to the skin depth effect on the feedback line metallization as highlighted previously in Figure 5.9. For the considered feedback line physical lengths, the simulated circuits present a matching level better than –20 dB in the NGD bandwidth.

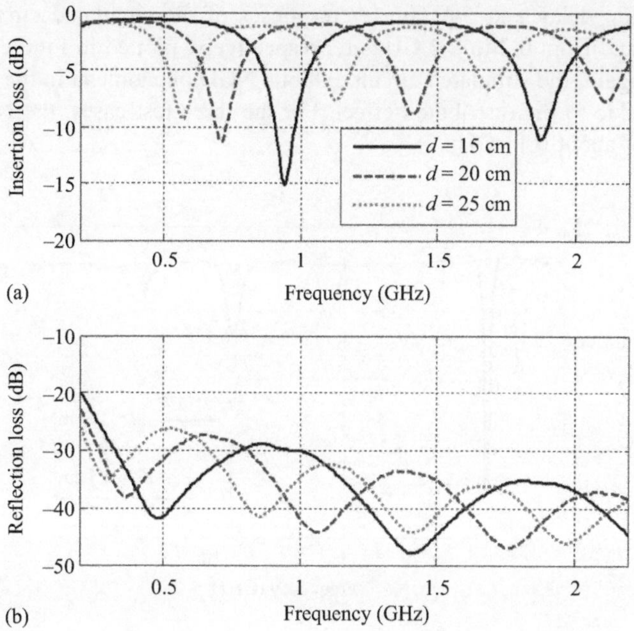

Figure 5.12 S-parameters of the structure shown in Figure 5.10

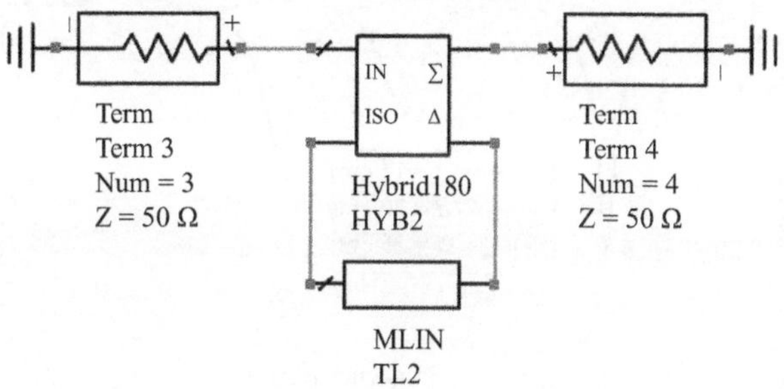

Figure 5.13 Schematic of the simulated circuit composed of 180° hybrid coupler with isolated and direct ways interconnected with a microstrip line

5.3.1.5 Application results with the NGD topology constituted by a 180° hybrid coupled with interconnected isolated and direct accesses

The schematic of the ADS simulated circuit is depicted in Figure 5.13. In this case, a 180° hybrid coupler was considered. It acts as fully distributed passive circuit. The in- and coupled-accesses constitute, respectively, the input and output main ports of the circuit. The isolated and output direct accesses are interconnected with the microstrip line used in the previous case.

The group delay and *S*-parameter responses of the simulated circuits in the frequency band from 0.2 to 2.2 GHz are, respectively, plotted in Figures 5.14 and 5.15. Once again, the simulated circuit exhibits NGD phenomena in the multiband frequencies due to the distributed effect. For the three test cases, the NGD bandwidths are of about 0.11 GHz.

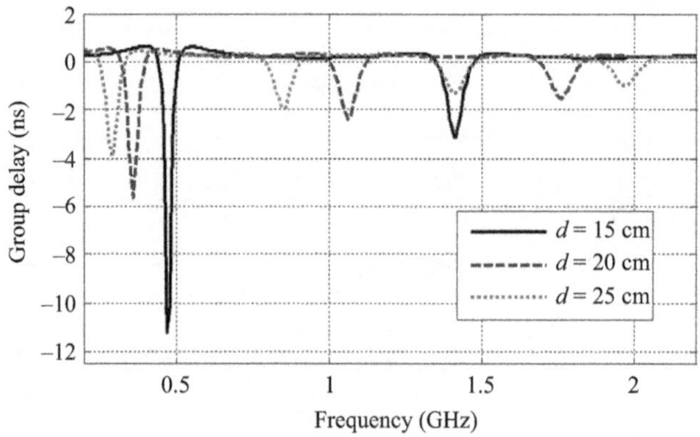

Figure 5.14 Group delay response of the structure shown in Figure 5.13

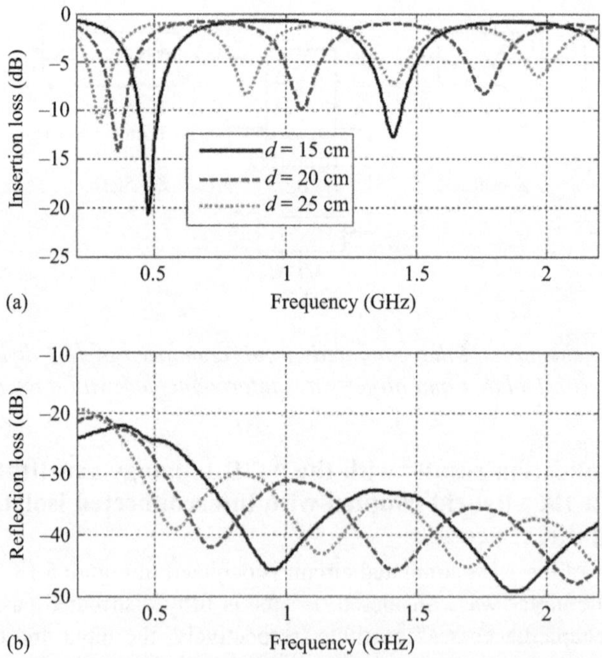

Figure 5.15 S-parameters of the structure shown in Figure 5.13

In this case, the NGD optimal values of about −3.9, −5.6, and −11 ns are achieved, respectively, at the fundamental center frequencies of 0.29, 0.36, and 0.47 GHz. However, due to the feedback line attenuation variation with the frequency as illustrated in Figure 5.9, the insertion losses are better at the multiple harmonic frequencies. The average insertion loss optimal values in the NGD frequency band are of about −11, −13, and −20 dB. Independently to the feedback line physical lengths, the simulated circuit presents a good matching level better than −15 dB.

5.4 Conclusion

A complete innovative circuit theory on the fully distributed NGD circuit is proposed. The NGD topology under study consists of a hybrid coupler with interconnected isolated and coupled- or direct-output ways in feedback. The feedback interconnection chain is constituted by a lossy TL. The *S*-parameter model of this original topology is analytically established. Then, the frequency responses and especially the group delay expression are derived in function of the feedback line delay and attenuation. It is deduced mathematically that the hybrid coupler-based topology is susceptible to exhibit the NGD phenomenon in function of the feedback line parameters. The hybrid coupler topology is characterized with the NGD level-bandwidth product. Then, the synthesis relation from the desired NGD level, which is systematically linked to the insertion loss, is established.

To verify the relevance of the theoretical NGD concept, illustrative examples based on the 90°- and 180°-hybrid couplers are analyzed. The verification analyses are performed with parametric analyses based on the variation of the feedback line attenuation and time delay. Therefore, as predicted in theory, low-pass NGD phenomena were realized with a feedback line having attenuation between −10 and −6 dB, and from 0.4 to 1.6 ns. Furthermore, it was found that the ADS simulations results are in good correlation with the theoretical approaches. It was pointed out that the optimal NGD level is inversely proportional to the feedback line attenuation. In addition, further application examples with a tens centimeter physical length feedback microstrip line are considered. Once again, NGDs with levels with minus some nanoseconds were occurred. With the application examples, the NGD bandwidths are in order of about 0.1 GHz. The obtained results illustrate the feasibility and the relevance of the NGD topology based on the classical hybrid coupler. The performed study is particularly promising for the application.

It is noteworthy that the present NGD topology can be fully implemented in distributed passive topology. Therefore, it can be potentially integrable as a microwave integrated circuit. Consequently, it is promising for wide range of applications notably to the field of telecommunication with microwave signal delay reduction and synchronization.

In the continuation of this work, an implementation of compact NGD circuit based on the coupled elements is in progress. Then, the NGD circuit applications can be potentially used for the signal delay reduction notably for the digital, microwave and mixed telecommunication devices.

References

[1] E. C. Heyde, "Theoretical methodology for describing active and passive recirculating delay line systems," Electronics Letters, Vol. 31, No. 23, Nov. 1995, pp. 2038–2039.

[2] I. M. Filanovsky, "One method of synthesis of time delay circuits," in Proc. of 42nd Midwest Symposium on Circuits and Systems, 1999, Vol. 2, Las Cruces, NM, 8–11 Aug. 1999, pp. 1032–1035.

[3] H. F. Li, S. C. Leung and P. N. Lam, "Optimised synthesis of delay-insensitive circuits using time-sharing," IEE Proceedings – Computers and Digital Techniques, Vol. 141, No. 2, Mar. 1994, pp. 111–118.

[4] S.-S. Myoung, B.-S. Kwon, Y.-H. Kim and J.-G. Yook, "Effect of group delay in RF BPF on impulse radio systems," IEICE Transactions on Communications, Vol. 90, No. 12, Dec. 2007, pp. 3514–3522.

[5] G. Groenewold, "Noise and group delay in active filters," IEEE Transactions on Circuits and Systems I: Regular Papers, Vol. 54, No. 7, Jul. 2007, pp. 1471–1480.

[6] M. Scalora, R. J. Flynn, S. B. Reinhardt and R. L. Fork, "Ultrashort pulse propagation at the photonic band edge: Large tunable group delay with minimal distortion and loss," Physical Review E, Vol. 54, No. 2, Aug. 1996, pp. 1078–1081.

[7] Y. Nabekawa and K. Midorikawa, "Group-delay-dispersion-matched sum-frequency mixing for the indirect phase control of deep ultraviolet pulses in the sub-20-fs regime," Applied Physics B, Vol. 78, No. 5, Mar. 2004, pp. 569–581.

[8] S. V. Narasimhan, M. Hazarathaiah and P. V. S. Giridhar, "Channel blind identification based on cyclostationarity and group delay," Signal Processing, Vol. 85, No. 7, Jul. 2005, pp. 1275–1286.

[9] J. Vemagiri, A. Chamarti, M. Agarwal and K. Varahramyan, "Transmission line delay-based radio frequency identification (RFID) tag," Microwave and Optical Technology Letters, Vol. 49, No. 8, Aug. 2007, pp. 1900–1904.

[10] C. Wijenayake, Y. Xu, A. Madanayake, L. Belostotski and L. T. Bruton, "RF analog beamforming fan filters using CMOS all-pass time delay approximations," IEEE Transactions on Circuits and Systems I: Regular Papers, Vol. 59, No. 5, May 2012, pp. 1061–1073.

[11] G. Agrawal, S. Aniruddhan and R. K. Ganti, "Multi-band RF time delay element based on frequency translation," in Proceedings of IEEE International Symposium on Circuits and Systems (ISCAS), 2014, Melbourne, VIC, 1–5 June 2014, pp. 1368–1371.

[12] M-E. Hwang, S-O. Jung and K. Roy, "Slope interconnect effort: Gate-interconnect interdependent delay modeling for early CMOS circuit simulation," IEEE Transactions on Circuits and Systems I, Vol. 56, No. 7, Jul. 2009, pp. 1428–1441.

[13] B. Ravelo, "Delay modelling of high-speed distributed interconnect for the signal integrity prediction," The European Physical Journal of Applied Physics, Vol. 57, No. 31002, Feb. 2012, pp. 1–8.

[14] F. R. Awwad, M. Nekili, V. Ramachandran and M. Sawan, "On modeling of parallel repeater-insertion methodologies for SoC interconnects," IEEE Transactions on Circuits and Systems I, Vol. 55, No. 1, Feb. 2008, pp. 322–336.

[15] B. Ravelo, "Recovery of microwave-digital signal integrity with NGD circuits," Photonics and Optoelectronics, Vol. 2, No. 1, Jan. 2013, pp. 8–16.

[16] O. F. Siddiqui, M. Mojahedi and G. V. Eleftheriades, "Periodically loaded transmission line with effective negative refractive index and negative group velocity," IEEE Transactions on Antennas and Propagation, Vol. 51, No. 10, Oct. 2003, pp. 2619–2625.

[17] L. Markley and G. V. Eleftheriades, "Quad-band negative-refractive-index transmission-line unit cell with reduced group delay," Electronics Letters, Vol. 46, No. 17, Aug. 2010, pp. 1206–1208.

[18] O. F. Siddiqui, S. J. Erickson, G. V. Eleftheriades and M. Mojahedi, "Time-domain measurement of negative-index transmission-line metamaterials," IEEE Transactions on Microwave Theory and Techniques, Vol. 52, No. 5, May 2004, pp. 1449–1453.

[19] N. S. Bukhman and S. V. Bukhman, "On the negative delay time of a narrow-band signal as i passes through the resonant filter of absorption," Radiophysics and Quantum Electronics, Vol. 47, No. 1, Jan. 2004, pp. 66–76.

[20] M. Kandic and G. E. Bridges, "Asymptotic limits of negative group delay in active resonator-based distributed circuits," IEEE Transactions on Circuits and Systems I: Regular Papers, Vol. 58, No. 8, Aug. 2011, pp. 1727–1735.

[21] M. Kandic and G. E. Bridges, "Bilateral gain-compensated negative group delay circuit," IEEE Microwave and Wireless Components Letters, Vol. 21, No. 6, May 2011, pp. 308–310.

[22] K-P Ahn, R. Ishikawa, A. Saitou and K. Honjo, "Synthesis for negative group delay circuits using distributed and second-order RC circuit configurations," IEICE Transactions on Electronics, Vol. E92-C, No. 9, Sep. 2009, pp. 1176–1181.

[23] B. Ravelo, "Methodology of elementary negative group delay active topologies identification," IET Circuits Devices Syst. (CDS), Vol. 7, No. 3, May 2013, pp. 105–113.

[24] B. Ravelo, "Similitude between the NGD function and filter gain behaviours," International Journal of Circuit Theory & Applications, Vol. 42, No. 10, Oct. 2014, pp. 1016–1032.

[25] W. Alomar and A. Mortazawi, "Method of generating negative group delay in phase arrays without using lossy circuits," in Proceedings of IEEE International Wireless Symposium (IWS), 2013, Beijing, China, 14–18 April 2013, pp. 1–3.

[26] W. Alomar and A. Mortazawi, "Elimination of beam squint in uniformly excited serially fed antenna arrays using negative group delay circuits," in Proceedings of IEEE Antennas and Propagation Society International Symposium (APSURSI), 2012, Chicago, IL, 8–14 July 2012, pp. 1–2.

[27] H. Mirzaei and G. V. Eleftheriades, "Realizing non-Foster reactances using negative-group-delay networks and applications to antennas," in Proceedings of IEEE Radio and Wireless Symposium (RWS), 2014, Newport Beach, CA, 19–23 Jan. 2014, pp. 58–60.
[28] C.-T. M. Wu, S. Gharavi and T. Itoh, "Negative group delay circuit based on a multisection asymmetrical directional coupler," in Proceedings of Asia-Pacific Microwave Conference Proceedings (APMC), 2013, Seoul, Korea, 5–8 Nov. 2013, pp. 333–334.
[29] G. Chaudhary and Yongchae Jeong, "Low signal-attenuation negative group-delay network topologies using coupled lines," IEEE Transactions on Microwave Theory and Techniques, Vol. 62, No. 10, Aug. 2014, pp. 2316–2324.
[30] G. Chaudhary and Y. Jeong, "Transmission-type negative group delay networks using coupled line doublet structure," IET Microwaves, Antennas & Propagation, Vol. 9, No. 8, Jun. 2015, pp. 748–754.
[31] V. K. Tripathi and Y. K. Chin, "Analysis of the general nonsymmetrical directional coupler with arbitrary terminations," IEE Proceedings H Microwaves, Optics and Antennas, Vol. 129, No. 6, Dec. 1982, pp. 360–362.
[32] G. Bogdanov and R. Ludwig, "Coupled microstrip line transverse electromagnetic resonator model for high-field magnetic resonance imaging," Magnetic Resonance in Medicine, Vol. 47, No. 3, Mar. 2002, pp. 579–593.
[33] J. Müller and A. F. Jacob, "Complex compensation of Coupled line structures in inhomogenous media," IEEE MTT-S International Microwave Symposium Digest, 2008, Atlanta, GA, 15–20 Jun. 2008, pp. 1007–1010.
[34] S. M. Musa and M. N. O. Sadiku, "Finite element method analysis of symmetrical coupled microstrip lines," International Journal of Computing And Digital Systems, Vol. 3, No. 3, Sep. 2014, pp. 189–195.

Chapter 6
NGD-based Hilbert filter

Blaise Ravelo[1]

Abstract

A particular application of the negative group delay (NGD) for a particular function of independent frequency phase shifter (PS) is developed in the present chapter. The independent frequency PS is a quadrature and unity magnitude PS which is known as Hilbert filter. An original circuit theory on analog Hilbert filter dedicated to the RF/microwave application is established in this chapter. The Hilbert filter can be assumed as an independent frequency quadrature PS with magnitude equal to unity. The identification methodology of the simplest lumped element-based topologies is presented. The proposed elementary cells are based on the NGD active cell constituted by a field effect transistor in cascade with a shunt RLC-series network. This NGD circuit is cascaded with a positive group delay (PGD) passive network in order to synthesize an Hilbert filter around the targeted operating frequency. The analytical expressions enable to establish the Hilbert filter designability characteristic condition between the topological parameters. The synthesis relations of the identified cells are established. The characteristics and properties are developed. Application example with the identified topology of LC- and NGD-cells in cascade is presented. The microwave Hilbert filter concept feasibility is demonstrated. Three proofs of concept (POCs) circuit operating around 2.45 GHz were synthesized and designed. After *S*-parameter simulations, transmission phase $90 \pm 1°$ flatness's are generated within the relative frequency better than 50% and ± 1 dB gain flatness's within 20% bandwidth. The identified Hilbert filter cells functioning are validated with simulations of SPICE environment designed circuits. As expected, typically narrow band Hilbert filters were obtained. The developed analog Hilbert filter topology is potentially useful for the RF/microwave transceiver system architecture designs notably for image intermediate frequency rejection.

Keywords

Negative group delay (NGD), Hilbert filter, phase shifter, identification methodology, microwave circuit, circuit theory, synthesis method, positive group delay (PGD) cell.

[1]ESIGELEC, France

6.1 Introduction

Nowadays, phase shifters (PSs) play an important role for the improvement of RF/ microwave and wireless communication terminal chains [1]. It enables to realize high resolution of beamformers in different scenario by using multibit phase-arrays [2]. The PSs are also adapted to new generation of hybrid beamformers with selection approach [3] and analog–digital mixed applications [4]. Various functions as the integration of FFT and IFFT operations were developed to increase the performances of such mixed transceivers [4]. Moreover, the PSs can be designed and fabricated for both narrow- and wide-band technologies with guarantee of low-cost fabrication process [5].

However, such operations are generally implemented in front- and back-end architectures with particular functions as Hilbert filters. The former ones can ensure in easy technical ways basic signal processing. It was emphasized that the Hilbert filters are promising one of ideal solutions to suppress image channels after down conversion mixer o RF/microwave wireless transceivers [6]. It highlights why, so far, these former ones constitute one of key elements enabling the improvement of hardware communication systems. Nevertheless, the Hilbert filter design technique remains an open challenge for the electronic and communication architecture engineers. So far, the Hilbert filters are usually designed and implemented based on programmable digital circuits [7,8]. Nonetheless, due to the technological design difficulties of analog-to-digital and digital-to-analog convertors, the Hilbert filter digital functions cannot operate with higher frequencies more than GHz, for example, in RF and microwave frequency L-band. To overcome this technical limitation, design of microwave Hilbert filters is proposed in the present chapter. As suggested in [9,10], the adjacent channels suppressor can be integrated in a complex active channel-select Hilbert filter. It was emphasized that the low intermediate frequency receiver design solution was performed with Hilbert filter built with passive RC polyphase filter [10]. But, this Hilbert filter topology is limited in terms of the operation frequency bandwidth, attenuation and design complexity.

For this reason, an original circuit topology of RF/ and microwave Hilbert filter is investigated in this chapter. The proposed Hilbert filter integrates a negative group delay (NGD) active cell investigated in [11]. It was reported in [12] that frequency-independent PS can be designed by cascading positive and NGD circuits. The feasibility of the NGD-based constant-phase PS is approved with several technologies [12–15]. An NGD PS implemented in microstrip technology is presented in [13]. A phase equalizer uses NGD times by reflection coefficients is introduced in [14]. Another design based of NGD circuit compensation true-time-delay (TTD) PS is developed in [15]. The TTD PSs were successfully designed in CMOS [16], millimeter wave [17], and LC delay line with gm-C cascading [18], technologies. But those TTD PSs [15–18] generate different time delays depending on the fabrication technology. For this reason, the NGD-based PS is proposed in the present chapter.

The compensation technique with typically NGD circuits was described in [19] by considering lossless double negative metamaterials. The double negative materials operate as negative group velocity somehow known as superluminal medium [20].

The first application of superluminal NGD circuit for delay cancellation with compensation was initiated in [21]. Then, the compensation technique was extended to the delay and electrical interconnect effect equalizations [22–26]. The delay equalization technique was validated in MMIC technology [22], in CMOS technology with metamaterial inspired cells [23], field effect transistor (FET)-based hybrid circuits for the printed circuit board electrical interconnect effect (RC-, LC-, RLC-, and RLCG-effects) reductions [24–26]. A similar principle is developed in the present paper to design Hilbert filter analog simple circuits with lumped elements.

In difference with the NGD-based PS developed in [12–15], the Hilbert filter topology under study is comprised of lumped LC-circuit instead of distributed transmission line. The innovative theoretical methodology is established in the next section to synthesize the Hilbert filter circuit elements.

6.2 Identification methodology of the NGD function-based Hilbert filter

The Hilbert filter is a system presenting transfer function equal to the pure imaginary number $j=\sqrt{-1}$. In other words, by denoting ω the angular frequency variable, for the input signal spectrum $X(j\omega)$, the output is given by

$$Y(j\omega)=j\cdot X(j\omega) \tag{6.1}$$

Figure 6.1 represents the synoptic diagram of Hilbert filter. The associated transfer function is assumed as a two-port system.

The next sections of this chapter are dedicated to the description of the proposal Hilbert filter design method by considering NGD concept. The Hilbert filter research identification methodology is based on the consideration of two blocks of two port networks in cascade as previously introduced in [12–15]. By considering the lumped element-based simplest topologies, the transfer matrices are proposed. Then, the associated *S*-parameters are extracted. The analysis of the transmission parameter enables to identify the basic simple topologies fulfilling the Hilbert filter transfer function.

6.2.1 General description of Hilbert filter basic topology

According to the circuit and system theory, by denoting $\tau(\omega)$ the group delay, the associated PS is defined by:

$$\varphi(\omega)=-\int\tau(\omega)d\omega \tag{6.2}$$

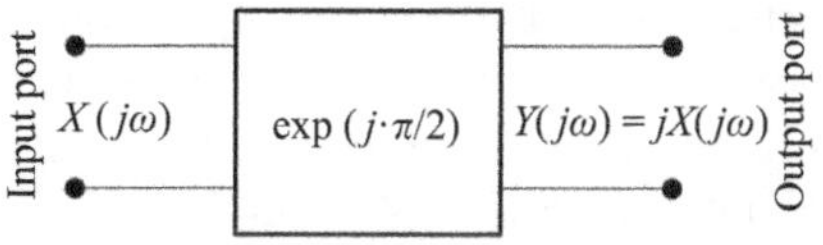

Figure 6.1 Hilbert filter transfer function

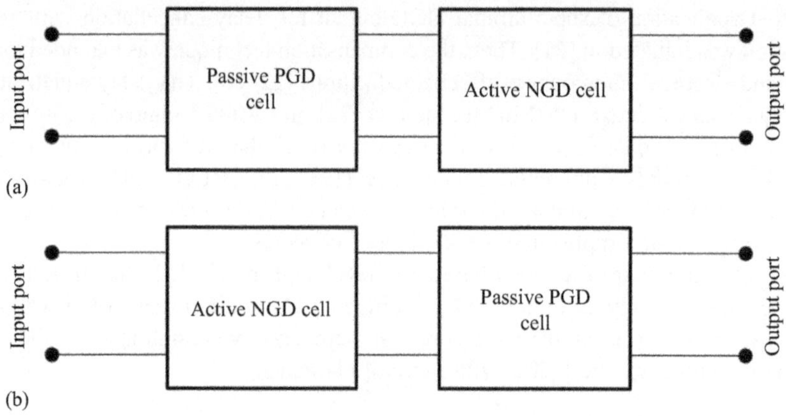

Figure 6.2 Topologies of Hilbert filter: (a) passive PGD cell cascaded with active NGD cell and (b) active NGD cell cascaded with PGD cell

If one takes the ideal case:

$$\tau(\omega) = \tau_{\text{NGD}} < 0 \tag{6.3}$$

we have:

$$\varphi(\omega) = -\tau_{\text{NGD}}\omega + \varphi_{\text{cst}} \tag{6.4}$$

with φ_{cst} is a real integration constant. It can be remarked from this simple demonstration that the NGD function is analytically equivalent to the positive slope phase shift in opposite with the classical ordinary circuit as the transmission lines.

By cascading positive and negative phase slope circuits, a constant phase or frequency-independent phase shift aspect can be generated. This principle is exploited in the present chapter for generating the unity magnitude and quadrature phase shift transmission coefficient circuit developed in this chapter. Meanwhile, an Hilbert filter topology can be synthesized by associating a PGD passive cell and NGD active cell.

Figure 6.2 represents the two topological architectures of PGD and NGD two-port networks associated in cascade. The PGD block is a two-port network passive circuit.

Acting as an ordinary PGD passive network, it should generate a transmission parameter magnitude lower than unity and negative slope negative phase shift. In order to compensate the loss and to equalize the phase shift, the other block must be an NGD active network.

The implementations of the PGD and NGD cells constitute the basic step for the present identification methodology. Different configurations of *Z-Y* network can be imagined. But the aim of this study is to identify the simplest elementary Hilbert filter topology. Table 6.1 addresses the possibility of the expected configurations. It indicates the a priori possibly successful "✓" and failed "×" combinations.

Based on these general topologies, the configurations of the resistor-, inductor-, and capacitor-lumped components-based cells are explored in Section 6.2.2.

Table 6.1 Configurations of Z-Y network

Z-Y network configuration	**Y = Resistor**	**Y = Inductor**	**Y = Capacitor**
Z = Resistor	×	✓	✓
Z = Inductor	✓	×	✓
Z = Capacitor	✓	✓	×

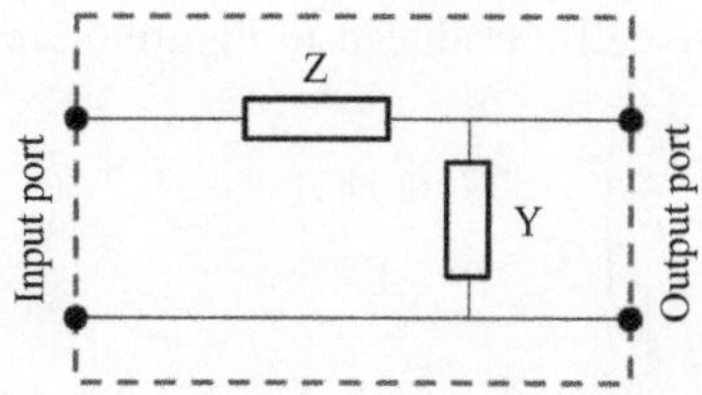

Figure 6.3 PGD passive cell configuration

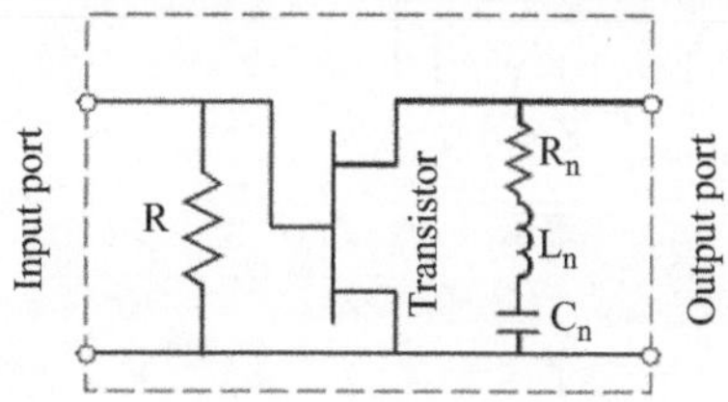

Figure 6.4 NGD active cell configuration

6.2.2 Basic configurations of PDG and NGD cells

The present chapter is focused on the PGD passive cell configured as an L-cell as shown in Figure 6.3. The cell is consisted of a series impedance Z associated with a parallel admittance Y.

The Z and Y elements can be implemented with a passive network constituted by a resistor, an inductor and a capacitor.

The NGD cell constituting the developed Hilbert filter topology consisted of a FET in cascade with an RLC-series shunt network as depicted in Figure 6.4. The RLC network connected between the FET drain and source is composed of the resistor R_n, inductor L_n, and the capacitor C_n. It is noteworthy that the circuit proposed in Figure 6.4 exhibits a group delay always negative at the resonance frequency:

$$\omega = \omega_n = 2\pi f_n = \frac{1}{\sqrt{L_n C_n}} \tag{6.5}$$

A shunt resistor R is connected between the FET Gate and Source for the NGD cell matching. In this chapter, for the sake of the mathematical simplicity, the FET is assumed as constituted by a voltage-controlled current source with transconductance g_m and a drain-source resistance R_{ds}. This model will be considered along the present chapter for the analytical investigation.

In the Section 6.2.3, the analytical characterization can be induced from these circuit configurations.

6.2.3 S-parameter analysis

The two-port PGD passive cell introduced in Figure 6.3 presents a transfer matrix written as:

$$[T_{PGD}] = \begin{bmatrix} 1+ZY & Z \\ Y & 1 \end{bmatrix} \tag{6.6}$$

The two-port NGD active cell one is obtained from the matrix product of the transfer matrices of the shunt resistance R, FET, and $R_nL_nC_n$ network:

$$[T_{NGD}] = \frac{-1}{g_m}\begin{bmatrix} \frac{1}{R_{ds}}+\frac{1}{Z_n} & 1 \\ \frac{1}{R}\left(\frac{1}{R_{ds}}+\frac{1}{Z_n}\right) & \frac{1}{R} \end{bmatrix} \tag{6.7}$$

and:

$$Z_n = R_n + j\left(L_n\omega - \frac{1}{C_n\omega}\right) \tag{6.8}$$

Consequently, the proposed Hilbert filter transfer matrix can be written as:

$$[T_{PGD\text{-}NGD}] = [T_{PGD}] \times [T_{NGD}] \tag{6.9}$$

or

$$[T_{NGD\text{-}PGD}] = [T_{NGD}] \times [T_{PGD}] \tag{6.10}$$

respectively, for the cells introduced in Figure 6.2(a) and (b). Along the chapter, the S-parameter characterization is associated with the reference impedance $Z_0 = 50\ \Omega$. The associated S-matrix:

$$[S] = \begin{bmatrix} S_{11} & S_{12} \\ S_{21} & S_{22} \end{bmatrix} \tag{6.11}$$

can be extracted from the transfer matrix:

$$[T] = \begin{bmatrix} T_{11} & T_{12} \\ T_{21} & T_{22} \end{bmatrix} \tag{6.12}$$

to the S-matrix transform [27]:

$$\begin{cases} S_{11} = \dfrac{Z_0(Z_0T_{21} + T_{22} - T_{11}) - T_{12}}{Z_0(Z_0T_{21} + T_{22} + T_{11}) + T_{12}} \\ S_{21} = \dfrac{2Z_0(T_{11}T_{22} - T_{12}T_{21})}{Z_0(Z_0T_{21} + T_{22} + T_{11}) + T_{12}} \\ S_{21} = \dfrac{T_{11}T_{12} + Z_0(2 + T_{11}T_{22} + Z_0T_{11}T_{21})}{Z_0(Z_0T_{21} + T_{22} + T_{11}) + T_{12}} \\ S_{22} = \dfrac{Z_0(Z_0T_{21} + T_{11} - T_{22}) - T_{12}}{Z_0(Z_0T_{21} + T_{22} + T_{11}) + T_{12}} \end{cases} \tag{6.13}$$

It is worth to recall that $S_{12} = 0$, because of the FET unilaterality. Accordingly, the S-parameters of the PGD-NGD cell are expressed as:

$$S_{11} = \frac{R - Z_0}{R + Z_0} \tag{6.14}$$

$$S_{21} = \frac{-2g_m R_{ds} Z_n Z_0 R/(R + Z_0)}{R_{ds}Z_0 + Z[R_{ds} + Z_0Y(R_{ds} + Z_n)] + Z_n[R_{ds} + Z_0(1 + R_{ds}Y)]} \tag{6.15}$$

$$S_{22} = \frac{Z_n[R_{ds} - Z_0Y(R_{ds} + Z_n)] + Z_n[R_{ds} - Z_0(1 + R_{ds}Y)] - R_{ds}Z_0}{R_{ds}Z_0 + Z[R_{ds} + Z_0Y(R_{ds} + Z_n)] + Z_n[R_{ds} + Z_0(1 + R_{ds}Y)]} \tag{6.16}$$

The S-parameters of the NGD-PGD cell are given by:

$$S_{11} = \frac{R + (Z_n - Z_0)(1 + RY)}{R + (Z_n + Z_0)(1 + RY)} \tag{6.17}$$

$$S_{21} = \frac{-2g_m R_{ds} Z_n Z_0 R}{[Z_0R_{ds} + Z_n(R_{ds} + Z_0)][R + Z_0 + (Z_n + Z_0)(1 + RY)]} \tag{6.18}$$

$$S_{22} = \frac{Z_nR_{ds} - Z_0(R_{ds} + Z_n)}{Z_nR_{ds} + Z_0(R_{ds} + Z_n)} \tag{6.19}$$

The analytical investigation on the associated PGD and NGD cell parameters can be established from these expressions.

6.2.4 *Identified Hilbert filter topologies implemented with PGD-NGD configuration*

By assuming, at the angular frequency $\omega = \omega_n$ (which implies spontaneously $Z_n = R_n$), the system behaves as an Hilbert filter if the transmission parameter is equal to j. So, the transmittance can be written as:

$$S_{21} = j = \exp(j \cdot \pi/2) \tag{6.20}$$

The analytical conditions corresponding to the Hilbert filter cell identification extracted from (6.15) are written as:

$$Y(j\omega_n) = \frac{\begin{array}{c}-Z_0R_n(Z_0+R) - Z_0^2(R_n+R_{ds}) - Z_0R_{ds}R- \\ Z_0R_{ds}R_n - R_{ds}R_nR + j[2g_mR_{ds}R_nZ_0R] \\ -[R_{ds}(R+Z_0)+R_n(Z_0+R)]Z(j\omega_n)\end{array}}{Z_0[R_nR_{ds}(R+Z_0)+[R_{ds}(Z_0+R)+R_n(R+Z_0)]Z(j\omega_n)]} \tag{6.21}$$

or

$$Z(j\omega_n) = \frac{\begin{array}{c}-Z_0^2(R_n+R_{ds}) - Z_0R(R_n+R_{ds}) - R_nR_{ds}R \\ +j[2g_mR_{ds}R_nZ_0R] - R_nR_{ds}Z_0(R+Z_0)Y(j\omega_n)\end{array}}{\begin{array}{c}R_nR + R_{ds}(R+Z_0) + Z_0R_n+ \\ Z_0[Z_0(R_n+R_{ds}) + (R_n+R_{ds})R]Y(j\omega_n)\end{array}} \tag{6.22}$$

6.2.4.1 Hilbert filter topology with Z = 0

The Hilbert filter topology constituted by PGD network with short-circuited impedance $Z = 0$ can be identified from the expression (6.21). In this case, the parallel admittance Y must be equal to:

$$Y(\omega_n) = \frac{\begin{array}{c}-Z_0R_n(Z_0+R) - Z_0^2(R_n+R_{ds}) - Z_0R_{ds}R \\ -Z_0R_{ds}R_n - R_{ds}R_nR + 2jg_mR_{ds}R_nZ_0R\end{array}}{Z_0R_nR_{ds}(R+Z_0)} \tag{6.23}$$

In other words, this admittance can be written as:

$$Y(j\omega_n) = G(\omega_n) + jB(\omega_n) \tag{6.24}$$

with:

$$G(\omega_n) = \frac{\begin{array}{c}R_{ds}R_nR + Z_0^2(R_n+R_{ds})+ \\ Z_0[R_n(Z_0+R) + R_{ds}(R+R_n)]\end{array}}{-Z_0R_nR_{ds}(R+Z_0)} \tag{6.25}$$

$$B(\omega_n) = \frac{2g_mR}{R+Z_0} \tag{6.26}$$

This topology will not be developed in this chapter because it requires a complex network with negative conductance $G(\omega_n) < 0$.

6.2.4.2 Hilbert filter with $Y = 0$

The Hilbert filter topology constituted by PGD network with open-circuit $Y = 0$ can be identified from the expression (6.22). In this case, the series impedance Z must be equal to:

$$Z(j\omega_n) = \frac{\begin{array}{c}-Z_0^2(R_n+R_{ds}) - Z_0R(R_n+R_{ds}) - R_nR_{ds}R \\ +j[2g_mR_{ds}R_nZ_0R]\end{array}}{R_nR + R_{ds}(R+Z_0) + Z_0R_n} \tag{6.27}$$

Similar to Section 6.2.4.1, this impedance can be rewritten as:

$$Z(j\omega_n) = R_z(\omega_n) + jX(\omega_n) \tag{6.28}$$

with:

$$R_z(\omega_n) = \frac{-\left[Z_0^2(R_n + R_{ds}) + Z_0R(R_n + R_{ds}) + R_nR_{ds}R\right]}{R_nR + R_{ds}(R + Z_0) + Z_0R_n} \tag{6.29}$$

$$X(\omega_n) = \frac{2g_mR_{ds}R_nZ_0R}{R_nR + R_{ds}(R + Z_0) + Z_0R_n} \tag{6.30}$$

This topology will not be developed in this chapter because it requires a complex network with negative resistance $R_z(\omega_n) < 0$.

6.2.4.3 Hilbert filter with $Z = R_p$

We assume that R_p is a resistor parameter. The Hilbert filter topology constituted by PGD network with a resistive impedance $Z = R_p$ can be identified from the expression (6.22). In this case, the parallel impedance Y must be equal to:

$$Y(j\omega_n) = \frac{\begin{array}{c}-Z_0R_n(Z_0 + R) - Z_0^2(R_n + R_{ds}) - Z_0R_{ds}R \\ -Z_0R_{ds}R_n - R_{ds}R_nR + j[2g_mR_{ds}R_nZ_0R] \\ -[R_{ds}(R + Z_0) + R_n(Z_0 + R)]R_p\end{array}}{Z_0\left[R_nR_{ds}(R + Z_0) + [R_{ds}(Z_0 + R) + R_n(R + Z_0)]R_p\right]} \tag{6.31}$$

In other words, this admittance can be written in (6.24) with:

$$G(\omega_n) = \frac{\begin{array}{c}-Z_0R_n(Z_0 + R) - Z_0^2(R_n + R_{ds}) - Z_0R_{ds}R- \\ Z_0R_{ds}R_n - R_{ds}R_nR - [R_{ds}(R + Z_0) + R_n(Z_0 + R)]R_p\end{array}}{Z_0\left[R_nR_{ds}(R + Z_0) + [R_{ds}(Z_0 + R) + R_n(R + Z_0)]R_p\right]} \tag{6.32}$$

$$B(\omega_n) = \frac{2g_mR_{ds}R_nZ_0R}{Z_0\left[R_nR_{ds}(R + Z_0) + [R_{ds}(Z_0 + R) + R_n(R + Z_0)]R_p\right]} \tag{6.33}$$

This topology will not be developed in this chapter because it requires a complex network with negative conductance $G(\omega_n) < 0$.

6.2.4.4 Hilbert filter with $Y = 1/R_p$

The Hilbert filter topology constituted by PGD network with conductance $Y = 1/R_p$ can be identified from the expression (6.23). In this case, the impedance Z must be equal to:

$$Z(j\omega_n) = \frac{\begin{array}{c}-Z_0^2(R_n + R_{ds}) - Z_0R(R_n + R_{ds}) - R_nR_{ds}R \\ +j[2g_mR_{ds}R_nZ_0R] - \dfrac{R_nR_{ds}Z_0(R + Z_0)}{R_p}\end{array}}{\begin{array}{c}R_nR + R_{ds}(R + Z_0) + Z_0R_n+ \\ \dfrac{Z_0}{R_p}[Z_0(R_n + R_{ds}) + (R_n + R_{ds})R]\end{array}} \tag{6.34}$$

In other words, this impedance can be written as in (6.28) with:

$$R_z(\omega_n) = \frac{-\begin{bmatrix} Z_0^2(R_n + R_{ds}) + Z_0R(R_n + R_{ds}) + R_nR_{ds}R \\ +\dfrac{R_nR_{ds}Z_0}{R_p}(R + Z_0) \end{bmatrix}}{\begin{array}{l} R_nR + R_{ds}(R + Z_0) + Z_0R_n + \\ \dfrac{Z_0}{R_p}[Z_0(R_n + R_{ds}) + (R_n + R_{ds})R] \end{array}} \tag{6.35}$$

$$X(\omega_n) = \frac{2g_mR_{ds}R_nZ_0R}{\begin{array}{l} R_nR + R_{ds}(R + Z_0) + Z_0R_n + \\ \dfrac{Z_0}{R_p}[Z_0(R_n + R_{ds}) + (R_n + R_{ds})R] \end{array}} \tag{6.36}$$

This topology will not be developed in this chapter because it requires a complex network with negative resistance $R_z(\omega_n) < 0$.

6.2.4.5 Hilbert filter with $Z = jX$

The Hilbert filter topology constituted by PGD network with reactive impedance $Z = jX$ can be identified from expression (6.23). As predicted in Table 6.1, there are two possible configurations which can be realized by assuming Z as an inductor or a capacitor.

In the case where $Z = jL\omega_n$, the parallel impedance Y must be equal to:

$$Y(j\omega_n) = \frac{\begin{array}{l} -Z_0R_n(Z_0 + R) - Z_0^2(R_n + R_{ds}) - \\ Z_0R_{ds}R - Z_0R_{ds}R_n - R_{ds}R_nR + \\ j[2g_mR_{ds}R_nZ_0R - L\omega_n[R_{ds}(R + Z_0) + R_n(Z_0 + R)]] \end{array}}{Z_0[R_nR_{ds}(R + Z_0) + jL\omega_n[R_{ds}(Z_0 + R) + R_n(R + Z_0)]]} \tag{6.37}$$

In the case where $Z = 1/(jC\omega_n)$, the parallel impedance Y must be equal to:

$$Y(\omega_n) = \frac{\begin{array}{l} -Z_0R_n(Z_0 + R) - Z_0^2(R_n + R_{ds}) - \\ Z_0R_{ds}R - Z_0R_{ds}R_n - R_{ds}R_nR + \\ j\left[2g_mR_{ds}R_nZ_0R + \dfrac{R_{ds}(R + Z_0) + R_n(Z_0 + R)}{C\omega_n}\right] \end{array}}{Z_0\left[R_nR_{ds}(R + Z_0) - \dfrac{j[R_{ds}(Z_0 + R) + R_n(R + Z_0)]}{C\omega_n}\right]} \tag{6.38}$$

Intuitively, the LC-circuit can be one of the simple solution which can satisfy these two last expressions. This topology will be developed in Section 6.2.5.

6.2.5 Identified Hilbert filter topologies implemented with NGD-PGD configuration

At the resonance $\omega=\omega_n$, the analytical conditions corresponding to the Hilbert filter cell identification extracted from (6.18) are written as:

$$Y(j\omega_n)=\frac{\begin{array}{c}-Z_0[R_n(R+Z_0+R_{\rm ds})+R_{ds}R]-Z_0^2(R_n+R_{\rm ds})-\\ R_{\rm ds}R_nR+2jg_mR_{\rm ds}R_nZ_0R-[Z_0(R_n+R_{\rm ds})+R_n+R_{\rm ds}]Z(j\omega_n)\end{array}}{R\{R_nZ_0(R_{\rm ds}+Z_0)+[R_nZ_0+R_{\rm ds}(Z_0+R_n)]Z(j\omega_n)\}} \tag{6.39}$$

or

$$Z(j\omega_n)=\frac{\begin{array}{c}-Z_0[R_n(R+Z_0+R_{\rm ds})+R_{\rm ds}(Z_0+R)]-R_{\rm ds}R_nR\\ +2jg_mR_{\rm ds}R_nZ_0R-Z_0R_{\rm ds}[Z_0(R_n+R)+R_nR]Y(j\omega_n)\end{array}}{R_{\rm ds}Z_0+R_n(R_{\rm ds}+Z_0)+R[R_nZ_0+R_{\rm ds}(Z_0+R_n)]Y(j\omega_n)} \tag{6.40}$$

6.2.5.1 Hilbert filter topology with $Z=0$

The Hilbert filter topology constituted by the PGD network with short-circuited impedance $Z=0$ can be identified from (6.39). In this case, the parallel admittance Y must be equal to:

$$Y(j\omega_n)=\frac{\begin{array}{c}-Z_0[R_n(R+Z_0+R_{\rm ds})+R_{\rm ds}R]-Z_0^2(R_n+R_{\rm ds})\\ -R_{\rm ds}R_nR+2jg_mR_{\rm ds}R_nZ_0R\end{array}}{R_nZ_0R(R_{\rm ds}+Z_0)} \tag{6.41}$$

In other words, this admittance can be written as in (6.24) with:

$$G(\omega_n)=\frac{-\left\{\begin{array}{c}Z_0[R_n(R+Z_0+R_{\rm ds})+R_{\rm ds}R]\\ +Z_0^2(R_n+R_{\rm ds})+R_{\rm ds}R_nR\end{array}\right\}}{R_nZ_0R(R_{\rm ds}+Z_0)} \tag{6.42}$$

$$B(\omega_n)=\frac{2g_mR_{\rm ds}}{R_{ds}+Z_0} \tag{6.43}$$

This topology will not be developed in this chapter because it requires a complex network with negative conductance $G(\omega_n)<0$.

6.2.5.2 Hilbert filter with $Y=0$

The Hilbert filter topology constituted by PGD network with open-circuit $Y=0$ can be identified from (6.40). In this case, the series impedance Z must be equal to:

$$Z(j\omega_n)=\frac{\begin{array}{c}-Z_0[R_n(R+Z_0+R_{\rm ds})+R_{\rm ds}(Z_0+R)]\\ -R_{\rm ds}R_nR+2jg_mR_{\rm ds}R_nZ_0R\end{array}}{R_{\rm ds}Z_0+R_n(R_{\rm ds}+Z_0)} \tag{6.44}$$

Similar to Section 6.2.5.1, this impedance can be rewritten as (6.28):

$$R_z(\omega_n) = \frac{-Z_0[R_n(R + Z_0 + R_{ds}) + R_{ds}(Z_0 + R)] - R_{ds}R_nR}{R_{ds}Z_0 + R_n(R_{ds} + Z_0)} \tag{6.45}$$

$$X(\omega_n) = \frac{2g_mR_{ds}R_nZ_0R}{R_{ds}Z_0 + R_n(R_{ds} + Z_0)} \tag{6.46}$$

This topology will not be developed in this chapter because it requires a complex network with negative resistance $R_z(\omega_n) < 0$.

6.2.5.3 Hilbert filter with $Z = R_p$

By denoting R_p the resistor parameter, the Hilbert filter topology constituted by PGD network with a resistive impedance $Z = R_p$ can be identified from (6.39). In this case, the parallel impedance Y must be equal to:

$$Y(j\omega_n) = \frac{\begin{array}{c}-Z_0[R_n(R + Z_0 + R_{ds}) + R_{ds}R] - Z_0^2(R_n + R_{ds}) - \\ R_{ds}R_nR + 2jg_mR_{ds}R_nZ_0R - [Z_0(R_n + R_{ds}) + R_n + R_{ds}]R_p\end{array}}{R\{R_nZ_0(R_{ds} + Z_0) + [R_nZ_0 + R_{ds}(Z_0 + R_n)]R_p\}} \tag{6.47}$$

In other words, this admittance can be written in (6.24) with:

$$G(\omega_n) = \frac{\begin{array}{c}-Z_0[R_n(R + Z_0 + R_{ds}) + R_{ds}R] - Z_0^2(R_n + R_{ds}) - \\ R_{ds}R_nR - [Z_0(R_n + R_{ds}) + R_n + R_{ds}]R_p\end{array}}{R\{R_nZ_0(R_{ds} + Z_0) + [R_nZ_0 + R_{ds}(Z_0 + R_n)]R_p\}} \tag{6.48}$$

$$B(\omega_n) = \frac{2g_mR_{ds}R_nZ_0}{R_nZ_0(R_{ds} + Z_0) + [R_nZ_0 + R_{ds}(Z_0 + R_n)]R_p} \tag{6.49}$$

This topology will not be developed in this chapter because it requires a complex network with negative conductance $G(\omega_n) < 0$.

6.2.5.4 Hilbert filter with $Y = 1/R_p$

The Hilbert filter topology constituted by PGD network with conductance $Y = 1/R_p$ can be identified from (6.40). In this case, the impedance Z must be equal to:

$$Z(j\omega_n) = \frac{\begin{array}{c}-Z_0[R_n(R + Z_0 + R_{ds}) + R_{ds}(Z_0 + R)] - R_{ds}R_nR \\ +2jg_mR_{ds}R_nZ_0R - \dfrac{Z_0R_{ds}}{R_p}[Z_0(R_n + R) + R_nR]\end{array}}{R_{ds}Z_0 + R_n(R_{ds} + Z_0) + \dfrac{R}{R_p}[R_nZ_0 + R_{ds}(Z_0 + R_n)]} \tag{6.50}$$

In other words, this impedance can be written as in (6.28) with:

$$R_z(\omega_n) = \frac{-\left\{\begin{array}{c}Z_0[R_n(R + Z_0 + R_{ds}) + R_{ds}(Z_0 + R)] \\ +R_{ds}R_nR + \dfrac{Z_0R_{ds}}{R_p}[Z_0(R_n + R) + R_nR]\end{array}\right\}}{R_{ds}Z_0 + R_n(R_{ds} + Z_0) + \dfrac{R}{R_p}[R_nZ_0 + R_{ds}(Z_0 + R_n)]} \tag{6.51}$$

$$X(\omega_n) = \frac{2g_m R_{ds} R_n Z_0 R}{R_{ds} Z_0 + R_n(R_{ds} + Z_0) + \frac{R}{R_p}[R_n Z_0 + R_{ds}(Z_0 + R_n)]} \tag{6.52}$$

This topology will not be developed in this chapter because it requires a complex network with negative resistance $R_z(\omega_n) < 0$.

6.2.5.5 Hilbert filter with $Z = jX$

The Hilbert filter topology constituted by PGD network with reactive impedance $Z = jX$ can be identified from (6.22). As predicted in Table 6.1, there are two possible configurations by assuming Z as an inductor or a capacitor.

In the case where $Z = jL\omega_n$, the parallel impedance Y must be equal to:

$$Y(j\omega_n) = \frac{\begin{array}{c}-Z_0[R_n(R + Z_0 + R_{ds}) + R_{ds}R] - Z_0^2(R_n + R_{ds}) - R_{ds}R_nR \\ +j\{2g_m R_{ds} R_n Z_0 R - L\omega_n[Z_0(R_n + R_{ds}) + R_n + R_{ds}]\}\end{array}}{R\{R_n Z_0(R_{ds} + Z_0) + jL\omega_n[R_n Z_0 + R_{ds}(Z_0 + R_n)]\}} \tag{6.53}$$

In the case where $Z = 1/(jC\omega_n)$, the parallel impedance Y must be equal to:

$$Y(j\omega_n) = \frac{\begin{array}{c}-Z_0[R_n(R + Z_0 + R_{ds}) + R_{ds}R] - Z_0^2(R_n + R_{ds}) - R_{ds}R_nR \\ +j\left(2g_m R_{ds} R_n Z_0 R + \dfrac{Z_0(R_n + R_{ds}) + R_n + R_{ds}}{C\omega_n}\right)\end{array}}{R\left[R_n Z_0(R_{ds} + Z_0) - j\dfrac{R_n Z_0 + R_{ds}(Z_0 + R_n)}{C\omega_n}\right]} \tag{6.54}$$

Intuitively, the LC-circuit can be one of the simple solution which can satisfy these two last expressions. This topology will be developed in Section 6.3.

6.3 Analysis of LC-NGD cell-based NGD Hilbert filter topology

The present section is focused on the investigation on the topology constituted by an LC-circuit. Figure 6.5 represents the schematic of the circuit. Based on the S-parameter analysis, the characteristics, properties, and synthesis relations will be formulated in Section 6.3.1.

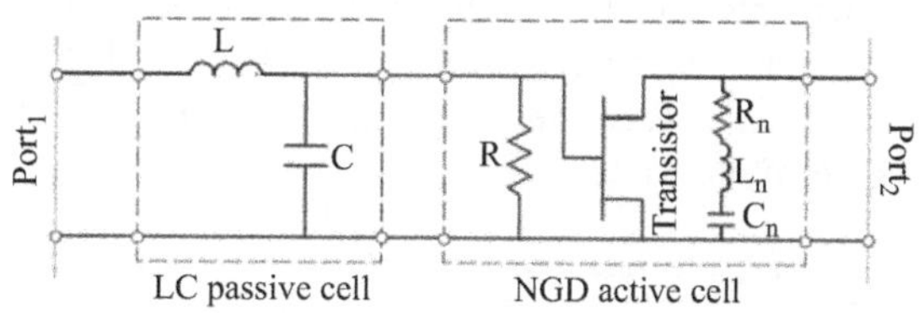

Figure 6.5 LC- and NGD-cells in cascade

6.3.1 S-parameter analysis

It is worth to emphasize that to operate as an Hilbert filter synthesizer, the proposed circuit transmission gain must be analytically equal to $S_{21}(j\omega) = \exp(j \cdot 90°)$ in the expected frequency band around the resonance frequency:

$$\omega_n - \frac{\Delta\omega}{2} \leq \omega \leq \omega_n + \frac{\Delta\omega}{2} \tag{6.55}$$

6.3.1.1 Analysis of the LC-NGD cell

We assumed that the passive LC network is characterized by the angular frequency:

$$\omega_1 = 2\pi f_1 = \frac{1}{\sqrt{LC}} \tag{6.56}$$

The calculated input reflection and transmission coefficients when the angular frequency $\omega = \omega_n$ are, respectively, expressed as:

$$S_{11}(j\omega_n) = \frac{1}{a}\left[\omega_n(Z_0^2 C + L(1 - \frac{\omega_n^2}{\omega_1^2})) + 2jZ_0(1 - \frac{\omega_n^2}{\omega_1^2})\right] \tag{6.57}$$

$$S_{21}(j\omega_n) = \frac{ja}{\omega_n\left[Z_0^2 C - L(1 - \frac{\omega_n^2}{\omega_1^2})\right]} \tag{6.58}$$

with

$$a = \frac{2Z_0^2 g_m R_{ds} R_n}{Z_0(R_{ds} + R_n) + R_{ds} R_n} \tag{6.59}$$

After analytical calculation, it is worth to emphasize that at the reference angular frequency $\omega = \omega_n$, the LC-NGD cell group delay is always equal to zero $\tau(\omega_n) = 0$. It means that the transmission phase should present a flatness aspect corresponding to the frequency independent function around $\omega = \omega_n$.

6.3.1.2 Synthesis formulation

We assume that the input reflection coefficient is equal to ρ_1 (as a positive real < 1):

$$|S_{11}(j\omega_n)| = \rho_1 \tag{6.60}$$

The LC-cell elements can be synthesized by inverting the magnitudes of reflection and transmission coefficients defined as (6.57) and (6.58), respectively, expressed in Section 6.3.1.1. Accordingly, it can be demonstrated that the capacitor and inductor can be formulated as:

$$C = \frac{2Z_0 + \sqrt{2\rho_1^2(2Z_0^2 + a^2) - a^2(\rho_1^4 + 1)}}{4Z_0^3\omega_n/a} \tag{6.61}$$

$$L = \frac{1 + \sqrt{1 + 4\omega_n C(\alpha - Z_0^2 \omega_n C)}}{2\omega_n^2 C} \tag{6.62}$$

Moreover, the matching resistance R and the NGD cell inductance L_n can be determined from the expected ideal Hilbert filter definition:

$$\angle S_{21}(j\omega_n) = 90^\circ \tag{6.63}$$

and

$$|S_{21}(j\omega_n)| = 1 \tag{6.64}$$

Then, the resistance R_n can be determined from the expected output reflection, ρ_2 (as a positive real < 1) with the equation:

$$|S_{22}(j\omega_n)| = \rho_2 \tag{6.65}$$

By inverting the output reflection coefficient expression, one can determine the output resistance:

$$R_n = \frac{Z_0 R_{\text{ds}}(1 - \rho_2)}{R_{\text{ds}} - Z_0 + \rho_2(R_{\text{ds}} + Z_0)} \tag{6.66}$$

The input matching resistance must be given by

$$R = Z_0 \frac{\omega_1}{\omega_n - \omega_1} \tag{6.67}$$

The NGD cell inductance can be obtained by inverting the transmission coefficient expression. Therefore, one gets

$$L_n = \frac{Z_0 R_n^2 (L + Z_0 RC)\left[Z_0 + R(1 + \omega_0^2/\omega_1^2)\right]}{\alpha\left[R^2\omega_0^4/\omega_1^4 + (R^2(Z_0^2 C^2 - 2/\omega_1^2) - 2L^2)\omega_0^2 + (R + Z_0)^2\right]} \tag{6.68}$$

The NGD cell capacitor can be calculated by inverting the resonance frequency defined in (6.4). It implies the equation:

$$C_n = \frac{1}{L_n \omega_n^2} \tag{6.69}$$

6.3.1.3 Properties and characteristics of LC-NGD cascaded cell

It is worth mentioning that the capacitor C maximal value is:

$$C_{\max} = \frac{\alpha + \sqrt{\alpha^2 + Z_0^2}}{2Z_0^2 \omega_n \alpha}, \tag{6.70}$$

and the minimal input reflection coefficient during the synthesis is:

$$\rho_{1_{\min}} = \sqrt{1 + \frac{2Z_0(Z_0 - \sqrt{\alpha^2 + Z_0^2})}{\alpha^2}} \leq \rho_1. \tag{6.71}$$

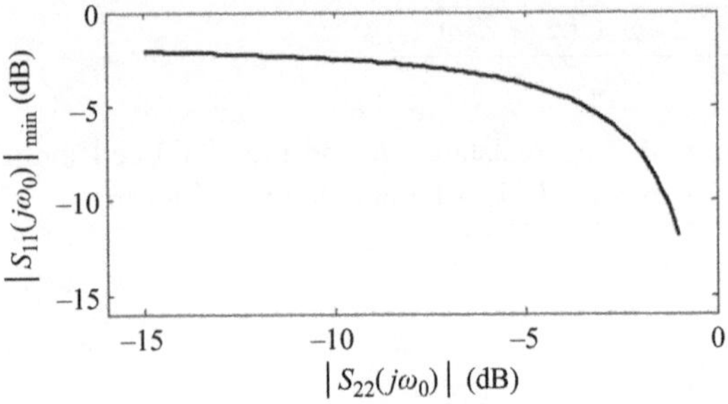

Figure 6.6 Proposed RF/microwave Hilbert filter $|S_{11}(j\omega_n)|_{min}$ *vs.* $|S_{22}(j\omega_n)|$

Table 6.2 RF/microwave Hilbert filter POC parameters

POC	PGD circuit parameters	NGD circuit parameters
POC_1	$R_1 = 534\ \Omega$	$R_{n_1} = 3.57\Omega$
	$L_1 = 9.8$ nH	$L_{n_1} = 4.35$ nH
	$C_1 = 0.47$ pF	$C_{n_1} = 9.7$pF
POC_2	$R_2 = 39\ \Omega$	$R_{n_2} = 35.1\ \Omega$
	$L_2 = 4.1$ nH	$L_{n_2} = 3.12$ nH
	$C_2 = 2.3$ pF	$C_{n_2} = 1.35$ pF
POC_3	$R_3 = 32\ \Omega$	$R_{n_3} = 53.6\ \Omega$
	$L_3 = 3.27$ nH	$L_{n_3} = 5.66$ nH
	$C_3 = 3.3$ pF	$C_{n_3} = 0.75$ pF

As highlighted in Figure 6.6, it can be emphasized that the input reflection coefficient $|S_{11}(j\omega_n)|_{dB,min}$ of the proposed RF/microwave Hilbert filter is inversely proportional to the output reflection coefficient $|S_{22}(j\omega_n)|_{dB}$. In function of the application specifications, the reflection coefficients can be improved with the integration of additional matching network.

6.3.2 Application examples

The present subsection is dedicated to the feasibility analysis of the developed Hilbert filter theory. To accomplish this, examples of lumped circuit will be synthesized, designed, and simulated. The proofs of concept (POC) are expected to operate in the frequency band around $f_n = 2.45$ GHz corresponding to the IEEE 802.11b Wi-Fi and ISM applications. One points out that the considered FET used in the NGD cell is characterized by the arbitrary chosen parameters $g_m = 100$ mS and $R_{ds} = 100\ \Omega$.

6.3.2.1 POC description

To confirm the feasibility of the established circuit theory, three POC cells are considered in this section. It acts a typically RF/microwave Hilbert filters. The POC circuits are designed in the electronic circuit and system SPICE environment. Table 6.2 addresses the POC circuit parameters.

6.3.2.2 Feasibility principle validation results

The simulated transmission phases and gains from 2 to 3 GHz are displayed in Figure 6.7.

As expected, around the operating frequency f_n, it can be seen that the transmission phase flatness is $\angle S_{21}(j\omega_n) = 90 \pm 1°$in the relative frequency band $\Delta f/f_n > 50\%$. In addition, the transmission gain is $|S_{21}(j\omega_n)| = \pm 1$dB within the relative frequency band $\Delta f/f_n > 23\%$.

6.4 Analysis of NGD-LC cell-based Hilbert filter topology

This section is focused on the investigation on the topology constituted by an NGD-LC cell. Figure 6.8 represents the circuit configuration.

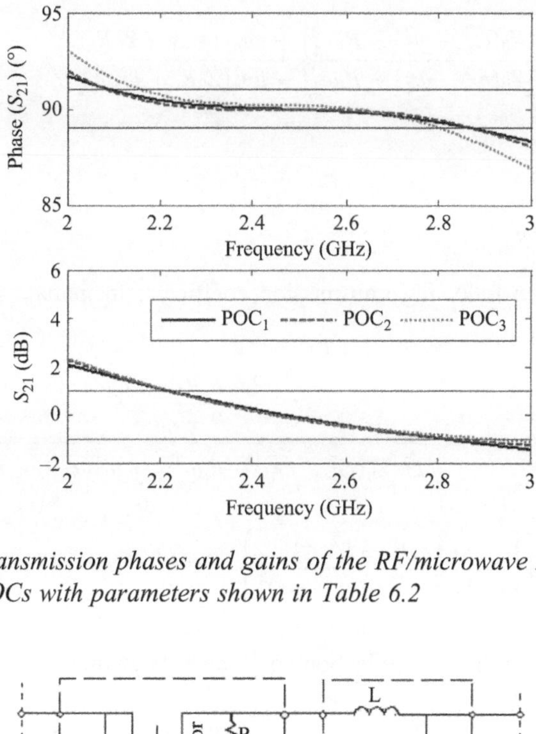

Figure 6.7 Transmission phases and gains of the RF/microwave Hilbert filter POCs with parameters shown in Table 6.2

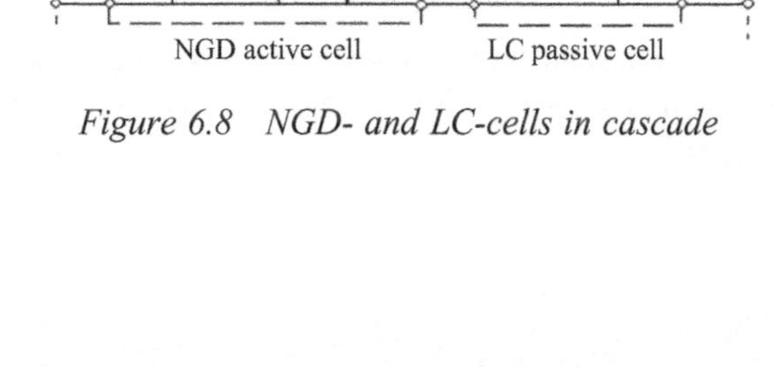

Figure 6.8 NGD- and LC-cells in cascade

Based on the *S*-parameter analysis, the characteristics, properties and synthesis relations will be formulated in Section 6.4.1.

6.4.1 S-parameter analysis

Similarly, to Section 6.3.1, the transmission gain must be analytically equal to $S_{21}(j\omega) = \exp(j \cdot 90°)$ in the expected frequency band around the resonance frequency as defined in (6.39).

6.4.1.1 Analysis of NGD-LC cell cascaded

The associated input reflection coefficient defined in (6.14) and transmission coefficients when $\omega = \omega_n$ are, respectively, expressed as:

$$S_{21}(j\omega_n) = \frac{2g_m R_t R Z_0 \omega_1^2 L}{(R+Z_0)\left\{L\left[Z_0(\omega_n^2 - \omega_1^2) - R_t\omega_1^2\right] - j\omega_n(Z_0 R_t + L^2\omega_1^2)\right\}} \tag{6.72}$$

$$S_{22}(j\omega_n) = \frac{L\left[Z_0(\omega_n^2 - \omega_1^2) - R_t\omega_1^2\right] + j\omega_n(L^2\omega_1^2 - Z_0 R_t)}{L\left[Z_0(\omega_n^2 - \omega_1^2) - R_t\omega_1^2\right] - j\omega_n(Z_0 R_t + L^2\omega_1^2)} \tag{6.73}$$

with

$$R_t = \frac{R_n R_{\mathrm{ds}}}{R_n + R_{\mathrm{ds}}} \tag{6.74}$$

It implies the following transmission coefficient magnitude and phase shift expressions:

$$S_{21}(\omega_n) = |S_{21}(j\omega_n)| = \frac{\dfrac{2g_m R_t R Z_0 \omega_1^2 L}{R+Z_0}}{\sqrt{L^2\left[Z_0(\omega_n^2 - \omega_1^2) - R_t\omega_1^2\right]^2 + \omega_n^2(Z_0 R_t + L^2\omega_1^2)^2}} \tag{6.75}$$

$$\varphi(\omega_n) = \arctan\left\{\frac{\omega_n(Z_0 R_t + L^2\omega_1^2)}{L\left[R_t\omega_1^2 + Z_0(\omega_1^2 - \omega_n^2)\right]}\right\} \tag{6.76}$$

It implies the following reflection coefficient magnitude:

$$S_{22}(\omega_n) = |S_{22}(j\omega_n)| = \frac{\sqrt{L^2\left[Z_0(\omega_n^2 - \omega_1^2) - R_t\omega_1^2\right] + \omega_n^2(L^2\omega_1^2 - Z_0 R_t)^2}}{\sqrt{L^2\left[Z_0(\omega_n^2 - \omega_1^2) - R_t\omega_1^2\right] + \omega_n^2(L^2\omega_1^2 + Z_0 R_t)^2}} \tag{6.77}$$

Similar to the LC-NGD cell analyzed in 6.3.1.1, it is worth to emphasize that at the reference angular frequency $\omega = \omega_n$, the NGD-LC cell group delay is always equal to zero $\tau(\omega_n) = 0$.

6.4.1.2 Synthesis formulation

The input matching resistance can be extracted by inverting the input reflection coefficient defined in (6.14):

$$R = Z_0 \cdot \begin{cases} \dfrac{1-\rho_1}{1+\rho_1} & \text{for} \quad R < Z_0 \\ \dfrac{1+\rho_1}{1-\rho_1} & \text{for} \quad R > Z_0 \end{cases} \tag{6.78}$$

In this case, the transmission coefficient magnitude can be written as:

$$S_{21}(\omega_n) = \frac{g_m R_t Z_0 \omega_n L(1+\rho_1)}{R_t^2 + Z_0 R_t + \omega_n^2 L^2} \tag{6.79}$$

The LC cell inductance L is determined from the combination of the Hilbert filter definition in (6.56) and the transmission coefficient expression:

$$L = \frac{1}{2\omega_n}\left[g_m Z_0 R_t(1+\rho_1) + \sqrt{g_m^2 R_t^2 Z_0^2 (1+\rho_1)^2 - 4R_t(Z_0 + R_t)}\right] \tag{6.80}$$

With the same way, the NGD cell resistance R_n can also be extracted:

$$R_t = \frac{1}{2}\left\{ Z_0[g_m\omega_n L(1+\rho_1) - 1] + \sqrt{\frac{[Z_0[1 - g_m\omega_n L(1+\rho_1)] - 2\omega_n L]}{[Z_0[g_m\omega_n L(1+\rho_1) - 1] + 2\omega_n L]}} \right\} \tag{6.81}$$

In parallel to this analytical calculation, with the obtained equation system of (6.20) and (6.76), the LC cell characteristic angular frequency can be determined. It implies the formula:

$$\omega_1 = \omega_n \sqrt{\frac{Z_0(R_n + R_{\mathrm{ds}})}{Z_0(R_n + R_{\mathrm{ds}}) + R_n R_{\mathrm{ds}}}} \tag{6.82}$$

The NGD cell capacitor can be calculated from (6.69). The LC-cell capacitor is determined from (6.56) and (6.80) via the relation:

$$C = \frac{1}{L\omega_1^2} \tag{6.83}$$

The properties and characteristics of the NGD-LC cell can be established from the mathematical analysis of the relations expressed in Section 6.4.1.1 combined with the physical requirement of each parameter.

First, the existence condition of this synthesis relation depends on the NGD cell resistor R_n, which must fulfill the following condition:

$$R_n > R_{n_{\min}} = \frac{4Z_0 R_{\mathrm{ds}}}{g_m^2 R_{\mathrm{ds}} Z_0^2 (1+\rho_1)^2 - 4(Z_0 + R_{\mathrm{ds}})} \tag{6.84}$$

Knowing that $R_{n_{\min}}$ must be positive, it means that the input reflection must satisfy the inequality:

$$(1+\rho_1)^2 > \frac{4(Z_0+R_{ds})}{g_m^2 R_{ds} Z_0^2} \quad \Rightarrow \quad \rho_1 > \rho_{1_{\min}} = \frac{2\sqrt{Z_0+R_{ds}}}{g_m Z_0 \sqrt{R_{ds}}} - 1 \tag{6.85}$$

The reflection coefficient definition $0 < \rho_1 < 1$ enables to describe the following FET characteristic:

$$0 < \frac{2\sqrt{Z_0+R_{ds}}}{g_m Z_0 \sqrt{R_{ds}}} - 1 < 1 \quad \Leftrightarrow \quad 1 < \frac{2\sqrt{Z_0+R_{ds}}}{g_m Z_0 \sqrt{R_{ds}}} < 2 \tag{6.86}$$

This inequality can be traduced with the minimal and maximal values of the transconductance:

$$g_{m_{\min}} = \frac{\sqrt{Z_0+R_{ds}}}{Z_0\sqrt{R_{ds}}} < g_m < g_{m_{\max}} = \frac{2\sqrt{Z_0+R_{ds}}}{Z_0\sqrt{R_{ds}}} \tag{6.87}$$

Figure 6.9 represents the graphical variation of these limits for R_{ds} increased from 0.1 to 200 Ω.

For example, if $R_{ds} = 100$ Ω, we have $g_{m_{\min}} = 24.5$ mS and $g_{m_{\max}} = 49$ mS. In this range of transconductance, the minimal reflection coefficient is plotted in Figure 6.10.

6.4.2 Application examples

The present subsection is dedicated to the feasibility analysis of the developed NGD-LC circuit theory. For this application, let us consider the FET drain-source resistance $R_{ds} = 100$ Ω.

6.4.2.1 Synthesis of the NGD-LC Hilbert filter prototype

Figure 6.11 illustrates the minimal values of the resistor R_n in order to realize an input reflection coefficient from −12 to −3 dB.

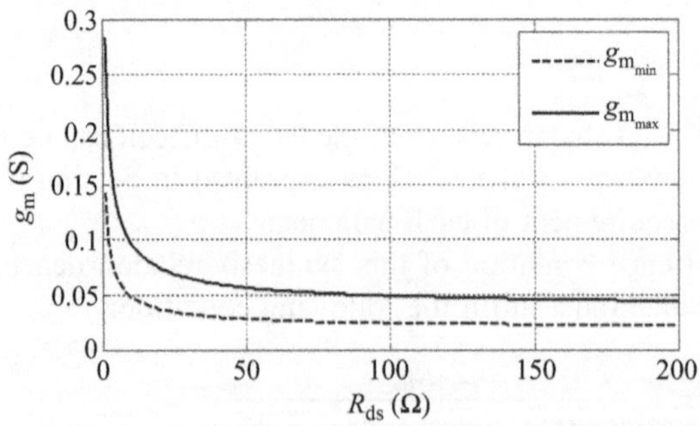

Figure 6.9 Limitation of the FET used to implement the NGD-LC cell

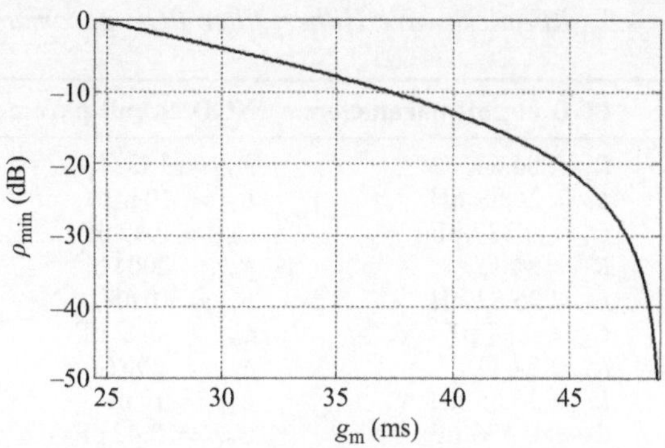

Figure 6.10 ρ_{min} *vs.* g_m *for* $R_{ds} = 100\ \Omega$

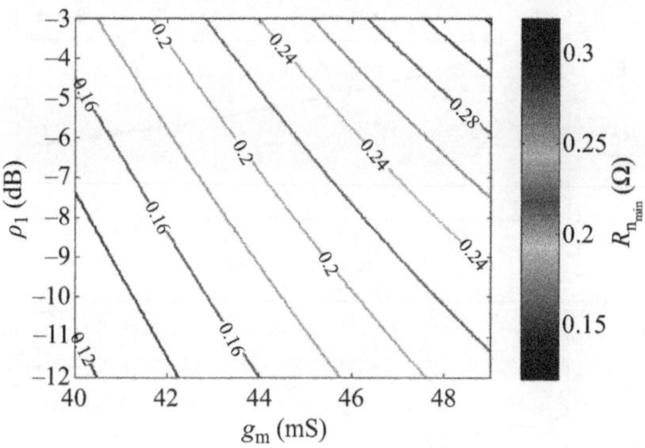

Figure 6.11 $R_{n_{min}}$ *minimal value vs.* (g_m, ρ_1)

Let us consider the three cases of $R_n = \{180, 200, \text{and } 220\ \Omega\}$ and an arbitrary transconductance $g_m = 100$ mS. Similar to the previous example, the operating frequency band is around $f_n = 2.45$ GHz. To achieve the input matching level $\rho_1 = -12$ dB, we need a matching resistor around $R = 84\ \Omega$. The synthesis relations established in Section 6.4.1.2 enables to find the results addressed in Table 6.3.

We will use these parameters to confirm the feasibility of the established circuit theory, three POC cells are considered in this section.

6.4.2.2 Illustrative validation results

Similar to the results presented in Section 6.3.2.2, the present POC circuits are designed in the SPICE environment. The simulated transmission phases and gains from 2 to 3 GHz are displayed in Figure 6.12. As expected, around the operating frequency f_n, it can be seen that the transmission phase flatness is $\angle S_{21}(j\omega_n) = 90 \pm 1°$in the relative frequency band $\Delta f/f_n > 60\%$. In addition, the transmission gain is $|S_{21}(j\omega_n)| = \pm 1$dB within the relative frequency band $\Delta f/f_n > 22\%$.

Table 6.3 RF/microwave Hilbert filter POC parameters

POC	PGD circuit parameters	NGD circuit parameters
POC_1	$R_1 = 84\ \Omega$	$R_{n_1} = 180\,\Omega$
	$L_1 = 24.88$ nH	$L_{n_1} = 10$ nH
	$C_1 = 0.387$ pF	$C_{n_1} = 0.42$ pF
POC_2	$R_2 = 84\ \Omega$	$R_{n_2} = 200\,\Omega$
	$L_2 = 25.82$ nH	$L_{n_2} = 10$ nH
	$C_2 = 0.38$ pF	$C_{n_2} = 0.42$ pF
POC_3	$R_3 = 84\ \Omega$	$R_{n_3} = 200\,\Omega$
	$L_3 = 26.65$ nH	$L_{n_3} = 10$ nH
	$C_3 = 0.376$ pF	$C_{n_3} = 0.42$ pF

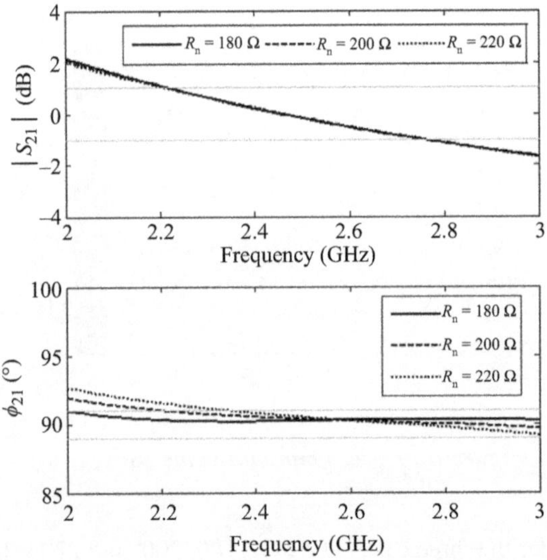

Figure 6.12 Transmission phases and gains of the RF/microwave Hilbert filter NGD-LC POCs with parameters shown in Table 6.3

6.5 Conclusion

An innovative circuit theory about the NGD concept-based Hilbert filter is developed in this chapter. The fundamental topology based on the PGD and NGD cells in cascade is analyzed. The identification methodology of the simplest topology composed of lumped components as resistor, inductor and capacitor is described. The analytical expression of the existence of Hilbert filter synthesizer is established. Based on the S-parameter theoretical modeling, the synthesis relations enabling to determine the circuit elements are formulated.

The feasibility of the established circuit theory is verified with RF/microwave lumped circuits. The POC circuits are designed with the SPICE environment electronic circuit simulator. The POC is simulated in frequency domain from 2 to 3 GHz.

In the future, this Hilbert filter topology opens new possibilities of the RF/microwave system design. Among the various applications, the proposed Hilbert filter can be implemented in downstream of dual down conversion block of RF/microwave transceiver chains to reject the second intermediate frequency.

References

[1] S. Gimenez, S. Roger, P. Baracca, *et al.*, "Performance evaluation of analog beamforming with hardware impairments for mmW massive MIMO communication in an urban scenario," Sensors, Vol. 16, No. 10, Sep. 2016, pp. 1–17.

[2] T. Shen, R. Zhang, and C. Shi, "A single-element of ka-band 0.13 μm CMOS all-RF phased-array receiver with 4-bit phase resolution," in Proceedings of 13th IEEE International Conference on Solid-State and Integrated Circuit Technology (ICSICT), 2016, Hangzhou, China, 25–28 Oct. 2016, pp. 1–5.

[3] S. Payami, M. Ghoraishi, and M. Dianati, "Hybrid beamforming for large antenna arrays with phase shifter selection," IEEE Transactions on Wireless Communications, Vol. 15, No. 11, Nov. 2016, pp. 7258–7271.

[4] T. E. Bogale, L. B. Le, A. Haghighat, and L. Vandendorpe, "On the number of RF chains and phase shifters, and scheduling design with hybrid analog-digital beamforming," IEEE Transactions on Wireless Communications, Vol. 15, No. 5, May 2016, pp. 3311–3326.

[5] J. C. Rudell, J.-J. Ou, T. B. Cho, *et al.*, "A 1.9 GHz wide-band IF double conversion CMOS integrated receiver for cordless telephone applications," in Proceedings of IEEE International Solid-State Circuits Conference Digest of Technical Papers, Vol. 476, San Francisco, CA, 1997, pp. 304–305.

[6] F. Behbahani, Y. Kishigami, J. Leete, and A. A. Abidi, "CMOS 10 MHz IF downconverter with on-chip broadband circuit for large image-suppression," in Proceedings of Symposium on VLSI Circuits Digest Technical Papers, Kyoto, Japan, 1999, pp. 83–86.

[7] S. Engelberg, "Part III-Chap 29: Implementing a Hilbert Filter," Digital signal processing – An experimental approach, Signals and Communication Technology, Ed. Springer, London, 2008, pp. 199–201.

[8] J. Ohm, Implementing a Hilbert Filter, Chap. 9: Part of the series Signals and Communication Technology, Ed. Springer London, 2008, pp. 199–201.

[9] J. Crois and M. Steyaert, "An analog integrated polyphase filter for a high performance low-IF receiver," in Proceedings Symposium on VLSI Circuits Digest of Technical Papers, 1995, Kyoto, Japan, 8–10 June 1995, pp. 87–88.

[10] J. Crois and M. Steyaert, "A fully integrated 900 MHz CMOS double quadrature downconverter," in Proceedings of IEEE International Solid-State Circuits Conference Digest Technical Papers, 1995, San Francisco, CA, 15–17 Feb. 1995, pp. 136–137.

[11] B. Ravelo, "Investigation on microwave negative group delay circuit," Electromagnetics, Vol. 31, No. 8, Nov. 2011, pp. 537–549.
[12] B. Ravelo, "Distributed NGD active circuit for RF-microwave communication," AEU- International Journal of Electronics And Communications, Vol. 68, No. 4, Apr. 2014 pp. 282–290.
[13] S. Keser and M. Mojahedi, "Broadband negative group delay microstrip phase shifter design," in Proceedings of IEEE Antennas and Propagation Society International Symposium (APSURSI), Charleston, SC, USA, Jun. 2009, pp. 1–4.
[14] S. Takeda and T. Anada, "Phase equalizer making use of negative group delay times by reflection coefficients," in Proceedings of 41st European Microwave Conference (EuMC 2011), Manchester, UK, 9–14 Oct. 2011, pp. 628–630.
[15] L. He, W. Li, N. Li, and J. Ren, "A 24-GHz novel true-time-delay phase shifter utilizing negative group delay compensation," in Proceedings of IEEE International Symposium on Phased Array Systems and Technology, Waltham, MA, USA, 2016, pp. 1–7.
[16] S. Park and S. Jeon, "A 15–40 GHz CMOS true-time delay circuit for UWB multi-antenna systems," IEEE Microwave and Wireless Components Letters, Vol. 23, No. 3, Mar. 2013, pp. 149–151.
[17] Q. Ma, D. Leenaerts, and R. Mahmoudi, "A 10–50 GHz true-time-delay phase shifter with max 3.9% delay variation," in Proceedings of IEEE Radio Frequency Integrated Circuits Symposium, Tampa, FL, USA, 2014, pp. 83–86.
[18] S. K. Garakoui, E. A. M. Klumperink, B. Nauta, and F. E. van Vliet, "Compact cascadable gm-C all-pass true time delay cell with reduced delay variation over frequency," IEEE J. Solid-State Circuits, Vol. 50, No. 3, Mar. 2015, pp. 693–703.
[19] S. S. Oh and L. Shafai, "Compensated circuit with characteristics of lossless double negative materials and its application to array antennas," IET Microwaves, Antennas & Propagation, Vol. 1, No. 1, Feb. 2007, pp. 29–38.
[20] L. Zhan, L. Zhang, J. Liu, Q. Shen, and Y. Xia, "Negative group velocity superluminal propagation in optical fibers using stimulated Brillouin scattering," in Proceedings of 9th International Conference on Optical Communications and Networks (ICOCN 2010), Nanjing, China, 24–27 Oct. 2010, pp. 245–248.
[21] D. Solli, R. Y. Chiao, and J. M. Hickmann, "Superluminal effects and negative delays in electronics, and their applications," Physical Review E, Vol. 66, No. 5 (056601), Nov. 2002, pp. 1–4.
[22] K.-P. Ahn, R. Ishikawa, and K. Honjo, "Group delay equalized UWB InGaP/GaAs HBT MMIC amplifier using negative group delay circuits," IEEE Transactions on Microwave Theory and Techniques, Vol. 57, No. 9, Sep. 2009, pp. 2139–2147.
[23] S. K. Podilchak, B. M. Frank, A. P. Freundorfer, and Y. M. M. Antar, "High speed metamaterial-inspired negative group delay circuits in CMOS for

delay equalization," in Proceedings of 2nd Microsystems and Nanoelectronics Research Conference (MNRC), Ottawa, ON, Canada, Oct. 2009, pp. 9–12.

[24] T. Eudes and B. Ravelo, "Cancellation of delays in the high-rate interconnects with UWB NGD active cells," Applied Physics Research, Vol. 3, No. 2, Nov. 2011, pp. 81–88.

[25] B. Ravelo, "Neutralization of LC- and RC-effects with left-handed and NGD circuits," Advanced Electromagnetics (AEM), Vol. 2, No. 1, Sept. 2013, pp. 73–84.

[26] B. Ravelo, "Recovery of microwave-digital signal integrity with NGD circuits," Photonics and Optoelectronics (P&O), Vol. 2, No. 1, Jan. 2013, pp. 8–16.

[27] D. A. Frickey, "Conversions between S, Z, Y, h, ABCD and T parameters which are valid for complex source and load impedances," IEEE Transactions on Microwave Theory and Techniques, Vol. 42, No. 2, Feb. 1994, pp. 205–216.

Chapter 7

Design technique on the microwave amplifier-based active lumped NGD devices

Fayu Wan[1] *and Junxiang Ge*[1]

Abstract

This chapter addresses a microwave circuit theory on the negative group delay (NGD) principle. The characterization methodology of this unfamiliar NGD function is described. The basic NGD topologies are typically active cells composed of RF low-noise amplifier (LNA). Typically, low-pass NGD topologies are identified and characterized. The low-pass NGD circuits consisted of simple first-order RL-series and RC-parallel networks. Similar to the filter theory, band-pass NGD circuits are introduced from low-pass to band-pass transform. The proposed NGD circuit theory is established on the *S*-parameter approach. The proposed NGD topology main characteristics and properties as the NGD level, cut-off frequencies or bandwidth, central frequency and figure-of-merit are established. Given the LNA *S*-parameter model, the NGD synthesis and design method of the low- and band-pass NGD cells in function of the specified NGD values is presented. The NGD cell parameter calculations in function of the expected NGD level, insertion loss and reflection coefficient are introduced. NGD proof of concept (PoC) are synthesized, designed and simulated in order to understand and validate the proposed NGD low-pass and band-pass functions. As expected, low-pass and band-pass NGD aspects are obtained. The low-pass NGD circuits present NGD level of about −2 ns over the bandwidth of about 70 MHz. Then, band-pass NGD circuits was synthesized to operate at about 0.5 GHz over the bandwidth of about 140 MHz with NGD level of about −4 ns. It is worth mentioning that the NGD circuits respect the requirements of RF and microwave specifications.

Keywords

Negative group delay (NGD), NGD characterization, synthesis method, microwave circuit theory, low-pass NGD circuit, band-pass NGD circuit.

[1]Electronic and Information Engineering College, Nanjing University of Information Science and Technology, China

7.1 Introduction

Since the 1990s, the negative group delay (NGD) phenomenon has been investigated theoretically and experimentally with low-frequency and RF/microwave electronic circuits [1–6]. In this period, the intriguing phenomenon was independently studied in two different ways. On the one hand, NGD function was synthesized with passive microwave circuits [1,2]. On the other hand, the NGD active circuits operating at very low frequencies below megahertz was identified [3–5]. This low-frequency NGD circuits are constituted by classical components as R, L, C and operational amplifier [3] was inspired from atomic medium in optical wavelengths [6]. The NGD effect was initially occurred with the propagation media presenting anomalous dispersion [7,8]. It was pointed out that the NGD effect is generated in abnormal media frequency band presenting negative refractive group index $n_g(\omega)$ in function of the angular frequency ω or the wavelength [9]. Because of the dispersion, this refractive group index can be negative was shown that at certain wavelengths. Meanwhile that the group velocity $v_g(\omega)$ can also be negative as analytically illustrated by

$$v_g(\omega) = \frac{c}{n_g(\omega)} \tag{7.1}$$

where c is the speed of light in the vacuum. It means that when using a propagation medium with geometrical length d, the group delay, definition given by:

$$\tau_g(\omega) = \frac{d}{v_g(\omega)} \tag{7.2}$$

can also become a negative quantity. Because of the inherent fact of the wave reshaping or the combination of the constructive and destructive interferences at the edge of the abnormal dispersive passive medium, the NGD phenomenon is systematically accompanied with significant losses at microwave wavelengths [10,11]. The NGD phenomenon was classified among the abnormal physical phenomenon as the superluminal signal propagating in anomalous dispersion media [7–12]. It was concluded from various experimentations that the NGD phenomenon means the possibility to generate an output signal with wave fronts propagating in advance of its input under certain conditions [13–15]. Inversely to the classical ordinary structure, in the case of NGD media, the time delay can be assumed as negative [13–15]. It should be emphasized that that this NGD effect does not forbid the causality principle [4,5].

In the early 2000s, the first tentative of NGD function tentative application for the signal delay cancellation was proposed in 2002 [16]. However, because of the NGD bandwidth limitations to some hundred kilohertz, the possibility of applications is technically restrained and was not expectedly successful. This limitation explains the main interest to develop NGD circuits capable to operate at RF and higher frequencies. Microwave NGD passive circuits based on resonant filter topologies presenting absorption phenomenon have been identified [17,18]. So, the applications of NGD passive circuits are still limited before the middle 2000s. In the same period, another NGD circuits inspired from left-handed metamaterial structures was also identified [19–21]. With metamaterial-based NGD circuits, the time-advance aspect was experimented in the time-domain at microwave wavelengths [21]. More recently, NGD distributed microwave circuits have been developed with coupling effects [22,23]. However, the passive NGD circuits are systematically absorbing or with outstanding reflections and limit the application possibilities.

To overcome this limitation, active microwave topologies based on the RF/microwave transistors have been developed [24–29]. A methodology to identify elementary field effect transistor (FET)-based NGD topologies has been introduced [27]. It was also emphasized that the intriguing NGD function presents similar behaviours as the classical NGD filter [28]. The electronic circuit NGD bandwidth can be defined as the frequency bands where the group delay is negative. Different types of NGD aspects as low-pass and band-pass NGD circuits have been categorized [28]. Despite the systematic limitation inherently to the NGD function [30,31], different applications notably in the RF and microwave engineering were proposed in the literature. One of the most published applications are the design of independent frequency NGD phase shifter [25,26,32]. Potential NGD function application for the high-performance feedforward amplifier design and bandwidth enhancement was also proposed [33–35]. Design of compact and performant antenna array was introduced [36,37]. One of the most obvious NGD applications on the propagation delay reduction based on the electrical interconnect equalization technique was also developed [38,39]. The feasibility of this delay cancellation was later extended to the signal recovery for the signal integrity enhancement. This delay reduction application technique can also be developed with microwave amplifier-based NGD circuits [40]. It can be pointed out that because of the transistor access-matching difficulties, the complexity of the biasing networks and the output voltage sign inversion compared to the input is found with certain NGD topologies as proposed in [24]. It means that somehow the use of the transistors is more and more sophisticated. To avoid the complexity of bias network and the issues related to the matching level, the replacement of the transistor by an integrated low-noise amplifier (LNA) was introduced in [41,42] for the base band applications. Despite this trend of the NGD theory democratization and exploration of applications, various ways of exploring the NGD concept are still needed for further understanding in academic and industrial point of view. Doing this, an alternative way to design NGD active RF/microwave circuits can be imagined by replacing the FET with LNA. This alternative can be a solution against the access matching, stability and bias circuit simplification. This aspect can be significantly important for the design simplicity.

In the present chapter, we are addressing a simple theoretical approach and easy to understand experimentation of NGD function using LNA. During the theoretical approach, this former one is assumed to be represented by its *S*-parameter touchstone model. This basic theoretical approach contains the basic principle and analysis enable to make the NGD concept familiar to electrical engineering students, researchers and industries. To accomplish this, we start with particularly simple low-pass topologies lumped element components based consisted of RL series network and RC parallel network. It acts as first-order circuits which can be investigated similarly to any electrical familiar function. The generality about the basic topology to be explored will be introduced in Section 7.2. Two types of NGD cells will be developed in the present paper. Section 7.3 is focused on the NGD analysis of the family of the shunt impedance type cell. Then, Section 7.4 describes the NGD analysis of the family of the series impedance type cell. In both sections, the theory presents also the basic NGD properties and synthesis methodology of the basic NGD cells. Then, by considering the familiarly known low-pass to band-pass circuit transform in filter theory, band-pass NGD cells are also investigated. The validity of the proposed NGD theory, simulations and experimental results are presented and explored in Section 7.5. Simulation validations of LNA-based lumped circuits will be described. Then, the conclusion of the paper will be presented in Section 7.6.

7.2 Generality on the active circuit topology under study

The present section is focused on the description of the general topologies to be developed in the whole paper. After the diagram description, the *S*-parameter model will be analytically expressed in function of the LNA and the constituting impedance of the circuit. Then, the NGD analysis principle will be introduced.

7.2.1 Basic topology introductory

The present subsection describes generally the elementary passive and active topologies of circuit to be developed in this paper. Like all classical electronic and electrical circuits, the simplest and elementary topologies are based on the two-port circuit as a simple impedance which can be denoted Z. As pointed out in [41,42], the basic topology is constituted by an LNA associated with a simple two-port passive network. Intuitively, the simplest passive network can be a parallel impedance Z_p or a series impedance Z_s. Along the paper, it is noteworthy that the analytical calculation will be developed with the LNA *S*-parameter model defined by

$$[S_{LNA}] = \begin{bmatrix} r & 0 \\ t & r \end{bmatrix} \tag{7.3}$$

with the real variables r and t assumed to be independent to the frequency in the operating frequency band are respectively the LNA access reflection coefficient and the insertion gain. The aim of the active circuit is to compensate the NGD circuit loss. This is very important for the time domain demonstration in order to show obviously the feasibility of the time advance aspect. This aspect can be visualized both with time domain simulations and time-domain measurements.

7.2.2 Shunt impedance-based topologies under study

The shunt impedance-based active cells investigated in the present paper are introduced in Figure 7.1. The three considered configurations depend on the position of the passive network placed in upstream or in downstream of the LNA. It should be underlined that in this paper R_0 represents the reference impedance equal to 50 Ω. Acting as a cascaded circuit configuration, the *S*-parameter analytical calculations can be obtained from the total circuit transfer matrix. According to the circuit and system theory, the total transfer matrix is equal to the non-commutative product of the cascaded constituting cells which are represented by the impedance and the LNA in the present case. By applying the *S*-parameter theory, the following expressions have been established:

- For the parallel impedance-based cell with LNA inserted in downstream as presented in Figure 7.1(a):

$$[S_{\text{p-LNA}}(j\omega)] = \begin{bmatrix} \dfrac{2rZ_p - (1+r)R_0}{2Z_p + (1+r)R_0} & 0 \\ \dfrac{2tZ_p}{2Z_p + (1+r)R_0} & r \end{bmatrix} \tag{7.4}$$

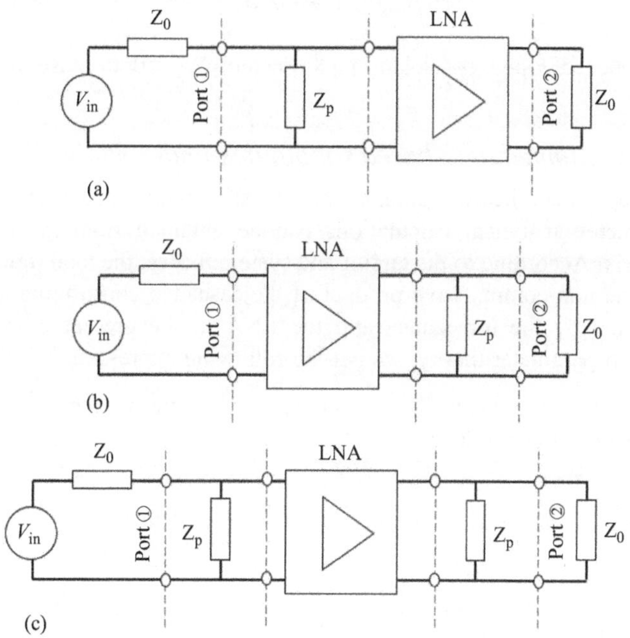

Figure 7.1 (a) Parallel impedance and LNA in cascade; (b) LNA and parallel impedance in cascade; and (c) active cell with double shunt impedance

- For the parallel impedance-based cell with LNA inserted in upstream as presented in Figure 7.1(b):

$$[S_{\text{LNA-p}}(j\omega)] = \begin{bmatrix} r & 0 \\ S_{\text{p-LNA}_{21}}(j\omega) & S_{\text{p-LNA}_{11}}(j\omega) \end{bmatrix}$$

$$= \begin{bmatrix} r & 0 \\ \dfrac{2tZ_p}{2Z_p + (1+r)R_0} & \dfrac{2rZ_p - (1+r)R_0}{2Z_p + (1+r)R_0} \end{bmatrix} \tag{7.5}$$

- For the parallel impedance-based cell with LNA inserted in upstream as presented in Figure 7.1(c):

$$[S_{\text{p-LNA-p}}(j\omega)] = \begin{bmatrix} S_{\text{p-LNA}_{11}}(j\omega) & 0 \\ \dfrac{S^2_{\text{p-LNA}_{21}}(j\omega)}{t} & S_{\text{p-LNA}_{11}}(j\omega) \end{bmatrix}$$

$$= \begin{bmatrix} \dfrac{2rZ_p - (1+r)R_0}{2Z_p + (1+r)R_0} & 0 \\ \dfrac{4Z_p^2 t}{[2Z_p + (1+r)R_0]^2} & \dfrac{2rZ_p - (1+r)R_0}{2Z_p + (1+r)R_0} \end{bmatrix} \tag{7.6}$$

The proposed topology isolation parameters $S_{12} = 0$ because of the LNA unilaterality.

7.2.3 Series impedance-based topologies under study

Figure 7.2 represents the configuration of the series impedance-based active cell. The *S*-parameter analytical calculations can be obtained from the total circuit transfer matrix. According to the circuit and system theory, the total transfer matrix is equal to the non-commutative product of the cascaded constituting cells which are represented by the impedance and the LNA in the present case. Then, by applying the *S*-parameter theory, we get the following expressions:

- For the series impedance-based cell with LNA inserted in downstream as presented in Figure 7.2(a):

$$[S_{\text{s-LNA}}(j\omega)] = \begin{bmatrix} r & 0 \\ \dfrac{2tR_0}{(1-r)Z_s + 2R_0} & \dfrac{(1-r)Z_s + 2rR_0}{(1-r)Z_s + 2R_0} \end{bmatrix}, \tag{7.7}$$

- For the series impedance-based cell with LNA inserted in upstream as presented in Figure 7.2(b):.

$$[S_{\text{LNA-s}}(j\omega)] = \begin{bmatrix} r & 0 \\ S_{\text{s-LNA}_{21}}(j\omega) & S_{\text{s-LNA}_{11}}(j\omega) \end{bmatrix} = \begin{bmatrix} \dfrac{(1-r)Z_s + 2rR_0}{(1-r)Z_s + 2R_0} & 0 \\ \dfrac{2tR_0}{(1-r)Z_s + 2R_0} & r \end{bmatrix} \tag{7.8}$$

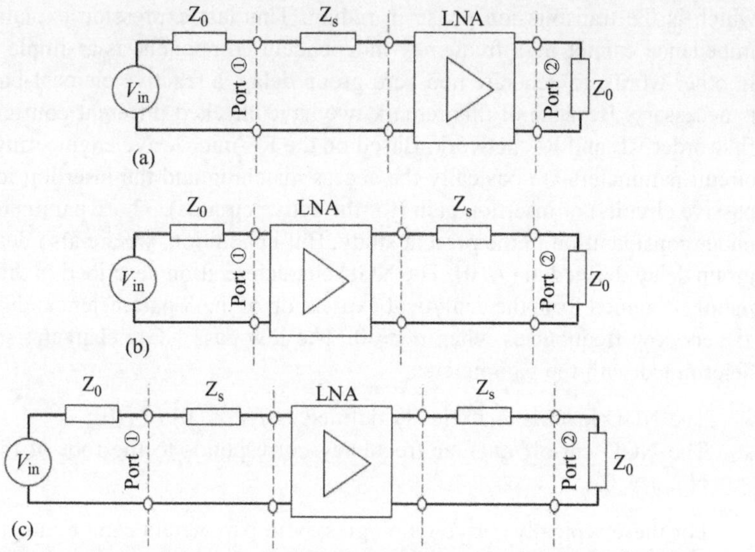

Figure 7.2 (a) Series impedance and LNA in cascade; (b) LNA and series impedance in cascade; and (c) active cell with double series impedance

- For the series impedance-based cell with LNA inserted in upstream as presented in Figure 7.2(c):

$$[S_{\text{LNA-s}}(j\omega)] = \begin{bmatrix} S_{\text{s-LNA}_{11}}(j\omega) & 0 \\ \dfrac{S^2_{\text{s-LNA}_{21}}(j\omega)}{t} & S_{\text{s-LNA}_{11}}(j\omega) \end{bmatrix}$$

$$= \begin{bmatrix} \dfrac{(1-r)Z_s + 2rR_0}{(1-r)Z_s + 2R_0} & 0 \\ \dfrac{4R_0^2 t}{[(1-r)Z_s + 2R_0]^2} & \dfrac{(1-r)Z_s + 2rR_0}{(1-r)Z_s + 2R_0} \end{bmatrix} \tag{7.9}$$

Again, it can be found that the isolation parameters $S_{12} = 0$ because of the LNA unilaterality.

7.2.4 Principle of NGD analysis

By denoting $j\omega$ the circuit angular frequency variable, according to the circuit and system theory, it can be recalled that the group delay is defined by

$$\tau(\omega) = -\frac{\partial \varphi(\omega)}{\partial \omega} \tag{7.10}$$

which is the transmission phase in radian. This last expression explains that our impedance cannot be a frequency independent component as a simple resistance. In other words, to generate non-zero group delay, a reactive element-based circuit is necessary. Because of this remark, we have checked different configurations of first-order RL and RC network. Based on the RF/microwave engineering, the main circuit parameters are basically the access matching and the insertion loss (for the passive circuits) or insertion gain (for the active circuits). Those parameters are still under consideration in the present study. But in addition, we are also dealing to the group delay defined in (7.10). The NGD characterization described in this section is mainly obtained from the analytical expression of the *S*-parameters and group delay at very low frequencies when $\omega \approx 0$. The low-pass NGD characteristics can be determined with the parameters:

- The NGD level is analytically defined as $\tau(\omega \approx 0) = \tau(0)$;
- The NGD cut-off angular frequency corresponds to the root of the equation $\tau(\omega_c) = 0$.

For these typically active topologies, both parameters can be integrated in the unified NGD figure-of-merit (FoM) which is defined by:

$$\mathrm{FoM_{NGD}} = \tau(0) \cdot \omega_c \tag{7.11}$$

Better is the NGD circuit, higher the FoM is.

7.2.5 NGD microwave circuit design method

The microwave circuit theory enabling to design low-pass and band-pass NGD circuits based on the familiar electronic components R, L and C will be presented in the two next sections. After exploring the NGD effect existence with lumped element-based two passive topologies, the possibility to compensate loss by inserting an LNA in cascade with the NGD passive network will be investigated. The theoretical characterization of the obtained active circuit will be described. It will be described how to compensate the NGD passive circuit loss attenuation $A < 1$, we can just insert an LNA with gain equal to $G = 1/A$. At the end of each step, synthesis expressions enabling to calculate easily the NGD circuit parameters in function of the NGD specifications will be formulated.

7.3 NGD analysis of the shunt impedance-based active cell

The analytical identification performed in [27] states that the only configuration of first-order or low-pass NGD circuit associated to the parallel impedance configuration is an RL-series network [43]:

$$Z_p(j\omega) = R + j\omega L \tag{7.12}$$

Knowing the LNA model defined in (7.3), the characterization and synthesis methods will be analytically developed.

7.3.1 Low-pass cell with LNA

Figure 7.3 represents the configuration of the RL-based low-pass NGD elementary cells [43].

7.3.1.1 Frequency-dependent S-parameter and group delay

Then, by applying the S-parameter theory, we get the following expressions for the parallel impedance-based cell with LNA inserted in downstream as presented in Figure 7.3(a):

$$[S_{pl}(j\omega)] = \begin{bmatrix} \dfrac{2r(R+j\omega L)-(1+r)R_0}{2(R+j\omega L)+(1+r)R_0} & 0 \\ \dfrac{4t(R+j\omega L)^2}{[2(R+j\omega L)+(1+r)R_0]^2} & \dfrac{2r(R+j\omega L)-(1+r)R_0}{2(R+j\omega L)+(1+r)R_0} \end{bmatrix} \quad (7.13)$$

7.3.1.2 NGD characterization and properties

At very low frequencies, the S-parameters of the parallel impedance-based circuit become:

$$[S_{pl}(0)] = \begin{bmatrix} \dfrac{2rR-(1+r)R_0}{2R+(1+r)R_0} & 0 \\ \dfrac{4R^2t}{[2R+(1+r)R_0]^2} & \dfrac{2rR-(1+r)R_0}{2R+(1+r)R_0} \end{bmatrix} \quad (7.14)$$

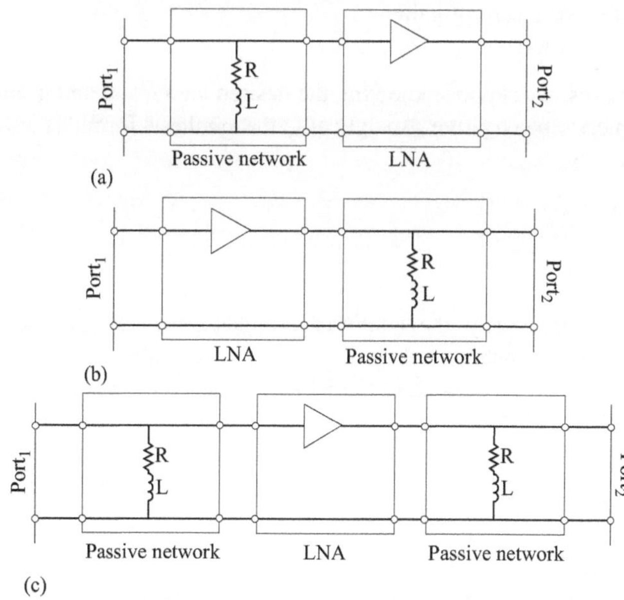

Figure 7.3 LNA-based low-pass NGD cell with RL-series shunt network in (a) upstream; (b) downstream; and (c) up- and down-downstream [43]

The associated group delay is given by

$$\tau_{\mathrm{pl}}(0) = \frac{-2(1+r)R_0 L}{R[2R+(1+r)R_0]} \tag{7.15}$$

It can be understood from the group delay expressed in (7.14) that the passive cell introduced in Figure 7.3(c) behaves as a low-pass NGD circuit. In addition to the previous characteristics, one of the most important properties of the NGD cell is the NGD cut-off frequency denoted which can be assumed as the root of the equation is given by

$$\omega_{pc} = \frac{\sqrt{2R[2R+(1+r)R_0]}}{2L} \tag{7.16}$$

Knowing all these, we can determine the NGD FoM by the relation:

$$\mathrm{FoM_p} = \tau_{\mathrm{pl}}(0)\cdot\omega_{\mathrm{pc}} = \frac{-2(1+r)R_0}{\sqrt{2R[2R+(1+r)R_0]}} \tag{7.17}$$

It can be underlined that in this case, the NGD FoM is inversely proportional to the resistance R.

7.3.1.3 NGD synthesis method

The synthesis method consists merely in the calculation of the NGD circuit parameters in function of the specified:

- NGD level τ_0, which is inversely linked to the NGD bandwidth
- Insertion loss or insertion gain
- Reflection coefficients.

In other words, we suppose knowing the desired gain $g > 0$ and group delay τ_0 which is assumed to be negative. Analytically, the synthesis formulas are generated from the equations:

$$\begin{cases} S_{\mathrm{pl}_{21}}(0) = g \\ \tau_{\mathrm{pl}}(0) = \tau_0 \end{cases} \tag{7.18}$$

As discussed previously, the synthesis formulas are merely the roots of (7.14) and (7.15). Parallel impedance-based cell: After calculations, we obtain the expressions of R and L associated to the RL-series cell. The synthesis relations for this low-pass NGD cell are

$$R = \frac{\sqrt{g}(1+r)}{2(\sqrt{t}-\sqrt{g})}R_0 \tag{7.19}$$

$$L = \tau_0 R\left[\frac{1}{2} + \frac{R}{(1+r)R_0}\right] \tag{7.20}$$

It can be understood from (7.19) that the synthesized resistance is realistic only if the existence condition $g < t$ is respected. Based on these synthesis formulas, the reflection coefficient can be rewritten in function of the desired gain as follows:

$$|S_{sl_{11}}(0)| = \left|1 - (1+r)\sqrt{\frac{g}{t}}\right| \tag{7.21}$$

It is worth mentioning that the resistance synthesis relation can be extracted also from the desired access matching $S_{11}(0) = m$.

7.3.2 *Band-pass NGD elementary cells*

Similar to the filter theory, the band-pass NGD cells can be synthesized easily from the low-pass ones via the low-pass to band-pass transform [28]. The present subsection is focused on the NGD characteristics of the band-pass NGD cells associated to the previous cells. In this case, to synthesis the associated band-pass NGD cell, we just need to replace the inductance L by an LC-series network. The associated capacitor can be calculated from the inductance and the expected NGD centre angular frequency which can be denoted ω_0. Therefore, we have the capacitor synthesis formula:

$$C_p = \frac{1}{\omega_0^2 L} \tag{7.22}$$

Figure 7.4 represents the configuration of the RLC-shunt network-based band-pass NGD elementary cells dual to the low-pass NGD cells shown in Figure 7.3 [43].

7.3.2.1 NGD characterization and properties

Then, by applying the S-parameter theory, we get the transmission gain, reflection coefficient and the group delay. NGD band-pass active cells: Then, the associated S-parameters at the centre frequency are expressed as

$$[S_{pb}(j\omega_0)] = [S_{pl}(0)] \tag{7.23}$$

The associated group delay is given by

$$\tau_{pb}(\omega_0) = \tau_{pl}(0) = \frac{-4(1+r)R_0 L}{R[2R + (1+r)R_0]} \tag{7.24}$$

7.3.2.2 NGD synthesis method

Similar to Section 7.3.2.1, the synthesis formulas of the active cell can be obtained from (7.23) and (7.24). For the considered parallel impedance-based cell, the expressions of R and L associated to the RL-series cell can be determined analytically. The synthesis relations for this band-pass NGD cell can be expressed in function of the resistance and inductance suggested in (7.19) and (7.20):

$$R_b = R. \tag{7.25}$$

$$L_b = L/2. \tag{7.26}$$

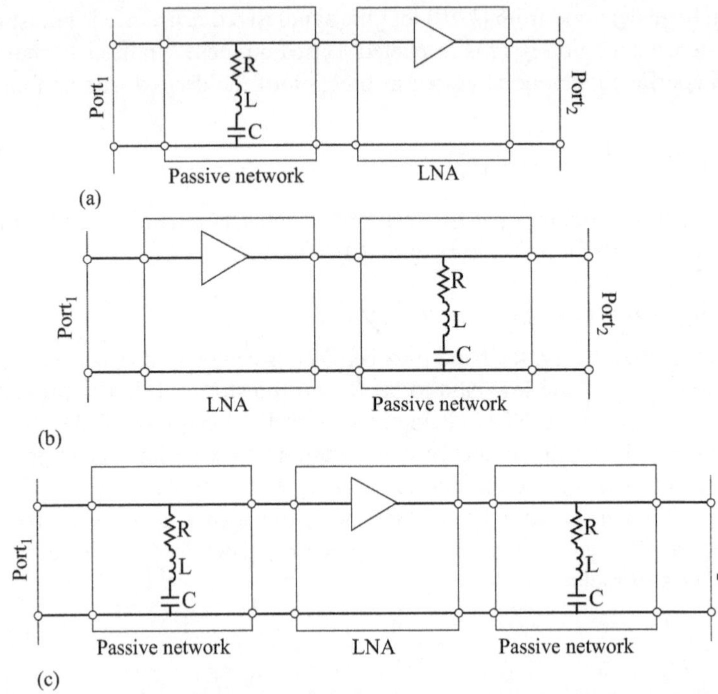

Figure 7.4 LNA-based band-pass NGD cell with RLC-series shunt network in (a) upstream; (b) upstream; and (c) up- and down-stream [43]

7.4 NGD analysis of the series impedance-based active cell

Based on the circuit theory, the simplest configuration of first-order or low-pass NGD circuit is intuitively associated to the series impedance configuration is an RC-parallel network:

$$Z_s(j\omega) = \frac{R}{1 + j\omega RC}. \tag{7.27}$$

7.4.1 Low-pass cell with LNA

Figure 7.5 represents the configuration of the RC-series network-based low-pass NGD elementary cells.

7.4.1.1 Frequency-dependent *S*-parameter and group delay

Then, by applying the *S*-parameter theory, we get the following expressions for the series impedance-based cell with LNA inserted in downstream as presented

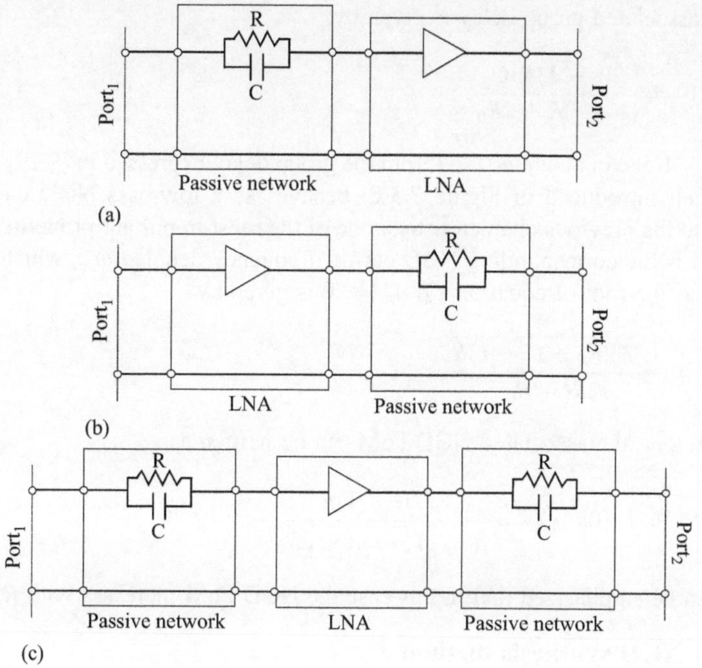

Figure 7.5 LNA-based low-pass NGD cell with RL-series shunt network in upstream

in Figure 7.4(a):

$$[S_{sl}(j\omega)] = \begin{bmatrix} \dfrac{(1-r)R + 2rR_0(1+j\omega RC)}{(1-r)R + 2R_0(1+j\omega RC)} & 0 \\ \dfrac{4tR_0^2(1+j\omega RC)^2}{[(1-r)R + 2R_0(1+j\omega RC)]^2} & \dfrac{(1-r)R + 2rR_0(1+j\omega RC)}{(1-r)R + 2R_0(1+j\omega RC)} \end{bmatrix} \tag{7.28}$$

7.4.1.2 NGD characterization and properties

At very low frequencies, the previous S-parameters of the parallel impedance-based circuit become:

$$[S_{sl}(0)] = \begin{bmatrix} \dfrac{(1-r)R + 2rR_0}{(1-r)R + 2R_0} & 0 \\ \dfrac{4tR_0^2}{[(1-r)R + 2R_0]^2} & \dfrac{(1-r)R + 2rR_0}{(1-r)R + 2R_0} \end{bmatrix} \tag{7.29}$$

The associated group delay is given by:

$$\tau_{sl}(0) = \frac{2(r-1)R^2C}{(1-r)R+2R_0}. \tag{7.30}$$

As $r < 1$, it can be understood from the group delay expressed in (7.29) that the passive cell introduced in Figure 7.5(c) behaves as a low-pass NGD circuit. In addition to the previous characteristics, one of the most important properties of the NGD cell is the corresponding NGD cut-off frequency denoted ω_{sc}, which can be assumed as the root of equation $\tau_{sl}(\omega_{sc}) = 0$ is given by

$$\omega_{sc} = \frac{\sqrt{2[2R_0+(1-r)R]}}{2RC\sqrt{R_0}} \tag{7.31}$$

Therefore, the associated NGD FoM can be written as

$$\mathrm{FoM}_s = \tau_{sl}(0)\cdot\omega_{sc} = \frac{\sqrt{2}(r-1)R}{R_0\sqrt{(1-r)R+2R_0}} \tag{7.32}$$

It can be emphasized that in this case the NGD FoM increases with R.

7.4.1.3 NGD synthesis method

Let suppose as targeted objective of the synthesis process the desired gain g and NGD level τ_0. Analytically, the synthesis formulas are generated from the equations $S_{21}(0) = g$ and $\tau_{sl}(0) = \tau_0$. Alternatively, the first relation can be replaced also by the synthesis from the desired access matching $S_{11}(0) = m$. As argued previously, the synthesis formulas are merely the roots of (7.29) and (7.30). For the configuration of the series impedance-based NGD cell, the expressions of R and C constituting the series impedance-based RC-parallel network can be formulated as follows:

$$R = \frac{2(\sqrt{t}-\sqrt{g})}{\sqrt{g}(1-r)}R_0 \tag{7.33}$$

$$C = \frac{-\tau_0}{R}\left[\frac{1}{2}+\frac{R_0}{(1-r)R}\right] \tag{7.34}$$

Based on these synthesis formulas, the reflection coefficient can be rewritten in function of the desired gain as follows:

$$|S_{sl11}(0)| = 1 \pm (1-r)\sqrt{\frac{g}{t}} \tag{7.35}$$

7.4.2 *Band-pass NGD elementary cells*

Similar to the filter theory, the band-pass NGD cells can be synthesized easily from the low-pass ones via the low-pass to band-pass transform. The present section is focused on the NGD characteristics of the band-pass NGD cells associated to the previous cells. In this case, to synthesis the associated band-pass NGD cell, we just need to replace the capacitor C by an LC-parallel network. The associated capacitor can be calculated from the inductance and the expected NGD centre angular frequency which can be denoted ω_0. Therefore, we have the inductance element synthesis formula

$$L_s = \frac{1}{\omega_0^2 C} \tag{7.36}$$

Figure 7.6 represents the configuration of the RLC-parallel series network-based band-pass NGD elementary cells associated family to the cells shown in Figure 7.5.

7.4.2.1 NGD characterization and properties

Similar to Section 7.3.2.1, the NGD band-pass active cell S-parameter and group delay can be inspired from the low-pass one introduced in Section 7.4.1. Accordingly, the associated S-parameters at the centre frequency ω_0 are expressed as

$$[S_{sb}(j\omega_0)] = [S_s(0)] \tag{7.37}$$

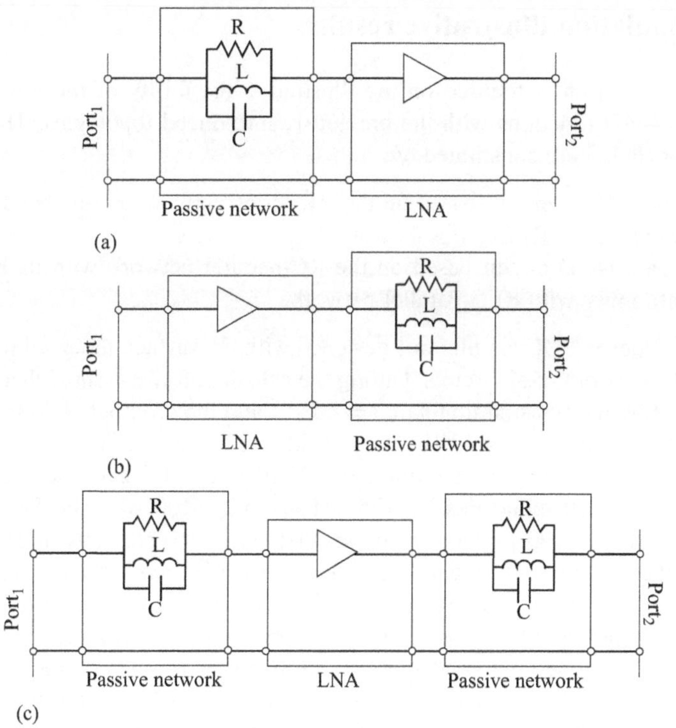

Figure 7.6 LNA-based band-pass NGD cell with RLC-series shunt network in (a) upstream; (b) downstream; and (c) up- and down-stream

The associated group delay is given by

$$\tau_{sb}(\omega_0) = 2\tau_{sl}(0) = \frac{2(r-1)R^2C}{(1-r)R + 2R_0} \tag{7.38}$$

7.4.2.2 NGD synthesis method

Similar to previous Section 7.3.2.2, the synthesis formulas of the active cell can be obtained from (7.37) and (7.38). In this case, the band-pass NGD cell synthesis relations can be expressed in function of the resistance and inductance suggested in (7.33) and (7.34) written as:

$$R_b = R. \tag{7.39}$$

$$C_b = C/2. \tag{7.40}$$

To check the effectiveness of the developed NGD circuit theory, the next section will be described the design, simulation and experimental application results. The proposed demonstrators are aimed to make the NGD concept as familiar possible to non-specialist.

7.5 Simulation illustrative results

The present section is focused on the illustrative feasibility of the low-pass and band-pass NGD functions with the previously introduced topologies. The proofs-of-concept (POC) are constituted by:

- Low-pass NGD circuit-based on the RL-series network and its band-pass circuit family with RLC-series network
- Low-pass NGD circuit based on the RC-parallel network with its band-pass circuit family with RLC-parallel network.

The proposed POC circuits was designed with the surface-mounted monolithic LNA LEE-9+ from mini-circuits. During the calculation, we assume that the LNA is specified by the transmission gain $t = 8.5$ dB and input/output reflection coefficient average value $r = -22$ dB. The considered NGD circuits was biased with $V_0 = 5V_{DC}$ power supply. In addition to the NGD circuit RF part, the bias network is consisted of self-inductance $L = 75$ nH and $C_p = 100$ pF. The DC blocking capacitors are $C = 100$ pF. As can be understood in the previous theory of Section 7.4, the NGD circuit under study can be designed similarly to classical and familiar electronic RF and microwave circuits. The POC modelled computed results are compared with simulations run in the ADS® environment of the electronic circuit designer and simulator. The proposed *S*-parameter simulations were performed from DC to 1 GHz. Comparisons of the *S*-parameter between the circuits integrating the LNA ideal model (defined in (7.3)) and touchstone *S*-parameter model provided by the manufacturer are presented. The obtained simulation results will be explored in the next sections.

7.5.1 Shunt impedance-type NGD topologies

This subsection is mainly focused on the validations of the shunt impedance-based NGD topologies explore previously in Section 7.3. Low-pass and band-pass NGD circuits are designed by considering the previous synthesis methods.

7.5.1.1 Synthesis approach

Table 7.1 summarizes the synthesized parameter of the shunt impedance-based low-pass cell introduced in Figure 7.3(c) and the band-pass cell introduced in Figure 7.4(c). The band-pass circuit was synthesized to operate at the centre frequency $f_0 = 0.5$ GHz.

7.5.1.2 Low-pass NGD simulations from the cell proposed in Figure 7.3(c)

The synthesis of the proposed low-pass NGD circuit based on the shunt impedance type circuit shown in Figure 7.3(c) is targeted to the specifications $g = 0$ dB and $\tau(0) = -2$ ns. Figure 7.7 represents the comparison between the simulated and measured S-parameters obtained from the shunt impedance-based low-pass NGD circuit configured in Figure 7.3(c). The ideal and real model-based LNA simulations are very well correlated. The transmission gain shown in Figure 7.7(b) is higher than 0 dB with flatness of about 2.5 dB from DC to 70 MHz. The reflection coefficients exposed in Figure 7.7(a) and (c) are slightly better than 10 dB.

Figure 7.8 displays the simulated group delays from the circuit shown in Figure 7.3(c). Despite the discrepancies and parasitic effects of the real LNA-based group delay, a good correlation between the ideal and real LNA-based circuits is confirmed. The low-pass NGD function is emphasized by the occurrence of NGD at very low frequencies up to the cut-off frequency of about $f_c = 70$ MHz and positive group delay when the frequency is higher than f_c. As expected, low-pass NGD level of about −2 ns is obtained. The NGD bandwidth is realized from DC to about 70 MHz.

7.5.1.3 Band-pass NGD simulations from the cell proposed in Figure 7.4(c)

The synthesis of the proposed band-pass NGD circuit based on the shunt impedance type circuit shown in Figure 7.4(c) is targeted to the specifications $g = 0$ dB and

Table 7.1 Synthesized shunt-based NGD circuit parameters given the expected transmission gain g and group delay $\tau_0 < 0$

g	τ_0	R	L	C (for the band-pass circuit)
0 dB	−2 ns	42.76 Ω	110 nH	0.92 pF
2 dB		59.47 Ω	190 nH	0.53 pF
0 dB	−5 ns	42.75 Ω	276 nH	0.37 pF
2 dB		59.47 Ω	476 nH	0.21 pF

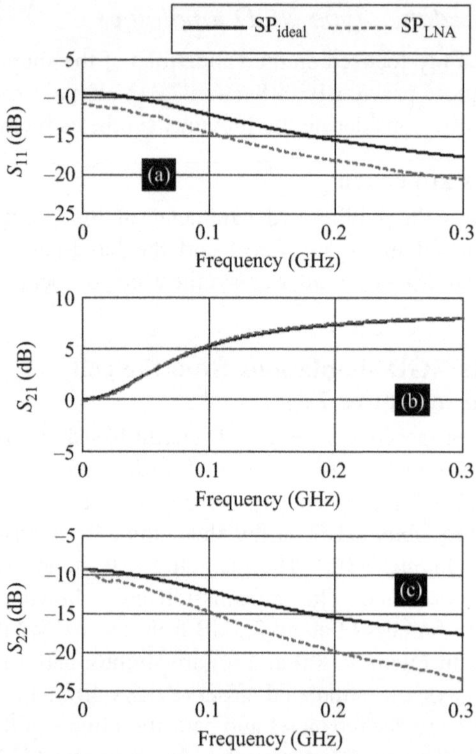

Figure 7.7 S-parameters of the shunt impedance type low-pass circuits shown in Figure 7.3(c)

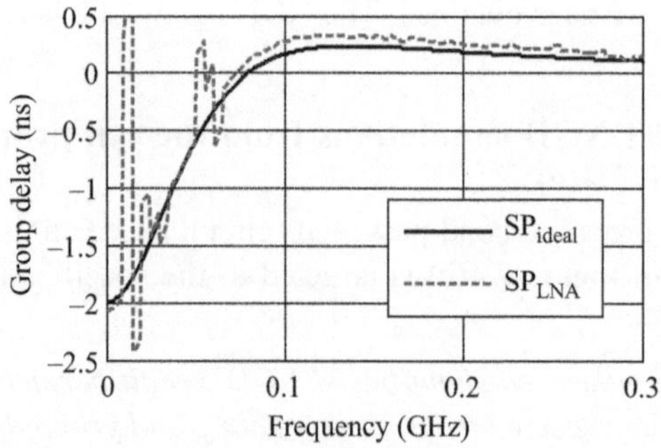

Figure 7.8 Group delays of the shunt impedance type low-pass circuits shown in Figure 7.3(c)

$\tau(f_0) = -4$ ns. Figure 7.9 introduces the comparisons between the ideal and real LNA-based NGD circuit *S*-parameters from the circuit shown in Figure 7.4(c). As expected, transmission gain above 0 dB is realized in addition to the reflection coefficient better than 10 dB.

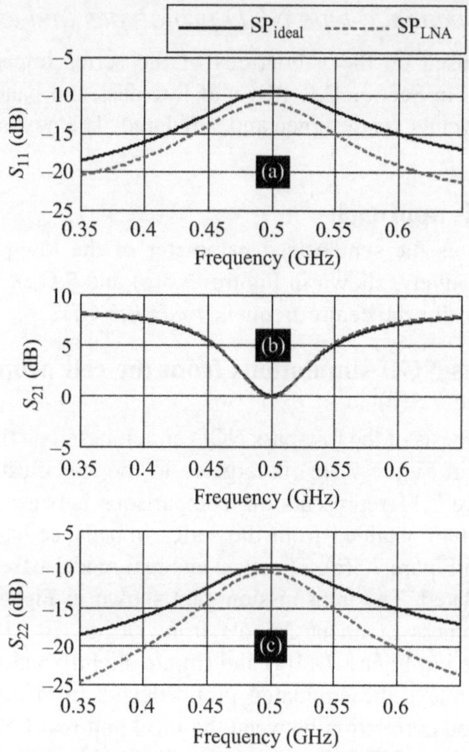

Figure 7.9 S-parameters of the shunt impedance type band-pass NGD circuits shown in Figure 7.4(c)

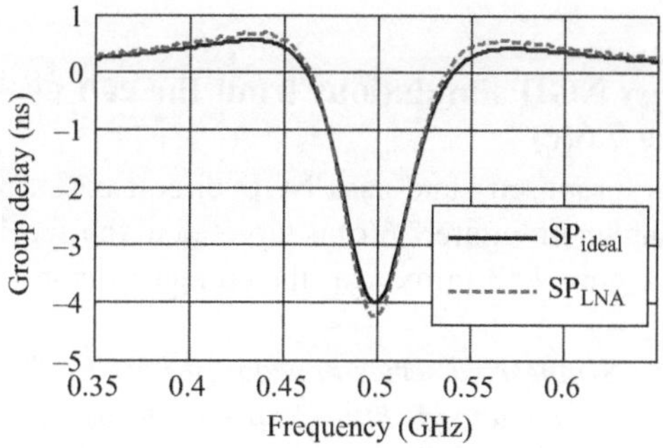

Figure 7.10 Group delays of the shunt impedance type band-pass NGD circuits shown in Figure 7.4(c)

Figure 7.10 displays the group delays of the circuit shown in Figure 7.4(c). The ideal and real LNA-based NGD circuit simulations are realized. As expected, a typically band-pass NGD behaviour centred at $f_0 = 0.5$ GHz is obtained. The NGD bandwidth is equal to $\Delta f = 2f_c$.

7.5.2 Series impedance-type NGD prototypes and experimentations

This section is focused on the validations of the series impedance-based NGD topologies proposed in Section 7.4. POCs of low-pass and band-pass NGD series impedance-based circuits are designed and simulated. The next sections present the obtained results.

7.5.2.1 Synthesis approach

Table 7.2 summarizes the synthesized parameter of the low-pass and band-pass NGD circuits respectively, shown in Figures 7.5(c) and 7.6(c). The band-pass circuit was synthesized for the centre frequency $f_0 = 0.5$ GHz.

7.5.2.2 Low-pass NGD simulations from the cell proposed in Figure 7.5(c)

In this case, the synthesis of the low-pass NGD circuit based on the shunt impedance type circuit shown in Figure 7.5(c) is targeted to the specifications $g = 0$ dB and $\tau(0) = -2$ ns. Figure 7.11 represents the comparisons between the simulated and measured S-parameters obtained from the series impedance-based low-pass NGD circuit configured in Figure 7.5(c). The ideal and real model-based LNA simulations are very well correlated. The transmission gain shown in Figure 7.11(b) is higher than 0 dB with flatness of about 2.5 dB from DC to 70 MHz. The reflection coefficients presented in Figure 7.11(a) and (b) are slightly better than 10 dB.

Figure 7.12 displays the simulated group delays from the circuit shown in Figure 7.5(c). A good correlation between the ideal and real LNA-based circuits is confirmed. The low-pass NGD function is emphasized by the occurrence of NGD at very low frequencies up to the cut-off frequency of about $f_c = 70$ MHz and positive group delay when the frequency is higher than f_c. As expected, low-pass NGD level of about −2 ns is obtained. The NGD bandwidth is realized from DC to about 70 MHz.

7.5.2.3 Band-pass NGD simulations from the cell proposed in Figure 7.6(c)

The synthesis of the proposed band-pass NGD circuit based on the shunt impedance type circuit shown in Figure 7.6(c) is targeted to the specifications $g = 0$ dB and $\tau(f_0) = -4$ ns. Figure 7.13 introduces the comparisons between the ideal and

Table 7.2 Synthesized series impedance-based NGD circuit parameters given the expected transmission gain g and group delay $\tau_0 < 0$

g	τ_0	R	C	L (for the band-pass circuit)
0 dB	−2 ns	68.56 Ω	0.38 pF	2.69 nH
2 dB		49.29 Ω	0.65 pF	1.56 nH
0 dB	−5 ns	68.56 Ω	94.23 pF	1.07 nH
2 dB		49.29 Ω	162.5 pF	0.62 nH

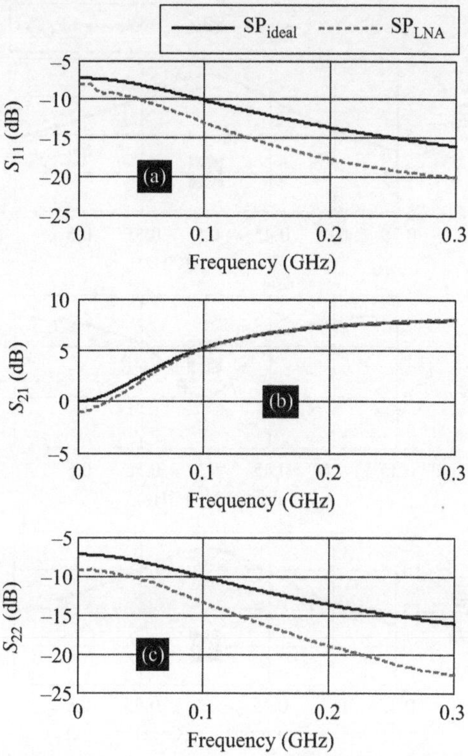

Figure 7.11 S-parameters of the shunt impedance type low-pass circuits shown in Figure 7.5(c)

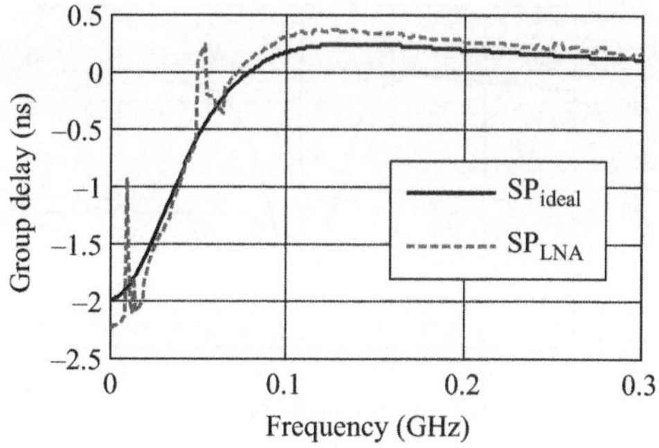

Figure 7.12 Group delays of the shunt impedance type low-pass circuits shown in Figure 7.5(c)

real LNA-based NGD circuit *S*-parameters from the circuit shown in Figure 7.6(c). As expected, transmission gain introduced in Figure 7.13(b) above 0 dB is realized. Moreover, the reflection coefficients shown in Figure 7.13(a) and (b) are better than 10 dB.

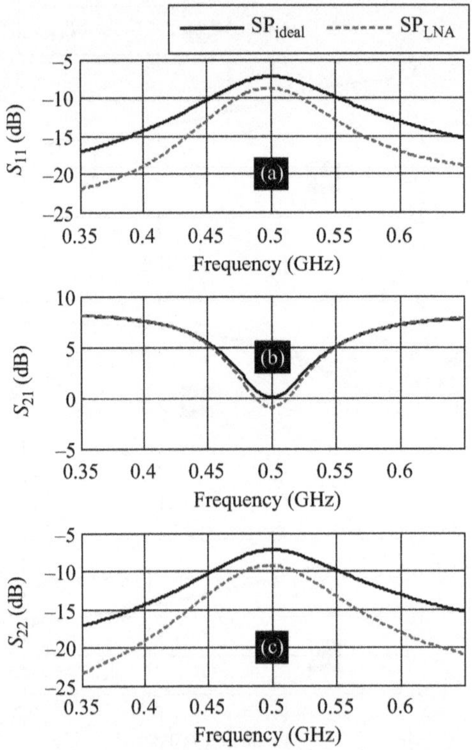

Figure 7.13 S-parameters of the shunt impedance type band-pass NGD circuits shown in Figure 7.6(c)

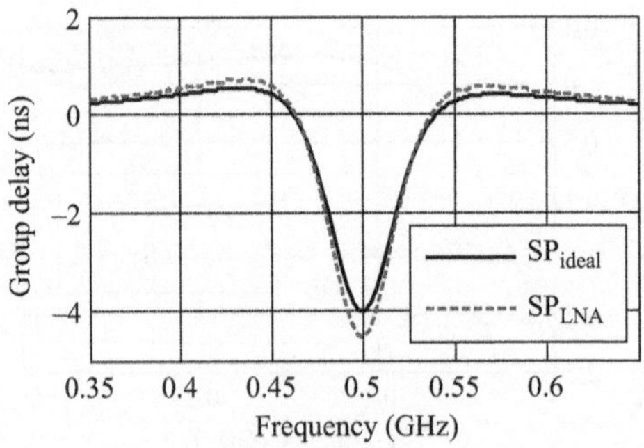

Figure 7.14 Group delays of the shunt impedance type band-pass NGD circuits shown in Figure 7.6(c)

Figure 7.14 displays the group delays of the circuit shown in Figure 7.6(c). The ideal and real LNA-based NGD circuit simulations are realized. As expected, a typically band-pass NGD behaviour centred at $f_0 = 0.5$ GHz with NGD level τ_0 $(f_0) = -4$ ns is obtained. The NGD bandwidth is equal to $\Delta f = 2f_c$.

7.6 Conclusion

A microwave circuit theory on the LNA-based active NGD elementary RF cells is developed. Two types of NGD cells based on the constituting shunt and series impedance configurations are explored. The general principle of the NGD analyses is presented. The principle was applied to the two proposed types of NGD circuit families. The analytical principle was established with *S*-parameter model. The meaning of low-pass and band-pass NGD functions is illustrated. The main properties of the NGD circuits are developed. The synthesis method enabling to calculate the NGD circuit parameters is proposed in function of the specified NGD level and transmission gain. The theoretical approach was validated with the two types of topologies. Low-pass NGD circuits are designed and simulated. As expected, low-pass NGD with level of about −5 ns from to about 70 MHz was obtained. In addition, band-pass NGD centred at 0.5 GHz with of about −10 ns from to about 140 MHz was obtained. The NGD circuits respect all the criteria of RF/microwave circuits.

References

[1] S. Lucyszyn, I. D. Robertson and A. H. Aghvami, "Negative group delay synthesizer," Electron. Lett., Vol. 29, No. 9, Apr. 1993, pp. 798–800.

[2] S. Lucyszyn and I. D. Robertson, "Analog reflection topology building blocks for adaptive microwave signal processing applications," IEEE Trans. Microwave Theory Techn., Vol. MTT-43, No. 3, Mar. 1995, pp. 601–611.

[3] R. Y. Chiao, E. L. Bolda, J. Bowie, J. Boyce and M. W. Mitchell, "Superluminality and amplifiers," Prog. Crystal Growth Charact. Mat., Vol. 33, Apr. 1996, pp. 319–325.

[4] M. W. Mitchell and R. Y. Chiao, "Causality and negative group delays in a simple bandpass amplifier," Am. J. Phys., Vol. 66, No. 14, Nov. 1998, pp. 14–19.

[5] M. W. Mitchell and R. Y. Chiao, "Negative group delay and 'Fronts' in a causal systems: An experiment with very low frequency bandpass amplifiers," Phys. Lett. A, Vol. 230, Jun. 1997, pp. 133–138.

[6] R. Y. Chiao, "Atomic coherence effects which produce superluminal (but causal) propagation of wavepackets," Quantum Optics: Journal of the European Optical Society Part B, Vol. 6, No. 4, Aug. 1994, pp. 359–369.

[7] A. Dogariu, A. Kuzmich and L. J. Wang, "Transparent anomalous dispersion and superluminal light pulse propagation at a negative group velocity," Phys. Rev. A, Vol. 63, No. 5, Apr. 2001, pp. 053806.1–053806.12.

[8] L. J. Wang, A. Kuzmich and A. Dogariu, "Gain-Assisted Superluminal Light Propagation," Nature, Vol. 406, No. 6, July 2000, pp. 277–279.

[9] J. F. Woodley and M. Mojahedi, "Negative group velocity and group delay in left-handed media," Phys. Rev. E, Vol. 70, No. 4, Oct. 2004, pp. 046603.1–046603.6.

[10] A. Dogariu, A. Kuzmich, H. Cao and L. J. Wang, "Superluminal light pulse propagation via rephasing in a transparent anomalously dispersive medium," Optics Express, Vol. 8, No. 6, Mar. 2001, pp. 344–350.

[11] H. Cao, A. Dogariu and L. J. Wang, "Negative group delay and pulse compression in superluminal pulse propagation," IEEE J. Sel. Top. Quant. Electron., Vol. 9, No. 1, Feb. 2003, pp. 52–58.

[12] R. Y. Chiao, "Faster than light propagations, negative group delays, and their applications," The Physics of Communication Proc. of the XXII Solvay Conference on Physics, DOI No. 10.1142, Nov. 2001, pp. 287–314.

[13] M. Kitano, T. Nakanishi and K. Sugiyama, "Negative group delay and superluminal propagation: An electronic circuit approach," IEEE J. Sel. Top. Quant. Electron., Vol. 9, No. 1, Apr. 2003, pp. 43–51.

[14] T. Nakanishi, K. Sugiyama and M. Kitano, "Demonstration of negative group delays in a simple electronic circuit," Am. J. Phys., Vol. 70, No. 11, Oct. 2002, pp. 1117–1121.

[15] J. N. Munday and R. H. Henderson, "Superluminal time advance of a complex audio signal," Appl. Phys. Lett., Vol. 85, July 2004, pp. 503–504.

[16] D. Solli, R. Y. Chiao and J. M. Hickmann, "Superluminal effects and negative delays in electronics, and their applications," Phys. Rev. E, Vol. 66, No. 5, Nov. 2002, pp. 056601.1–056601.4.

[17] C. D. Broomfield and J. K. A. Everard, "Broadband negative group delay networks for compensation of oscillators using feedforward amplifiers," Electron. Lett., Vol. 36, No. 23, Sept. 2000, pp. 1710–1711.

[18] N. S. Bukhman and S. V. Bukhman, "On the negative delay time of a narrow-band signal as it passes through the resonant filter of absorption," Radiophys. Quan. Electron., Vol. 47, No. 1, 2004, pp. 66–76.

[19] G. V. Eleftheriades, O. Siddiqui and A. K. Iyer, "Transmission line models for negative refractive index media and associated implementations without excess resonators," IEEE Microw. Wireless Comp. Lett., Vol. 13, No. 2, Feb. 2003, pp. 51–53.

[20] O. F. Siddiqui, M. Mojahedi and G. V. Eleftheriades, "Periodically loaded transmission line with effective negative refractive index and negative group velocity," IEEE Trans. Antennas Propagat., Vol. 51, No. 10, Oct. 2003, pp. 2619–2625.

[21] O. F. Siddiqui, S. J. Erickson, G. V. Eleftheriades and M. Mojahedi, "Time-domain measurement of negative group delay in negative-refractive-index transmission-line metamaterials," IEEE Trans. Microwave Theory Techn., Vol. 52, No. 5, May 2004, pp. 1449–1454.

[22] G. Chaudhary and Y. Jeong, "Tunable center frequency negative group delay filter using coupling matrix approach," IEEE Microw. Wireless Compon. Lett., Vol. 27, No. 1, 2017, pp. 37–39.

[23] G. Liu and J. Xu, "Compact transmission-type negative group delay circuit with low attenuation," Electron. Lett., Vol. 53, No. 7, Mar. 2017, pp. 476–478.

[24] B. Ravelo, "Investigation on microwave negative group delay circuit," Electromagnetics, Vol. 31, No. 8, Nov. 2011, pp. 537–549.
[25] B. Ravelo, Étude des circuits analogiques numériques NGD et leurs applications: Théorie fondamentale et expérimentations des circuits analogiques et numériques à temps négatif et leurs applications, (in French), Éditions Universitaires Européennes (EUE), Ed. by M. Fabre, 432 Pages, 13 Jan. 2012.
[26] B. Ravelo, Theory and Design of Analogue and Numerical Elementary NGD Circuits: Theoretical Characterization of Analogue and Numerical NGD Circuits, LAP Lambert Academic Publishing, Germany, 352 pages, 14 Mar. 2012.
[27] B. Ravelo, "Methodology of elementary negative group delay active topologies identification," IET Circuits Devices Syst. (CDS), Vol. 7, No. 3, May 2013, pp. 105–113.
[28] B. Ravelo, "Similitude between the NGD function and filter gain behaviours," Int. J. Circ. Theor. Appl., Vol. 42, No. 10, Oct. 2014, pp. 1016–1032.
[29] C.-T. M. Wu and T. Itoh, "Maximally flat negative group delay circuit: A microwave transversal filter approach," IEEE Trans. on Microwave Theory and Techniques, Vol. 62, No. 6, June 2014, pp. 1330–1342.
[30] M. Kandic and G. E. Bridges, "Asymptotic limits of negative group delay in active resonator-based distributed circuits," IEEE Transactions on Circuits and Systems I: Regular Papers, Vol. 58, No. 8, Aug. 2011, pp. 1727–1735.
[31] M. Kandic and G. E. Bridges, "Transient-imposed limitations of negative group delay circuits," in Proc. of 14th International Symposium on Antenna Technology and Applied Electromagnetics & the American Electromagnetics Conference (ANTEM-AMEREM), 2010, Ottawa, ON, Canada, 5–8 July 2010, pp. 1–4.
[32] S. Keser and M. Mojahedi, "Broadband negative group delay microstrip phase shifter design," in Proc. of IEEE Ant. Prop. Soc. Int. Symp. 2009 (APSURSI'09), Charleston, SC, 1–5 June 2009, pp. 1–4.
[33] H. Noto, K. Yamauchi, M. Nakayama and Y. Isota, "Negative group delay circuit for feed-forward amplifier," IEEE Int. Microw. Symp. Dig., Honolulu, Hawaii, Jun. 2007, pp. 1103–1106.
[34] H. Choi, Y. Jeong, C. D. Kim and J. S. Kenney, "Efficiency enhancement of feedforward amplifiers by employing a negative group delay circuit," IEEE Trans. Microw. Theory Tech., Vol. 58, No. 5, May 2010, pp. 1116–1125.
[35] H. Choi, Y. Jeong, C. D. Kim and J. S. Kenney, "Bandwidth enhancement of an analog feedback amplifier by employing a negative group delay circuit," Prog. Electromag. Res., Vol. 105, 2010, pp. 253–272.
[36] S. S. Oh and L. Shafai, "Compensated circuit with characteristics of lossless double negative materials and its application to array antennas," IET Microw. Antennas Propag., Vol. 1, No. 1, Feb. 2007, pp. 29–38.
[37] W. Alomar, and A. Mortazawi, "Method of generating negative group delay in phase arrays without using lossy circuits," Proc. IEEE Int. Wireless Symp. (IWS) 2013, Beijing, China, 14–18 Apr. 2013, pp. 1–4.

[38] T. Eudes and B. Ravelo, "Cancellation of delays in the high-rate interconnects with UWB NGD active cells," Applied Physics Research, Vol. 3, No. 2, Nov. 2011, pp. 81–88.

[39] B. Ravelo, "Recovery of microwave-digital signal integrity with NGD circuits," Photonics and Optoelectronics (P&O), Vol. 2, No. 1, Jan. 2013, pp. 8–16.

[40] B. Ravelo and A. K. Jastrzebski, "NGD circuit using a microwave amplifier for the signal integrity improvement," Proc. of 2012 Asia-Pacific Symposium on Electromagnetic Compatibility (APEMC 2012), Singapore, 21–24 May 2012, pp. 1–4.

[41] B. Ravelo, "Baseband NGD circuit with RF amplifier," Electron. Lett., Vol. 47, No. 13, June 2011, pp. 752–754.

[42] B. Ravelo, "Synthesis of RF Circuits with Negative Time Delay by Using LNA," Adv. Electromagn. (AEM), Vol. 2, No. 1, Feb. 2013, pp. 44–54.

[43] F. Wan, N. Li, B. Ravelo, Q. Ji, B. Li and J. Ge, "The design method of the active negative group delay circuits based on a microwave amplifier and an RL-series network," IEEE ACCESS (Early Access), 2018, pp. 1–10.

Chapter 8
Limits of passive and active negative group delay distributed circuits

Miodrag Kandic[1] *and Greg E. Bridges*[2]

Abstract

The chapter examines the negative group delay (NGD) limits of active and passive distributed element transmission line circuits. The first section illustrates NGD characteristic examples in both frequency and time domains, and pertinent figure of merits and trade-offs which can be used to evaluate performance of different NGD devices. The relationship between the time-domain transients and out-of-band gain will be analyzed and discussed in the context of modulated signals. The second section introduces resonator-based circuit NGD topologies and shows how active elements are employed for gain compensation. In the third section, the asymptotic limits of these circuits are then derived while considering issues of gain, bandwidth, in-band distortion, and out-of-band gain. The final section will extend the theory to NGD phenomena in general linear causal media, derived from Kramers–Kronig relations. Experimental results of fabricated designs demonstrating the concepts will be put forward throughout the chapter.

Keywords

Distributed circuit, NGD limitation, passive and active circuits.

8.1 Introduction

An example of input and output waveforms of an NGD-exhibiting network is illustrated in Figure 8.1(a), whereas the network itself is represented by its frequency-domain *S*-parameters, as shown in Figure 8.1(b). In this example, the input waveform is an amplitude modulated signal, where the modulation, or the waveform envelope, is a Gaussian pulse with a temporal standard deviation significantly larger than the period of the sinusoidal carrier. The NGD effect is represented by

[1]AC/DC System Planning Group, Manitoba Hydro, Canada
[2]Department of Electrical and Computer Engineering, University of Manitoba, Canada

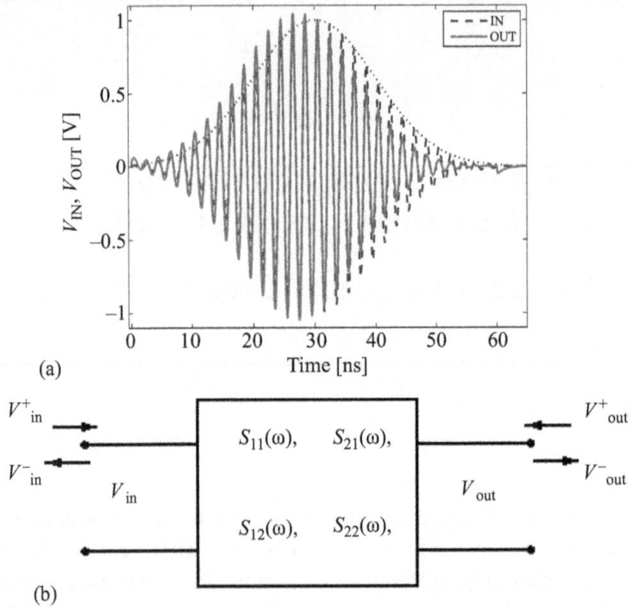

Figure 8.1 (a) An example of waveforms at the input and output of a network exhibiting NGD and (b) network's S-parameter representation

the time-advance of the peak of the output waveform envelope, relative to the input waveform envelope peak. The phase advance or delay of the sinusoidal carrier has no consequence on the NGD.

8.1.1 Frequency-domain analysis

Typical frequency domain transfer function (parameter $S_{21}(\omega)$ amplitude, phase, group delay) plots of a medium exhibiting NGD around a center frequency ω_0 are illustrated in Figure 8.2.

Generally, any transfer function exhibiting NGD has its amplitude characteristic minimum at the center frequency [1–3], and the particular example shown in Figure 8.2 is fully gain-compensated so its minimum value is $A = 1$, or 0 dB. On the other hand, a passive, noncompensated NGD medium would have an amplitude characteristic offset to the negative decibel region ($A_0 = -20$ dB, -40 dB, -60 dB for the three examples shown), but the phase and group delay characteristics would ideally remain the same. Further, an NGD medium-phase characteristic has a positive slope at the center frequency as evident from Figure 8.2(b), and consequently the group delay is negative at this frequency, as evident from Figure 8.2(c).

The particular transfer function illustrated in Figure 8.2 is a rational second-order function given by

$$S_{21}(\omega) = \frac{V^-_{out}}{V^+_{in}} = A_0 A \frac{\omega^2 - j\Delta\omega \times \omega - \omega_0^2}{\omega^2 - jA\Delta\omega \times \omega - \omega_0^2}, \tag{8.1}$$

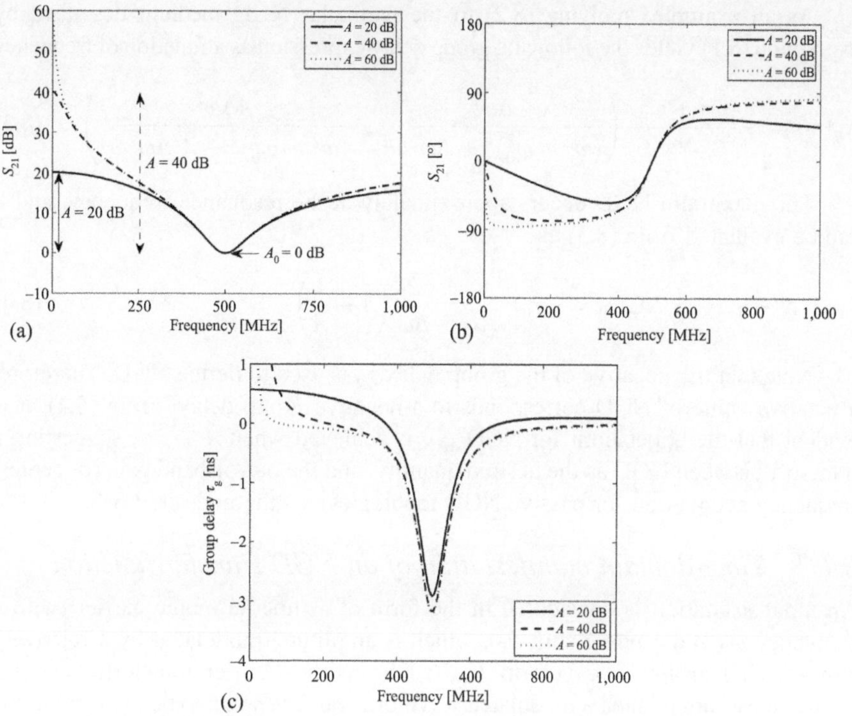

Figure 8.2 Amplitude (a), phase (b), and group delay (c) transfer function characteristics of an NGD medium with a center frequency of 500 MHz, 3-dB bandwidth of 100 MHz, 0 dB center frequency attenuation, and out-of-band gain values of 20, 40, and 60 dB

where A is the out-of-band gain (given in magnitude, not dB), $\Delta\omega$ is approximately the 3-dB bandwidth, ω_0 is the center frequency, A_0 is the center-frequency magnitude, and V^{-}_{out} and V^{+}_{in} are the output transmitted and input reflected waves, respectively, corresponding to S-parameters labeling in Figure 8.1(b). This is also the simplest possible NGD transfer function around a nonzero center frequency, whereas more complex examples (which would also exhibit larger NGD values) would be higher-order rational functions, or an exponential function describing a continuous or distributed physical medium, such as Lorentzian medium [4].

The phase characteristic and the corresponding group delay for any transfer function are given by [5]

$$\phi(\omega) = \arctan\left(\frac{\text{Im}(S_{21})}{\text{Re}(S_{21})}\right), \quad \tau_g(\omega) = -\frac{d\phi(\omega)}{d\omega}. \tag{8.2}$$

As an example, applying (8.2) to the particular NGD medium described by expression (8.1) yields the following group delay expression as a function of frequency

$$\tau_g(\omega) = -\frac{(\omega^2+\omega_0^2)}{\Delta\omega}\left(\frac{\Delta\omega^2}{(\omega^2-\omega_0^2)^2+\Delta\omega^2\omega^2} - \frac{A\Delta\omega^2}{(\omega^2-\omega_0^2)^2+A^2\Delta\omega^2\omega^2}\right). \tag{8.3}$$

The maximum NGD occurs approximately at the resonance frequency, and it can be evaluated from (8.3) as

$$\mathrm{NGD_{MAX}} \approx \mathrm{NGD_0} = -\tau_g\Big|_{\omega=\omega_0} = \frac{2}{\Delta\omega}\left(1-\frac{1}{A}\right). \tag{8.4}$$

Note that the negative of the group delay τ_g is used to define NGD. Therefore, a positive value of NGD corresponds to a negative group delay. From (8.4), it is evident that the upper limit for $\mathrm{NGD_{MAX}}$ is achieved when $A \to \infty$, suggesting a trade-off between NGD as the desired quantity, and the out-of-band gain (or center-frequency attenuation for passive NGD topologies), as the undesired one.

8.1.2 Time-domain manifestation of an NGD transfer function

An input excitation is considered in the form of a time-harmonic carrier (with a frequency ω_0 and a phase shift θ_0), which is amplitude-modulated by a relatively slow-varying arbitrary waveform $U_{\mathrm{env}}(t)$. From the Fourier transform, the frequency spectrum of such a modulated waveform can be shown to be identical to the "baseband" spectrum of the modulating waveform $U_{\mathrm{env}}(t)$, except that it is shifted to the carrier frequency ω_0. The input waveform can then be expressed by inverse-Fourier transform integral as

$$u_{IN}(t) = U_{env}(t)\cos(\omega_0 t+\theta_0) = \frac{1}{\pi}\mathrm{Re}\left\{\int_0^\infty U_{env}(\omega-\omega_0)e^{j\theta_0}e^{j\omega\cdot t}d\omega\right\}. \tag{8.5}$$

Further, an assumption is made that within the frequency bandwidth of interest the phase characteristic changes approximately linearly with frequency, and the amplitude characteristic is close to constant, as

$$\phi(\omega) \approx \phi_0 + (\omega-\omega_0)\frac{d\phi(\omega)}{d\omega}\Big|_{\omega=\omega_0}, \qquad A(\omega) \approx A(\omega_0) = \mathrm{const.} \tag{8.6}$$

The output waveform can now be described by

$$u_{OUT}(t) \approx A(\omega_0)\frac{1}{\pi}\mathrm{Re}\left\{\int_0^\infty \left[U_{env}(\omega-\omega_0)e^{j\theta_0}e^{j\phi_0-j\omega_0(d\phi/d\omega)|_0}\right]e^{j\omega(d\phi/d\omega)|_0}e^{j\omega\cdot t}d\omega\right\}, \tag{8.7}$$

which yields

$$u_{OUT}(t) \approx A(\omega_0)U_{env}\left(t+\frac{d\phi(\omega)}{d\omega}\Big|_{\omega=\omega_0}\right)\cos\left(\omega_0\left(t+\frac{\phi_0}{\omega_0}\right)+\theta_0\right). \tag{8.8}$$

It is evident from (8.8) that the output waveform retains the amplitude-modulated form, with its time-harmonic carrier shifted by the phase delay ($t_{ph} = -\phi_0/\omega_0$), while the modulating "envelope" part of the waveform is shifted by the group delay ($\tau_g = -d\phi/d\omega$, evaluated at ω_0). For baseband waveforms ($\omega_0 = 0$, and $\theta_0 = 0$ since phase shift has no meaning in this case), (8.8) is still valid and in this case $U_{env}(t) = U_{IN}(t)$.

Negative group delay phenomenon is demonstrated on an example in the time domain, as shown in Figure 8.3. The idealized amplitude characteristic of the medium is considered approximately constant within the bandwidth of interest, and the phase characteristic is considered approximately linear, as given by (8.6). The waveform shown in Figure 8.3 is a Gaussian pulse modulated sinusoidal carrier. The envelope of the output waveform seems to be advanced in time relative to the input envelope, thus exhibiting an NGD. The output sinusoidal carrier on the other hand is positively delayed by value t_{ph}, thus exhibiting a positive phase delay. The Gaussian envelope peak forms at the output even before the input envelope peak enters the medium. This phenomenon seems to be counter-intuitive and that it violates causality. However, the input waveform shown in Figure 8.3 does not have a definite turn-on time, so it presumably originated infinitely back in time and therefore does not carry any information [1–3].

Signals of practical interest, however, have a defined turn-on time. The information is contained in nonanalytical parts of the waveform [3], such as turn-on point or any point where the waveform value or any of its derivatives is not

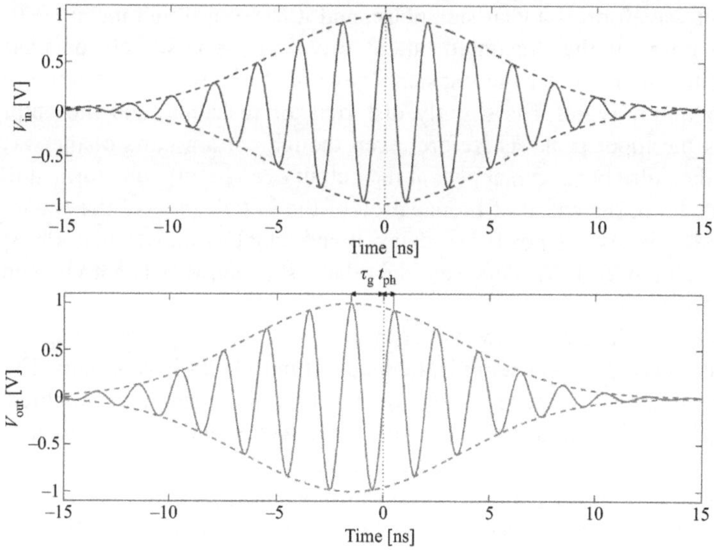

Figure 8.3 An idealized NGD-exhibiting output waveform for an amplitude modulated input waveform

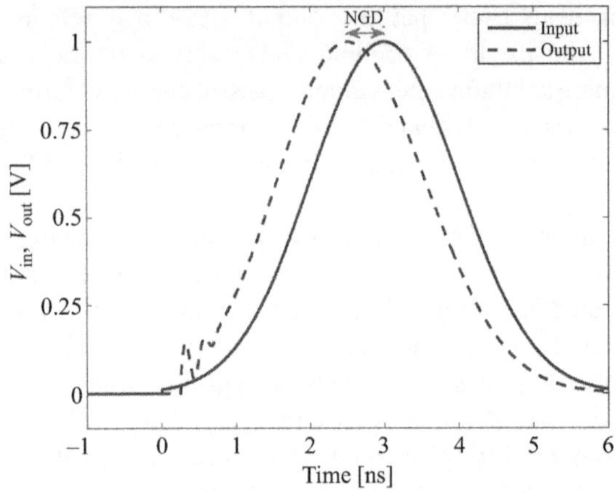

Figure 8.4 An illustration of input and output waveform envelopes in an NGD medium

continuous. Theoretically, the entire analytical part of a waveform could be determined when the value of the waveform and its derivatives are known at any point, by employing Taylor series. Therefore, waveforms of practical interest have nonanalytical points, such as turn-on/off points. Figure 8.4 shows an exaggerated illustration of such a practical waveform (or waveform envelope, if an amplitude modulated waveform are considered) at the input and output of an NGD medium. The input waveform is a Gaussian pulse, and it has a turn-on time at $t = 0$.

The onset of the illustrated output waveform is positively delayed and it contains transients. After the transients have decayed, the rest (steady-state part) of the output waveform is slightly distorted, or reshaped, and the output peak precedes the input peak. Therefore, even though certain parts of the waveform, such as the pulse peak, can appear at the output even before they form at the input of the medium, the speed of propagation of the initial onset of the waveform, or the "front," is always positively delayed and equal or slower than the speed of light in vacuum, c [1–3]. This way, the relativistic causality is not violated in this type of a medium.

In general, the output waveform can have a different amplitude magnitude than the input waveform, depending if the medium introduces loss or gain. The output waveform shape also depends on the level of the medium's amplitude/phase characteristics distortion within the NGD bandwidth, which translates into distortion of the steady-state part of the waveform.

It is also intuitive that the waveform propagating through an NGD medium needs to be distorted to a degree, if one part of the waveform is shifted in time by one value (turn-on point is delayed positively in the example above according to the "front" velocity), and another part of the waveform is delayed by a different value (peak delayed in the opposite direction in this case).

8.1.3 NGD metrics and trade-offs

In order to evaluate and effectively compare performances of different media or circuits exhibiting NGD, the following frequency domain characteristic properties of the medium need to be determined:

- **Amplitude-bandwidth (Δf)** of the medium is taken as the 3-dB-bandwidth of the frequency-domain amplitude characteristic, to ensure acceptable in-band amplitude distortion (which is proportional to time-domain distortion of the steady-state part of a pulse propagated through the medium). Note that the frequency band over which the group delay characteristic is negative could be larger than this 3-dB amplitude bandwidth. However, in that case the overall useful bandwidth over which the waveform distortion is acceptable is still the 3-dB amplitude bandwidth.
- **NGD** is the negative group delay value at center frequency (derivative of the frequency-domain phase characteristic, evaluated at center frequency), which is proportional to time-domain advance of the peak of a pulse (or the pulse envelope for amplitude modulated pulses—nonzero carrier/center frequency) propagated through the medium.
- **NGD-bandwidth product ($NGD\cdot\Delta f$)** represents a normalized NGD, which is proportional to time-domain advance of the peak of a pulse, relative to temporal duration/width of the pulse. This quantity is a representative measure of the magnitude of achieved NGD phenomenon in a particular medium, and typically it is the quantity that is set to be maximized in the NGD media design. For example, if a Gaussian pulse is propagated through the medium, with temporal standard deviation chosen such that its 6 standard deviations in frequency domain fits within the bandwidth Δf, and the pulse duration is 6 standard deviations in time (3 away from the peak on each side), then $\sigma_t = 1/\sigma_\omega = 1/(2\pi\sigma_f) = 6/(2\pi\Delta f) = T/6$, and $1/\Delta f = T/(18/\pi) \approx T/5.73$, and $NGD/T \approx NGD\cdot\Delta f/5.73$, which shows that $NGD\cdot\Delta f$ is indeed proportional to the relative pulse peak advancement in the time-domain. The proportional factor depends on the particular waveform pulse applied.
- **Out-of-band gain (A_{max})** is an undesired, trade-off quantity, which is inherent to all media exhibiting NGD. It is the ratio of the maximum out-of-band value, and the center frequency value of the amplitude characteristic (inherently, $A_{max} > \sqrt{2}$). If the amplitude characteristic is given in dB, then the out-of-band gain is determined as a difference of those values, instead of the ratio (inherently, $A_{max} > 3$ dB). This trade-off quantity is generally set to be minimized in a design, since, as it will be shown, it is proportional to the amplitude of time-domain transients, when a pulse with finite duration (defined turn-on/off times) is propagated through the medium.
- **In-band phase (and therefore group delay) distortion** is proportional to the time-domain distortion of the steady-state part of a pulse propagated through the medium. Note that for linear causal media, the phase characteristic is not independent of the amplitude characteristic. Instead, they are linked through

Kramers–Kronig relations [3,4], and this will be discussed further and made use of in Section 8.4. One way to quantify this distortion is the product of the bandwidth and the maximum in-band group delay variation ($\Delta\tau_{\text{in-max}}\cdot\Delta f$). For example, an acceptable level of phase distortion can be selected to match the value for a first-order low-pass filter ($\Delta\tau_{\text{in-max}}\cdot\Delta f \leq 1/(2\pi)$, or $\Delta\tau_{\text{in-max}}\cdot\Delta\omega \leq 1$, since for this filter $\tau(\omega) = \omega_c/(\omega_c^2 + \omega^2)$, and $\Delta\omega = 2\omega_c$, so $\Delta\tau_{\text{in-max}} = \tau(0) - \tau(\omega_c) = 1/(2\omega_c) = 1/\Delta\omega$). Ultimately, the distortion level in the time-domain is the true test, and it will depend on the particular waveform used and its frequency spectral density. Due to the nonexact nature of quantifying an acceptable level of in-band phase distortion in the frequency domain, phase/group distortion of NGD topologies presented in this chapter will not be examined in detail. It is noted however, that all topologies presented in this chapter satisfy the above defined $\Delta\tau_{\text{in-max}}\cdot\Delta\omega \leq 1$ criterion, and by visual inspection of time-domain results obtained for Gaussian waveforms, the 3-dB bandwidth in-band phase distortion seems acceptable (amplitude characteristic distortion was dominant, and since it is kept under 3-dB, phase distortion had no significant additional impact).

With the parameters above determined, a frequency domain performance figure of merit of an NGD medium can be evaluated as the ratio of desired, and the trade-off quantities as

$$\text{NGD}_{\text{merit}} = \frac{\text{NGD} \times \Delta f}{\text{A}_{\text{max-dB}}}, \tag{8.9}$$

with units for this figure of merit expressed in (s·Hz/dB) or just (1/dB). In-band phase distortion is a requirement that needs to be met just like 3-dB amplitude distortion, so its actual value is not used in the merit figure of calculation (8.9).

8.1.3.1 NGD figure of merit for media with same out-of-band gains, but different NGD

This example is given for a baseband waveform, for simplicity. The baseband versions of various nonzero center-frequency transfer functions exhibiting NGD will be discussed in more detail in subsequent sections. In this example, medium 1 is a first-order high-pass filter, and medium 2 is comprised of five buffered stages of identical first-order high-pass filters. More on NGD-exhibiting circuit topologies will be discussed in subsequent sections. Frequency-domain amplitude and group delay characteristics for the two media, as well as time-domain Gaussian pulse propagation example, respectively, are illustrated in Figure 8.5.

Frequency-domain figure of merits and trade-offs for these two examples are given in Table 8.1, whereas pertinent time-domain quantities are given in Table 8.2.

Both media have out-of-band gain $A_{\max} = 35$ dB and $\Delta f = \Delta\omega/(2\pi) = 2 \cdot 1/(2\pi) = 1/\pi$ (Hz), in the frequency domain. The input signal used in the time domain is a Gaussian pulse with $T = 6 \cdot \sigma_t = 6/\sigma_\omega = 6/(2\pi\sigma_f) = 6/(2\pi(\Delta f/6)) = 18$ s. However, medium 2 (a fifth-order high-pass filter) outperforms medium 1 (a first-order high-pass filter) since for the same out-of-band gain (undesired trade-off quantity), it achieves a higher NGD-bandwidth product.

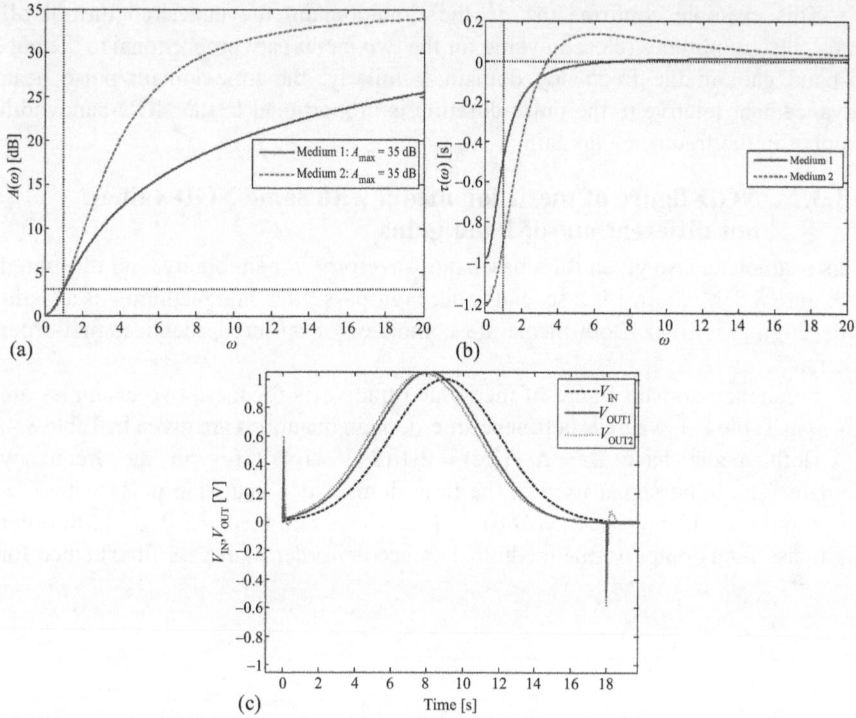

Figure 8.5 Time-domain example for NGD media with the same out-of-band gain, but different NGD

Table 8.1 Frequency-domain figure of merits and trade-offs, for example, for NGD media with the same out-of-band gain, but different NGD

Medium	**NGD_{freq} (s)**	**$NGD_{freq} \cdot \Delta f$**	**A_{max} (dB)**	**Merit (1/dB)**
Medium 1	0.983	0.313	35	0.0089
Medium 2	1.215	0.387	35	0.0111

Table 8.2 Time-domain figure of merits and trade-offs, for example, for NGD media with the same out-of-band gain, but different NGD

Medium	**NGD_{time} (s)**	**NGD_{time}/T_{pulse}**	**Transient amplitude**
Medium 1	0.886	0.0492	0.521
Medium 2	1.174	0.0652	0.592

This example confirms that in the time-domain, the undesired turn-on/off transients' amplitudes (close in value for the two media) are proportional to the out-of-band gain in the frequency domain. Similarly, the time-domain pulse peak advancement relative to the pulse duration is proportional to the NGD-bandwidth product in the frequency domain.

8.1.3.2 NGD figure of merit for media with same NGD values, but different out-of-band gains

This example is also given for a baseband waveform, for simplicity, and illustrated in Figure 8.6. Medium 1 is a second-order high-pass filter, and medium 2 is a sixth-order high-pass filter. Both media are comprised of buffered, identical first-order filters.

Frequency-domain figure of merits and trade-offs for these two examples are given in Table 8.3, whereas pertinent time-domain quantities are given in Table 8.4.

Both media have $\Delta f = \Delta\omega/(2\pi) = 2\cdot 1/(2\pi) = 1/\pi$ (Hz), in the frequency domain. The input signal used in the time domain is a Gaussian pulse with $T = 6\cdot\sigma_t = 6/\sigma_\omega = 6/(2\pi\sigma_f) = 6/(2\pi(\Delta f/6)) = 18$s. However, medium 2 (a sixth-order high-pass filter) outperforms medium 1 (a second-order high-pass filter) since for

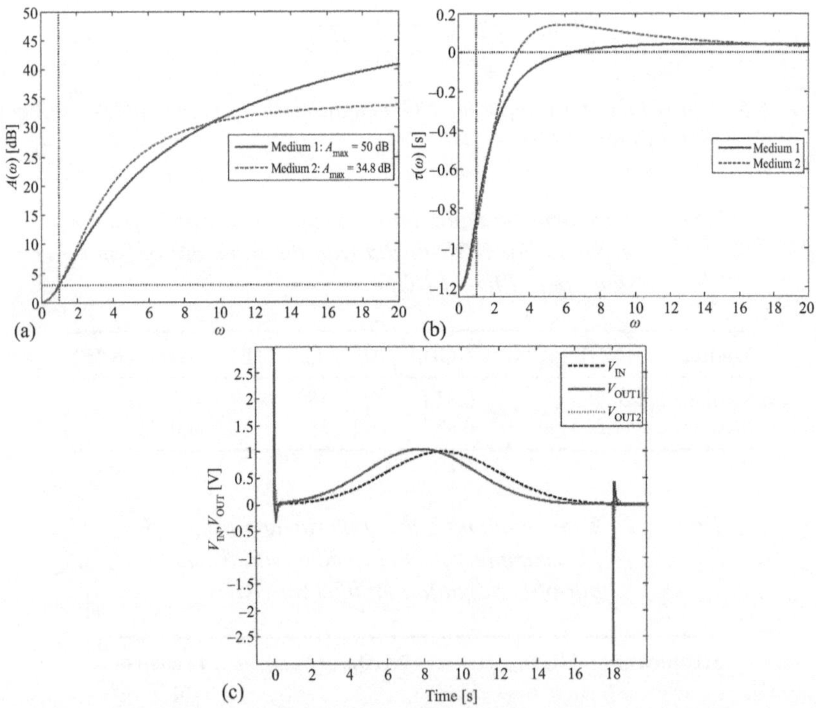

Figure 8.6 Time-domain example for NGD media with the same NGD values, but different out-of-band gains

Table 8.3 Frequency-domain figure of merits and trade-offs, for example, for NGD media with the same NGD values, but different out-of-band gains

Medium	NGD_{freq} (s)	$NGD_{freq} \cdot \Delta f$	A_{max} (dB)	Merit (1/dB)
Medium 1	1.218	0.388	50	0.0078
Medium 2	1.218	0.388	34.8	0.0111

Table 8.4 Time-domain figure of merits and trade-offs, for example, for NGD media with the same NGD values, but different out-of-band gains

Medium	NGD_{time} (s)	NGD_{time}/T_{pulse}	Transient amplitude
Medium 1	1.159	0.0644	2.946
Medium 2	1.174	0.0652	0.579

the same NGD-bandwidth product it has a lower out-of-band gain (undesired trade-off quantity).

This example also confirms that in the time-domain, the undesired turn-on/off transients' amplitudes are proportional to the out-of-band gain in the frequency domain: $20 \times \log(v_{\mathrm{trans}_1}/v_{\mathrm{trans}_2}) \approx 14.1$ dB, and $A_{\max_1}/A_{\max_2} = 50$ dB $-$ 34.8 dB $=$ 15.2 dB. Similarly, the time-domain pulse peak advancement relative to the pulse duration (close values for two media) is proportional to the NGD-bandwidth product in the frequency domain.

8.2 Single-stage resonator-based NGD circuit topologies

In the previous section, a second-order rational transfer function exhibiting NGD was introduced as given by (8.1). In this section, different circuit topologies that can achieve this transfer function will be examined. Further, impedance matching of these topologies and cascading issues for multi-stage designs will be discussed as well.

Circuit topologies comprised of a single resonator and arbitrary resistive matching networks can be described by the same function (8.1), given by [6]

$$S_{21}(\omega) = A_0 A \frac{\omega^2 - \omega_0^2 - j\frac{1}{Q}\omega_0\omega}{\omega^2 - \omega_0^2 - j\frac{1}{Q}A\omega_0\omega}, \tag{8.10}$$

where Q is the ratio of center frequency and the 3-dB amplitude bandwidth ($Q = \omega_0/\Delta\omega$), A is the ratio of the maximum and minimum values of the transfer function amplitude (out-of-band gain), and A_0 is a the amplitude characteristic value minimum, which occurs at the center frequency, $S_{21}(\omega_0) = A_0$. As shown in Section 8.1, the maximum NGD occurs approximately at the resonance/center frequency, and it is given by

$$\mathrm{NGD_{MAX}} \approx \mathrm{NGD_0} = -\tau_g\Big|_{\omega=\omega_0} = \frac{2Q}{\omega_0}\left(1-\frac{1}{A}\right) = \frac{1}{\pi}\frac{1}{\Delta f}\left(1-\frac{1}{A}\right). \quad (8.11)$$

Different resonator-based circuit topologies will be analyzed next, showing that those are all particular cases of the generic expression given by (8.1).

8.2.1 Unmatched parallel RLC resonator circuit

This circuit topology is assembled by simply connecting a parallel RLC resonator to a voltage source with impedance Z_0 on one end, and to a load on the other, as shown in Figure 8.7.

Applying basic circuit theory, the input reflection coefficient for the circuit shown in Figure 8.7, and assuming that the load is matched to the source ($Z_0 = R_L$), can be expressed as

$$S_{11}(\omega) = \frac{Z_{res}(\omega)}{2Z_0 + Z_{res}(\omega)}, \quad (8.12)$$

where $Z_{res}(\omega)$ is the frequency-dependent parallel RLC resonator impedance. The circuit's transfer function, or transmission coefficient $S_{21}(\omega)$, can be shown to fit the generic form (8.10), with parameters A_0 and A given by

$$A_0 = \frac{2Z_0}{2Z_0 + R}, \qquad A = 1 + \frac{R}{2Z_0} = \frac{1}{A_0}, \qquad Q = \frac{R}{\sqrt{L/C}}. \quad (8.13)$$

The transmission and reflection coefficient magnitude plots for a selected case with $Q = 5$ and $R = 250\ \Omega$, as well as the transmission phase plot, are shown in Figure 8.8.

The minimum transmission magnitude (maximum loss), A_0, occurs at the resonance frequency, when the resonator has a maximum impedance of R. At the same time, the largest negative group delay (tangent slope on the phase

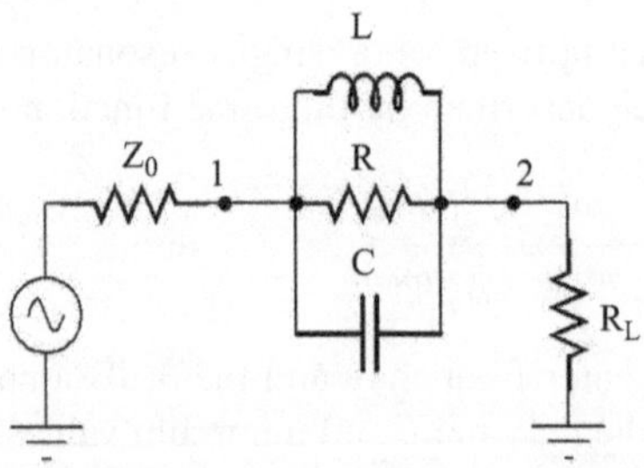

Figure 8.7 An unmatched parallel RLC-resonator NGD circuit

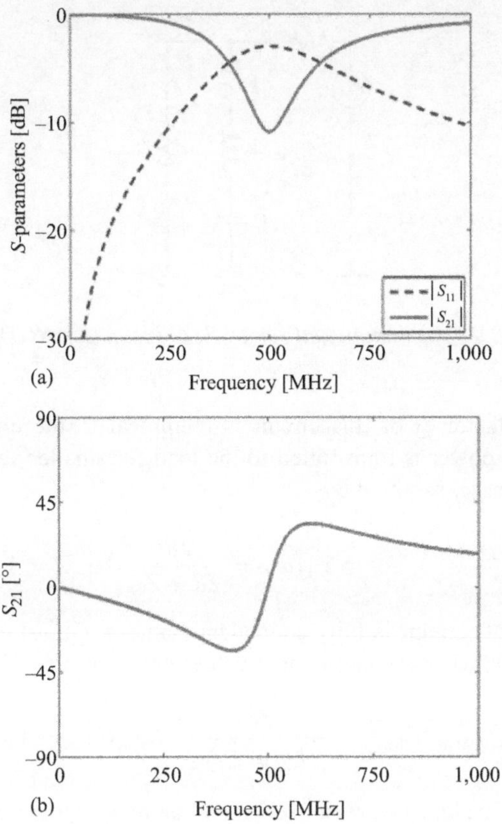

Figure 8.8 Unmatched parallel RLC circuit (a) input reflection and transmission magnitude and (b) transmission phase plots for $Q = 5$, $R = 250\ \Omega$

characteristic) occurs approximately at the resonance frequency as well, and can be determined from (8.11). From this expression, it is clear that there is a trade-off between the negative group delay, and the power dissipated in the resonator. The closer the transmitted power at resonance, A_0, is to zero (or to negative infinity in decibels), the larger the NGD is. This result is expected since the resonator is connected in series with the source and load impedances, so the higher the resonator impedance is (more loss), the more it will dominate the series connection and consequently yield more NGD. The output waveform envelope will therefore be attenuated, in addition to being advanced in time. The extent to how much NGD can be pushed is limited by how much loss can be tolerated (or compensated for through a cascaded amplifier), in a given application.

8.2.2 Unmatched series RLC resonator circuit

This circuit topology has a series RLC circuit connected in parallel with the load and the source, as shown in Figure 8.9.

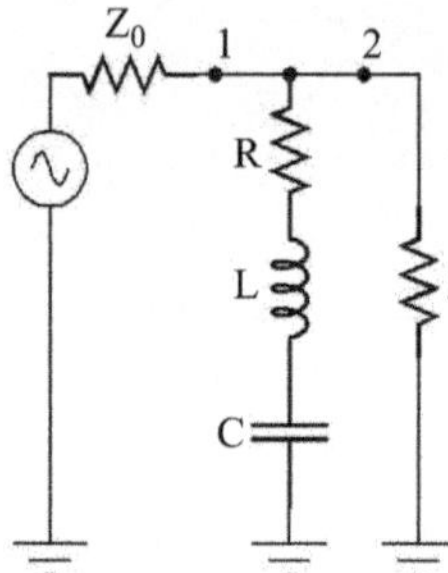

Figure 8.9 An unmatched series RLC-resonator NGD circuit

The quality factor Q of this circuit is reciprocal to the one of the parallel circuit. Also, less power is transmitted to the load for smaller values of resonator resistance in this case, as given by

$$Q_{SER} = \frac{\sqrt{L/C}}{R}, \qquad A_0 = S_{21}(\omega_0) = \frac{2R}{2R + Z_0}, \qquad A = 1 + \frac{Z_0}{2R}. \tag{8.14}$$

The series RLC circuit is fully equivalent (same A, Q, ω_0) to the parallel one from Section 8.2.1, when the component values are chosen as

$$R_{SER} = \frac{Z_0^2}{R_{PAR}}, \ \sqrt{L_{SER}/C_{SER}} = \frac{Z_0^2}{\sqrt{L_{PAR}/C_{PAR}}}, \ L_{SER}C_{SER} = L_{PAR}C_{PAR}. \tag{8.15}$$

The same equivalency between series and parallel topologies is observed for the examples to follow, and therefore only the parallel RLC circuit versions will be presented, to avoid repetition.

8.2.3 Resistive-matched RLC resonator circuit

The unmatched parallel or series topologies presented in Sections 8.2.1 and 8.2.2, respectively, cannot be effectively cascaded to produce higher-order transfer functions and proportionally cumulative NGD. For example, a circuit with two cascaded, unmatched, identical parallel resonators with component values R, L, and C is equivalent to a single parallel resonator circuit with values $2R$, $2L$ and $C/2$, which has the same Q as individual resonators, and approximately twice the out-of-band gain, $2A$. The factor of increase in NGD is $(1 - 1/(2A))/(1 - 1/A)$ in this case according to (8.4), which is not significant for larger values of A. For example, for an individual resonator $R = 18Z_0 = 900\ \Omega$, which yields $A = 10$, two cascaded unmatched resonators would have an increase in NGD by a factor of only $0.95/0.9 = 1.056$, at the expense of doubling the out-of-band gain.

In order to achieve effective cascading of RLC resonators to yield a more substantial NGD increase, a matching of individual stages to the system impedance is needed. This will be analyzed in more detail in Section 8.3. In this section,

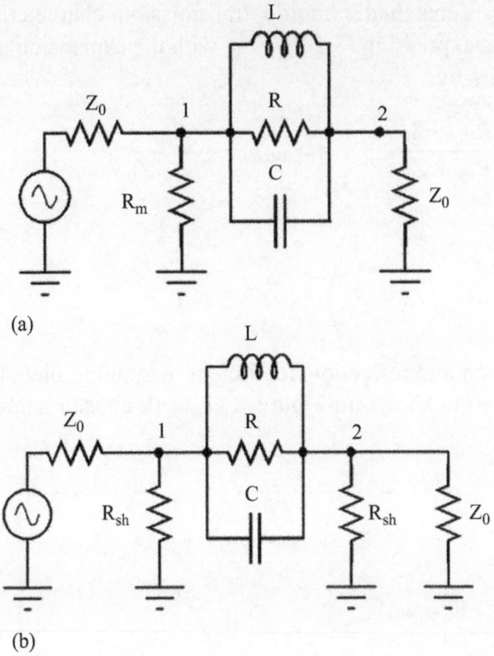

Figure 8.10 Parallel RLC-resonator NGD circuits with resistive matching, (a) at the source end, R-matched, and (b) at both ends, π-matched (©2011 IEEE. Reprinted, with permission, from [6])

two different approaches for resistive matching will be discussed, R-matching and π–matching.

A circuit consisting of the same parallel resonator from Section 8.2.1 with the addition of a matching resistor R_m at the input, will be referred to as R-matched topology here, and is shown in Figure 8.10(a). Similar topology, but having matching resistors at both ends will be referred to as π-matched, and is shown in Figure 8.10(b). The π-matched topology is symmetric, and two-port matched at the resonance frequency.

The R-matched circuit's transmission characteristic is described by the generic expression (8.10), with the expression parameters related to circuit parameters by

$$A_{0\_\text{R\_match}} = S_{21}(\omega_0) = \frac{Z_0}{Z_0 + R}, \qquad A_{\text{R\_match}} = \frac{2}{Z_0} \frac{(R + Z_0)^2}{3R + 2}, \tag{8.16a}$$

and the required matching resistor value is given by

$$R_m = Z_0\left(1 + \frac{Z_0}{R}\right). \tag{8.16b}$$

Similarly, the π-matched circuit's transmission characteristic is described by the same generic expression (8.10) again, with the expression parameters related to circuit parameters by

$$A_{0\_\pi\_\text{match}} = \frac{R_{\text{sh}} - Z_0}{R_{\text{sh}} + Z_0}, \qquad A_{\pi\_\text{match}} = \frac{R_{\text{sh}}}{R_{\text{sh}} - Z_0}, \tag{8.17a}$$

where

$$R_{\text{sh}} = \frac{Z_0}{R}\left(Z_0 + \sqrt{Z_0^2 + R^2}\right). \tag{8.17b}$$

The transmission and reflection coefficient magnitude plots for both resistive-matched topologies are shown in Figure 8.11, with chosen parameters $Q = 5$ and

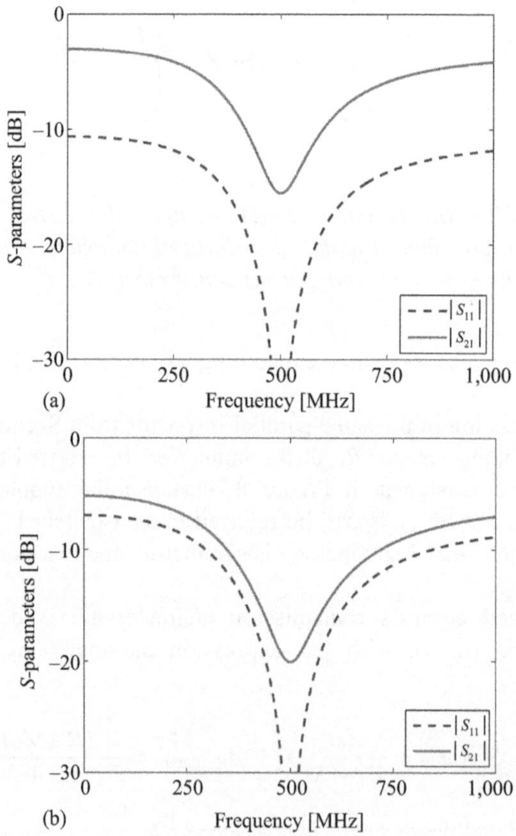

Figure 8.11 Input reflection S_{11} and transmission S_{21} magnitude plots for parallel RLC circuit with $Q = 5$, $R = 250\ \Omega$. (a) R-matched topology with $R_m = 60\ \Omega$ and (b) π-matched topology with $R_{sh} \approx 61\ \Omega$

$R = 250\ \Omega$. Resulting matching resistors for the two topologies are $R_m = 60\ \Omega$ and $R_{sh} \approx 61\ \Omega$, obtained from (8.16b) and (8.17b), respectively.

It can be noted that the transmission coefficient magnitude value at low and high frequencies (when the resonator impedance is essentially zero) is not 100% (or zero dB), as in the unmatched topology, due to the power to ground dissipation through the matching resistor(s). The limit value of the low- and high-frequency end-transmission coefficient value, for a large out-of-band gain, is given by

$$\lim_{A\to\infty} S_{21}(0) = \lim_{A\to\infty} S_{21}(\infty) = \lim_{A\to\infty} (A_0 A) = \begin{cases} 1, & \text{for unmatched} \\ \frac{2}{3}, & \text{for R} - \text{matched} \\ \frac{1}{2}, & \text{for } \pi\text{−matched} \end{cases}. \tag{8.18}$$

Comparing the resistive-matched circuits with the unmatched one, it can be noted that even though the two circuits are matched at the input, so no power is reflected back to the source at the resonance frequency, overall less power is transmitted to the load for the same value of the resonator resistor *R*. This is due to more power being lost in the circuit itself, through the matching shunt to ground resistor(s). On the other hand, unlike the unmatched topology from Section 8.2.1, this topology allows for a cascaded design with a compounded increasing NGD effect, as it will be discussed in Section 8.3.

8.2.4 Gain-compensated RLC resonator circuit

Resonator-based circuits discussed in Sections 8.2.2 and 8.2.3 are passive and contain resistive elements. Therefore, they exhibit attenuation, which is largest at the center frequency. In general, passive media exhibiting anomalous wave propagation such as superluminal or NGD propagation is accompanied by an attenuation/loss at the NGD center frequency [1–3].

As depicted in Figure 8.12, an amplifier can be cascaded with a resonator-based circuit to gain-compensate for this attenuation. As a result, the amplitude characteristic will ideally be scaled (or upshifted if given in decibels) over the entire frequency range, such that at the center frequency the amplification is 1 (or 0 dB), and the highest out-of-band amplification will equal *A*. Some examples of gain-compensated NGD circuits are reported in [6–8]. As it will be discussed in Section 8.3, gain-compensating amplifiers can also serve as buffers for cascaded multi-staged designs, and an example of actual gain-compensated circuit will be presented.

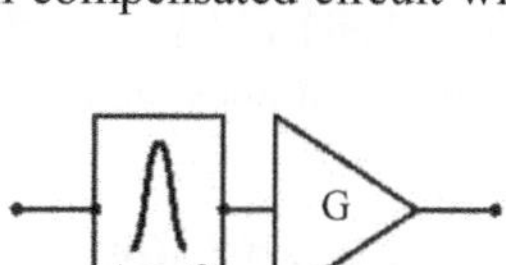

Figure 8.12 Schematic of a generic resonator-based gain-compensated NGD circuit

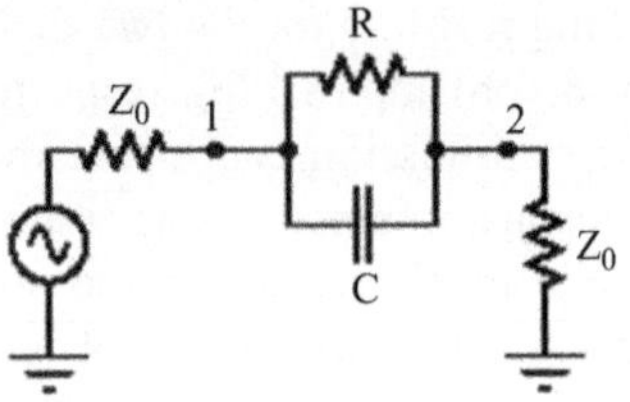

Figure 8.13 *Unmatched RC-based high-pass filter exhibiting NGD in the baseband*

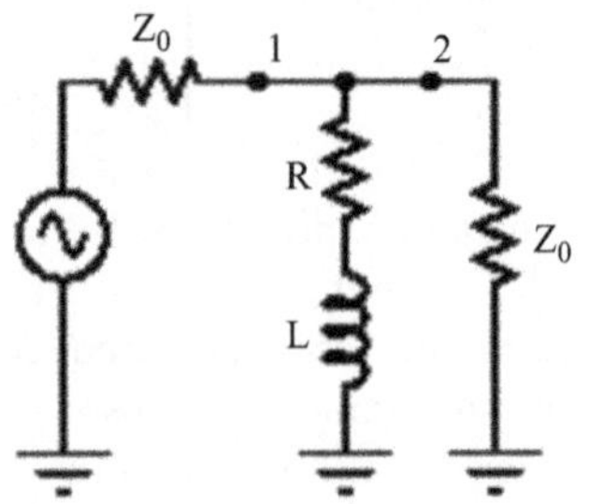

Figure 8.14 *Unmatched RL-based high-pass filter exhibiting NGD in the baseband*

8.2.5 Baseband RC and RL high-pass circuits

Resonator-based NGD topologies discussed so far can be modified to exhibit similar characteristics at baseband (center frequency of zero), by replacing the RLC resonators by either parallel RC, or series RL elements, as shown in Figures 8.13 and 8.14, respectively. Some examples of baseband NGD circuits are reported in [9–11].

The transmission coefficient of the single-stage baseband NGD circuits is a first-order variation of (8.10), and it can be derived as

$$S_{21}(\omega) = A_0 A \frac{\omega - j\dfrac{\Delta\omega}{2}}{\omega - j \cdot A\dfrac{\Delta\omega}{2}}, \tag{8.19}$$

where $\Delta\omega$ is the bandwidth given by either $\Delta\omega = 1/(RC)$ or $\Delta\omega = R/L$ for the circuits in Figures 8.13 and 8.14, respectively. All expressions, derived for A and A_0 for unmatched, R-matched, and π-matched baseband topologies, are identical to those of the corresponding resonator-based topologies presented in Sections 8.2.2–8.2.4.

8.3 NGD distributed circuits and limits

A single-stage NGD circuit can be represented by a two-port electrical network, which in turn can be described by a number of different parameters, such as

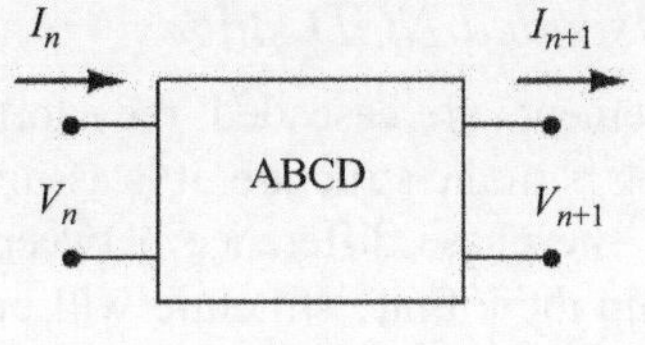

Figure 8.15 ABCD matrix representation of a network

impedance (*Z*), admittance (*Y*), scattering (*S*), transmission (ABCD), parameters/matrices, and others [5].

The ABCD matrix, for example, relates the input port voltage and current to the output voltage and current, which is illustrated in Figure 8.15 and given by [5]

$$\begin{bmatrix} V_n \\ I_n \end{bmatrix} = \begin{bmatrix} A & B \\ C & D \end{bmatrix} \cdot \begin{bmatrix} V_{n+1} \\ I_{n+1} \end{bmatrix} \tag{8.20}$$

The ABCD matrix analysis is preferred for cascaded networks, since the overall matrix of the system is obtained by a simple multiplication of individual cascaded element matrices. The overall cascaded design matrix can then easily be converted into any other network parameters, such as *S*-parameters, through conversion expressions. For example, the transmission coefficient characteristic S_{21} is obtained [5] as

$$S_{21} = \frac{2}{A + B + C + D}. \tag{8.21}$$

The generic transmission characteristic (frequency-dependent complex ratio of output and input steady-state sinusoidal waveforms) for circuit topologies comprised of a single resonator and arbitrary resistive matching networks is given by [6]:

$$S_{21}(\omega) = K \frac{\omega^2 - \omega_0^2 - j\frac{1}{Q}\omega_0\omega}{\omega^2 - \omega_0^2 - j\frac{1}{Q}A\omega_0\omega} \tag{8.22}$$

where *A* is the ratio of the maximum and minimum values of the transfer function amplitude (out-of-band gain), and *K* is a constant which takes into account optional gain-compensation ($S_{21}(\omega_0) = K/A$). As shown in Section 8.2, the maximum NGD occurs approximately at the resonance frequency, and it is given by

$$\mathrm{NGD_{MAX}} \approx \mathrm{NGD_0} = -\tau_g\Big|_{\omega=\omega_0} = \frac{2Q}{\omega_0}\left(1 - \frac{1}{A}\right) = \frac{1}{\pi}\frac{1}{\Delta f}\left(1 - \frac{1}{A}\right) \tag{8.23}$$

From (8.23), it is evident that an upper limit for $\mathrm{NGD_{MAX}}$ is achieved when $A \to \infty$, and the NGD-bandwidth product of a single-stage resonator circuit has an upper limit of

$$\mathrm{NGD_{MAX}} \cdot \Delta f \leq \frac{1}{\pi}. \tag{8.24}$$

8.3.1 Nonbuffered cascaded NGD stages

When identical network elements are cascaded, they form a periodic structure. In the case of an infinitely long periodic structure, it is assumed that for a steady-state sinusoidal applied voltage, the phase difference between the input and output of any cascaded element within the infinite structure will be the same [5]. Similarly, the input to output voltage amplitude ratio will be the same for every element, as well as voltage to current ratio at the input or output of each element, which is referred to as the Bloch impedance (according to Bloch's theorem). Applying the described conditions to (8.20), it can be shown [5] that the transmission coefficient for an N-element block of an infinitely long periodic structure is given by

$$e^{N\gamma} = \left[e^{\alpha+j\beta}\right]^N = \left[\frac{A+D}{2} \pm \frac{1}{2}\sqrt{(A-D)^2+4BC}\right]^N, \tag{8.25}$$

$$Z_B = Z_0\frac{B}{e^{\alpha+j\beta}-A} = Z_0\frac{e^{\alpha+j\beta}-D}{C}, \tag{8.26}$$

where A, B, C, and D are transmission matrix parameters of individual (and identical) cascaded elements, and Z_B is the Bloch's impedance of the infinite structure. In essence, through expressions (8.25) and (8.26), an infinitely long periodic structure is described as a medium with a propagation constant γ and a characteristic impedance Z_B.

For example, applying (8.25) and (8.26) to an infinitely long periodic structure comprised of π-matched NGD elements presented in Section 8.2.3 yields propagation parameters given by

$$e^{\gamma} = A - \sqrt{BC} = 1 + \frac{Z}{R_{sh}} - \sqrt{\frac{2Z}{R_{sh}} + \frac{Z^2}{R_{sh}^2}}, \qquad Z_B = Z\Big/\sqrt{\frac{2Z}{R_{sh}} + \frac{Z^2}{R_{sh}^2}}, \tag{8.27}$$

where Z is the resonator frequency-dependent impedance, and R_{sh} is the shunt matching resistance for this configuration, given by (8.17b).

If a periodic structure has a finite number of cascaded elements, N, and it is connected at both ends (source and load) to a system with characteristic impedance Z_0, which in a general case is different from the structure's Bloch's impedance, Z_B, the transmission coefficient of the overall network is then given by [5]

$$S_{21} = \frac{(1-\Gamma^2)e^{N\gamma}}{1-\Gamma^2 e^{N\gamma}}, \qquad \Gamma = \frac{Z_B - Z_0}{Z_B + Z_0}, \tag{8.28}$$

where Γ is the incident reflection coefficient between the outside system and the periodic structure. If the system impedance is matched to the periodic structure impedance (the incident reflection coefficient $\Gamma = 0$), then the overall transmission coefficient S_{21} in (8.28) reduces to $e^{N\gamma}$, as in the case of an infinite structure.

How well the periodic structure's impedance Z_B is matched to the system impedance Z_0 over a frequency band, depends on the structure topology. An example of the impedance mismatch of a periodic structure comprised of π-matched NGD elements is shown in Figure 8.16.

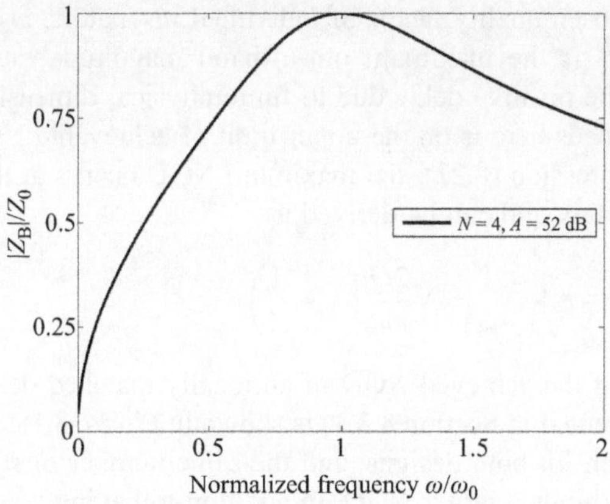

Figure 8.16 Bloch impedance of a cascaded four-stage π-matched NGD circuit with overall A = 52 dB

This particular structure's impedance is always matched at the resonance, and the mismatch increases away from this frequency. How small the mismatch is over the entire frequency band will depend on the number of stages, as well as the overall out-of-band gain.

8.3.2 Buffered cascaded NGD stages and limits

As discussed in Section 8.2, any resistive matched resonator-based NGD circuit will have the same second-order transfer function form given by (8.10). In order to ensure ideal matching between cascaded NGD elements, an ideal amplifier can be cascaded with each element, as depicted in Figure 8.17. In other words, the amplifier needs to provide an ideal output-to-input isolation (ideally unilateral), and needs to have an output impedance matched to the system impedance within the entire frequency band.

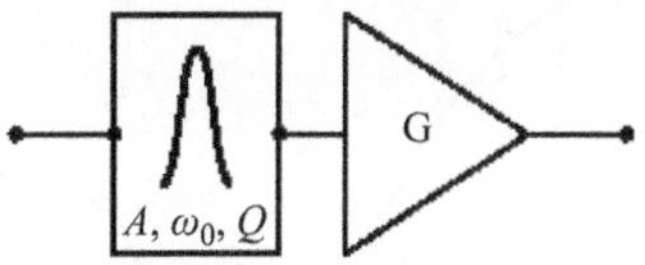

Figure 8.17 Schematic of a generic resonator-based gain-compensated NGD circuit

For this generic, ideally matched gain-compensated case, the transmission coefficient of a cascaded N-stage design is given [6] by

$$S_{21}(\omega) = A^N \left[\frac{\omega^2 - \omega_0^2 - j\frac{1}{Q}\omega_0\omega}{\omega^2 - \omega_0^2 - j\frac{1}{Q}A\omega_0\omega} \right]^N, \tag{8.29}$$

where Q is the chosen quality factor of individual resonators, ω_0 is the resonance frequency, and A^N is the maximum out-of-band magnitude ratio of the overall N-stage design. The positive delay due to finite physical dimensions is ignored in (8.29), since the focus here is on the upper limit of achievable NGD.

Similar to expression (8.22), the maximum NGD occurs at the resonance frequency approximately and can be derived as

$$\mathrm{NGD_{MAX}} = -\tau_g\Big|_{\omega=\omega_0} = N\frac{2Q}{\omega_0}\left(1 - \frac{1}{A}\right). \tag{8.30}$$

Comparison of the achieved NGD of an ideally matched design and the non-buffered one (presented in Section 8.3.1) is shown in Figure 8.18. The same out-of-band gain is chosen for both designs, and the same number of stages. The ideally matched design exhibits a larger NGD, in addition to having a simpler expression

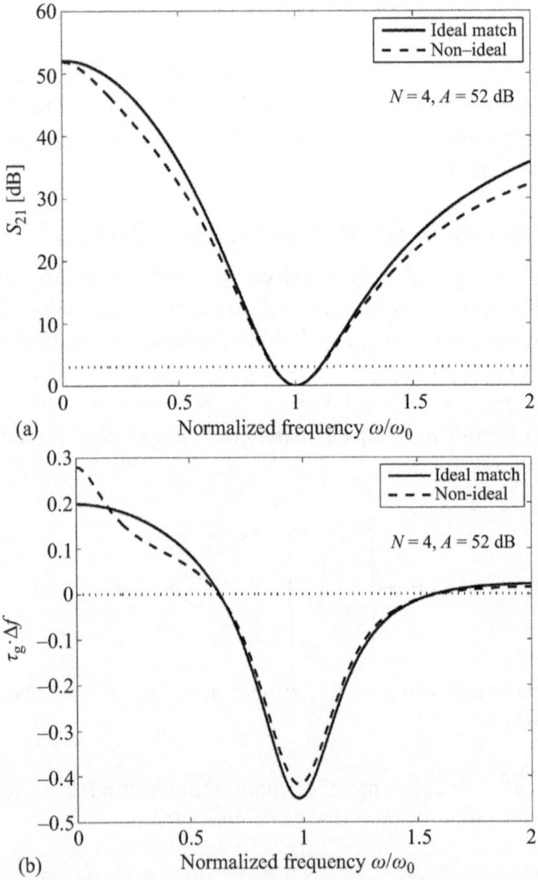

Figure 8.18 (a) Transmission coefficient and (b) group delay comparison of buffered and nonbuffered matched cascaded NGD circuits

for the transmission coefficient. Therefore, since one of the questions addressed in this chapter deals with establishing the upper limit of achievable NGD for a given out-of-band gain, only the ideally matched case will be considered for the remainder of this section.

In the multistage design, if the Q's of individual resonators are kept constant as the number of stages increases, the overall Q will increase (and thus the 3-dB bandwidth will decrease). This effect is demonstrated in Figure 8.19, for ideally matched/buffered cascaded designs with individual stages having $A = 20$ dB and $Q = 5$.

The achieved NGD is linearly increasing with the number of stages in this case (Figure 8.19(c)), but the bandwidth is decreasing (Figure 8.19(a)). It will be shown that the NGD is increasing at the higher rate than the rate of the bandwidth decrease, so the overall NGD-bandwidth product still increases with number of stages, but at a slower than linear rate.

In order to maintain a constant bandwidth for the overall N-stage design, the individual stage Qs need to be adjusted properly. As evident from Figure 8.19, the individual stage Qs will need to be lower than the overall Q-factor, which will

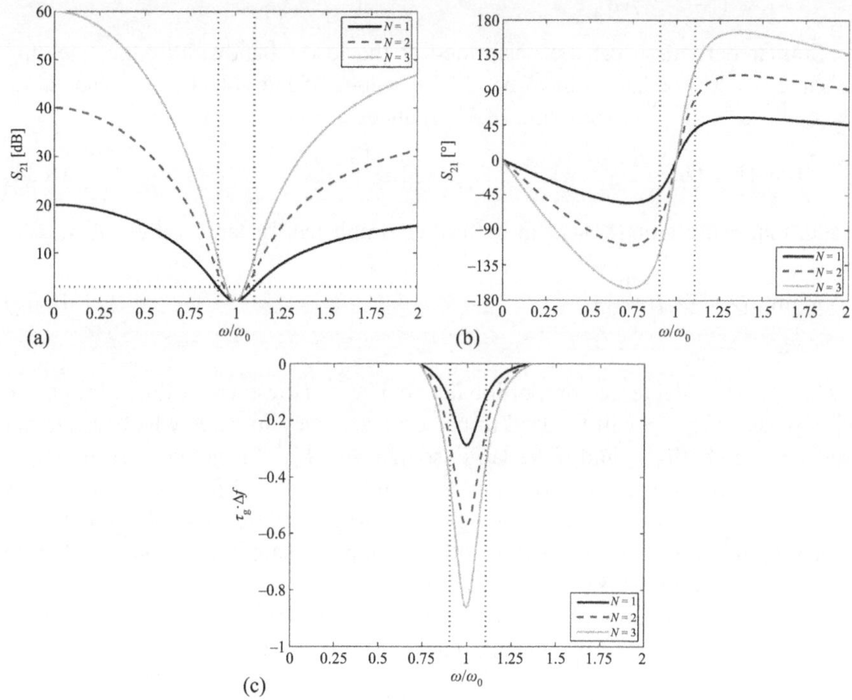

Figure 8.19 (a) Transmission coefficient; (b) transmission phase; and (c) group delay for ideally matched cascaded designs with N = 1, 2, and 3 stages, with individual stage A = 20 dB and individual Q-factors unadjusted (Q = 5)

be called Q_{tot}. The condition to maintain a 3 dB N-stage transmission coefficient ratio from the upper band-edge to the center frequency, can be written as [6]

$$\left|S_{21}\left(\omega_0+\frac{\Delta\omega}{2}\right)\Big/S_{21}(\omega_0)\right|^2=\left|S_{21}\left(\omega_0+\frac{\omega_0}{2Q_{tot}}\right)\Big/S_{21}(\omega_0)\right|^2=2. \quad (8.31)$$

Substituting S_{21} from (8.29) into (8.31), and assuming that the overall N-stage bandwidth is much smaller than the center frequency, $1/(2Q_{tot}) << 1$, yields [6]

$$\frac{1+\left(\dfrac{Q}{Q_{tot}}\right)^2}{1+\dfrac{1}{A^2}\left(\dfrac{Q}{Q_{tot}}\right)^2}\approx 2^{1/N} \quad (8.32)$$

Finally, by introducing a variable for the overall N-stage out-of-band gain, $A_{tot}=A^N$, and solving (8.32), the approximate expression for individual stage Q-factor adjustment required to maintain the overall bandwidth can be obtained as [6]

$$Q\approx\frac{Q_{tot}\sqrt{2^{1/N}-1}}{\sqrt{1-2^{1/N}/A_{tot}^{2/N}}}. \quad (8.33)$$

Similar derivation can be performed at the lower bandwidth edge, yielding the same result. For large out-of-band gain values of individual stage resonators, $A=A_{tot}^{1/N}>>2^{1/(2N)}$, expression (8.33) reduces to

$$\lim_{A\to\infty}Q=Q_{tot}\sqrt{2^{1/N}-1}. \quad (8.34)$$

In addition, expression (8.34) can be further simplified for large values of N, as

$$\lim_{\substack{A\to\infty\\ N\gg 1}}Q\approx Q_{tot}\frac{\sqrt{\ln 2}}{\sqrt{N}}, \quad (8.35)$$

in which case Q decreases proportionally to $1/N^{1/2}$. However, in this section, the full expression (8.33) will be used to preserve accuracy in cases where A_{tot} is not large, or when both A_{tot} and N are large, so that $A=A_{tot}^{1/N}$ is not necessarily large.

In order to validate the derived expression (8.33), the Q-factors of three cases from Figure 8.19 are adjusted accordingly, and the results are shown in Figure 8.20. Evidently, the overall Q_{tot} and bandwidths are approximately the same for all three cases, thus validating (8.33).

The N-stage maximum NGD-bandwidth product for the maintained bandwidth case can now be obtained by substituting (8.33) into (8.30), as [6]

$$\mathrm{NGD_{MAX}}\Delta f=\frac{1}{\pi}N\sqrt{2^{1/N}-1}\frac{1-1/A_{tot}^{1/N}}{\sqrt{1-2^{1/N}/A_{tot}^{2/N}}}. \quad (8.36)$$

For $N=1$ and $A\to\infty$, expression (8.36) yields the same limit of $1/\pi$, as given in Section 8.2. For validation purposes, the NGD-bandwidth values of designs

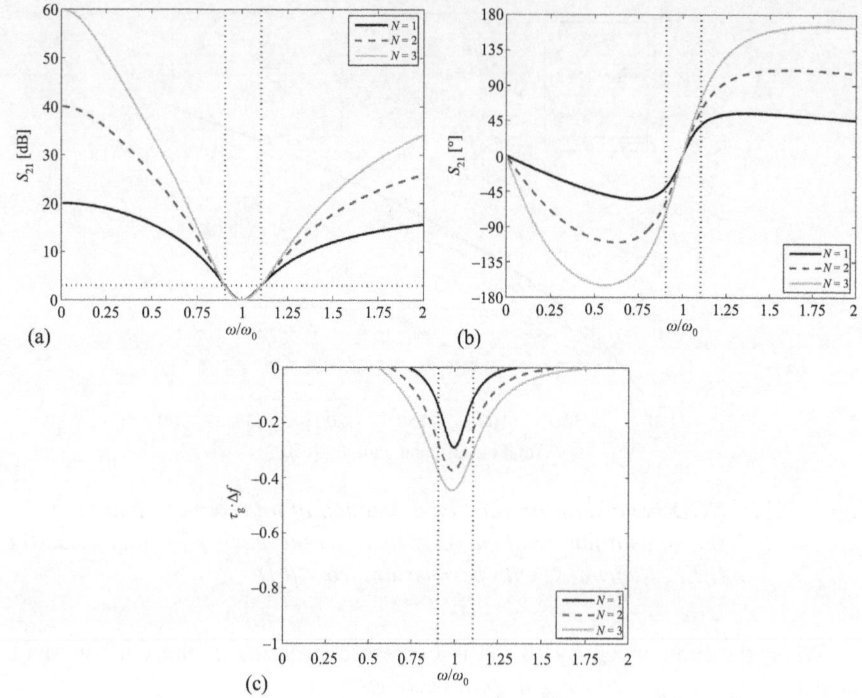

Figure 8.20 (a) Transmission coefficient; (b) transmission phase; and (c) group delay for ideally matched cascaded designs with $N = 1$, 2, and 3 stages, with individual stage $A = 20$ dB and individual Q-factors adjusted to yield same overall $Q_{tot} = 5$

reported in [7,8], for example, have been extracted and found to fit within the upper limit given by (8.36).

The maximum NGD-bandwidth product as a function of total out-of-band gain A_{tot} is plotted in Figure 8.21, as a family of curves with different number of stages N [6]. As expected, NGD is a monotonically increasing function of both A_{tot} and N. The out-of-band gain A_{tot} is an undesirable quantity since it is directly related to transient amplitudes of the circuit response to finite duration signals, as discussed in Section 8.1. Therefore, there is a trade-off between the increasing NGD and A_{tot}.

The family of curves in Figure 8.21 has an asymptotic limit for NGD-bandwidth product, as the number of stages $N \to \infty$, and the total out-of-band gain A_{tot} is kept as a finite variable. The asymptotic limit for this infinitely distributed resonator-based medium can be derived from (8.36), as [6]

$$\lim_{N\to\infty} \mathrm{NGD_{MAX}}\Delta f = \frac{1}{\pi}\sqrt{\ln 2}\frac{\ln(A_{\mathrm{tot}})}{\sqrt{2\ln(A_{\mathrm{tot}}) - \ln 2}} \approx \frac{1}{\pi}\frac{\sqrt{\ln 2}}{\sqrt{2}}\sqrt{\ln(A_{\mathrm{tot}})}, \qquad (8.37)$$

where the final expression in (8.37) was further simplified for large out-of-band gain values, $A_{\mathrm{tot}} >> \sqrt{2}$.

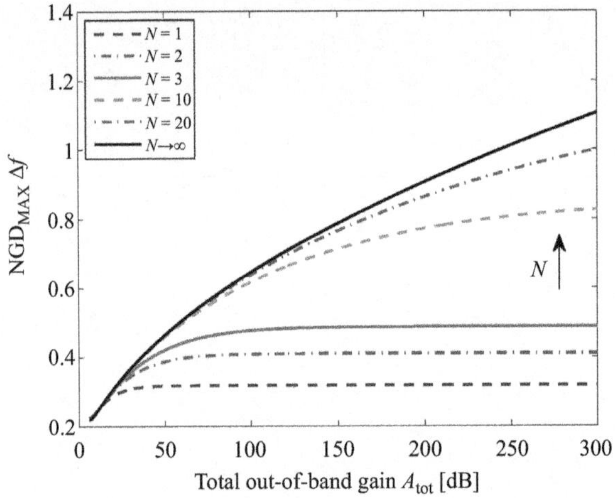

Figure 8.21 NGD-bandwidth product as a function of total out-of-band gain, shown as a family of curves with different number of stages (©2011 IEEE. Reprinted, with permission, from [6])

When the limit given by (8.37) is expressed in terms of the out-band-gain decibel value, $A_{\text{tot-dB}} = 20 \cdot \log(A_{\text{tot}})$, it becomes

$$\lim_{N\to\infty} \text{NGD}_{\text{MAX}}\Delta f \approx \frac{1}{\pi}\frac{\sqrt{\ln 2 \cdot \ln 10}}{\sqrt{40}}\sqrt{A_{\text{tot-dB}}} \approx 0.0636\sqrt{A_{\text{tot-dB}}}. \tag{8.38}$$

Similarly, an asymptotic limit as a function of number of stages can be derived from (8.36), by keeping N finite, and letting $A_{\text{tot}} \to \infty$, as

$$\lim_{A_{\text{tot}}\to\infty} \text{NGD}_{\text{MAX}}\Delta f = \frac{1}{\pi}N\sqrt{2^{1/N}-1} \approx \frac{\sqrt{\ln 2}}{\pi}\sqrt{N}, \tag{8.39}$$

where the final expression in (8.39) was further simplified for cases with large number of stages, $N >> 1$.

From (8.38) and (8.39), it is evident that achievable NGD-bandwidth product is proportional to the square root of $\ln(A_{\text{tot}})$, or square root of the out-of-band decibel value $A_{\text{tot-dB}}$, as well as to the square root of number of stages, respectively. It can alternatively be said that the magnitude of the total out-of-band gain, A_{tot}, increases exponentially with the square of NGD (A_{tot} is a reciprocal Gaussian function of NGD). This particular formulation can come in handy when discussing transients for Gaussian modulated signals with finite duration (defined turn-on/off times) [6].

From Figure 8.21, it can be noted that for a given NGD value, cases with larger number of stages, N, yield smaller out-of-band gain values, A_{tot}. This is also depicted in the frequency domain plots in Figure 8.22, for a few selected cases. This result shows that having a larger number of stages (distributed case) is preferred, since it minimizes the undesirable out-of-band gain trade-off, for a given NGD-bandwidth product.

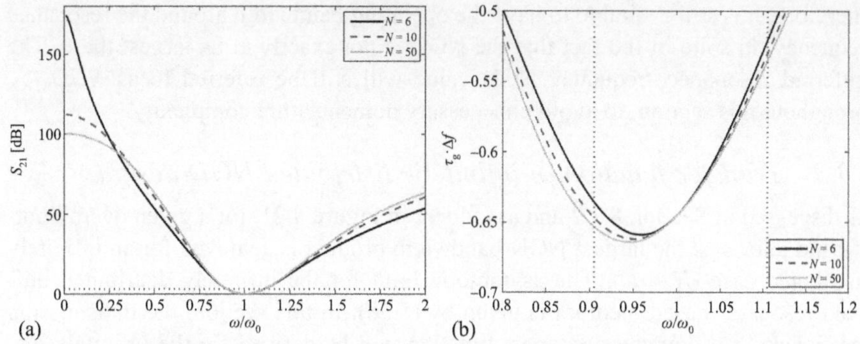

Figure 8.22 (a) Transmission coefficient and (b) group delay plots for a fixed NGD-bandwidth product value, and different number of stages N

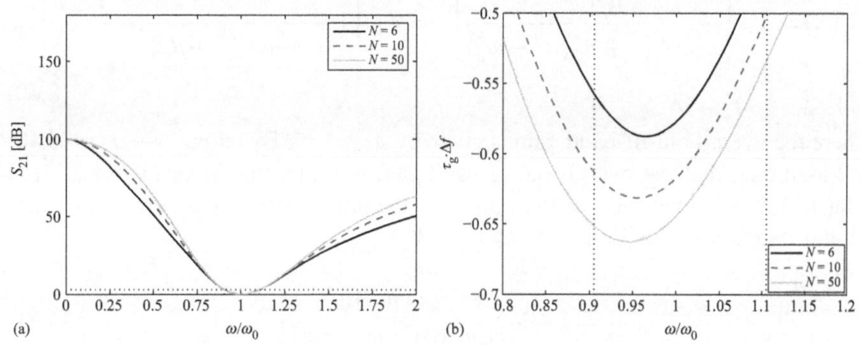

Figure 8.23 (a) Transmission coefficient and (b) group delay plots for a fixed A_{tot} value, and number of stages N

Similarly, the preference for larger number of stages can be demonstrated for a case when the out-of-band value is fixed, and the achieved NGD is to be maximized. From Figure 8.21 it can be observed again that the larger N values achieve this, which is also depicted in the frequency domain plots in Figure 8.23, for a few selected cases. As expected, a larger number of stages (distributed case) maximizes the NGD-bandwidth, for a given out-of-band gain.

From Figures 8.22(b) and 8.23(b), it can be noted that the maximum NGD does not occur exactly at the resonance frequency, but at progressively lower frequency values as the number of stages increases. The minimum transmission magnitude value, however, stays at the resonance frequency for all cases. The drifting NGD maximum is a consequence of resonator's asymmetric frequency response, which gets more pronounced for lower Q values of individual resonators, which is exactly the case with increasing number of stages, as given by (8.35). However, trying to operate at frequencies where the NGD is exactly at its largest value would correspond to regions of larger transmission magnitude swings, as evident from Figures 8.22(a) and 8.23(a), which would considerably reduce the 3 dB bandwidth.

Therefore, it is still desirable to have the operating bandwidth around the resonance frequency, in spite of the fact that the NGD is not exactly at its largest there. The preferred resonance frequency NGD value will still be referred to as NGD_{MAX} throughout this section, to avoid unnecessary nomenclature complexity.

8.3.3 Transfer function of infinitely distributed NGD design

As discussed in Section 8.3.2 and as evident in Figure 8.21, for a given overall out-of-band gain A_{tot}, the largest NGD-bandwidth product is achieved for an infinitely distributed case ($N \to \infty$). The asymptotic limit for the infinitely distributed buffered resonator-based medium is given by (8.38). In this section, the transmission coefficient $S_{21}(\omega)$ expression approximation will be derived for the infinitely distributed medium from Section 8.3.2. First, the original transfer function for the N-stage circuit (8.29) can be rewritten as

$$S_{21}(\omega) = A_{tot}\left[\frac{\omega_0\omega + jQ(\omega^2 - \omega_0^2)}{\omega_0\omega A_{tot}^{1/N} + jQ(\omega^2 - \omega_0^2)}\right]^N = A_{tot}\left[1 + \frac{\omega_0\omega\left(1 - A_{tot}^{1/N}\right)}{\omega_0\omega A_{tot}^{1/N} + jQ(\omega^2 - \omega_0^2)}\right]^N, \tag{8.40}$$

where the overall out-of-band gain is given by $A_{tot} = A^N$. By letting $N \to \infty$ in (8.40) and assuming that the overall out-of-band gain is finite, the factor $(1 - {A_{tot}}^{1/N}) \approx -\ln(A_{tot})/N$ becomes infinitesimal and the overall expression becomes an exponential function in this limit case as given by

$$\lim_{N\to\infty} S_{21}(\omega) = \lim_{N\to\infty} A_{tot} \exp\left[\frac{-\omega_0\omega \ln(A_{tot})}{\omega_0\omega + jQ(\omega^2 - \omega_0^2)}\right]. \tag{8.41}$$

The individual stage Q-factor in (8.41) is now substituted by (8.33) and its limit expression found by letting $N \to \infty$, which yields [6]

$$\lim_{N\to\infty} S_{21}(\omega) = A_{tot} \exp\left[\ln(A_{tot})\frac{jb\omega}{\omega^2 - jb\omega - \omega_0^2}\right], \tag{8.42a}$$

where

$$b = \Delta\omega\sqrt{\frac{2\ln(A_{tot})}{\ln 2} - 1}, \tag{8.42b}$$

and the overall distributed circuit bandwidth is given by $\Delta\omega = \omega_0/Q_{tot}$. Expression (8.42) is a connection between an NGD multistage buffered resonator-based discrete circuit model and its continuous medium equivalent in the limit case. The phase characteristic can be extracted from (8.42a) as the imaginary part of the exponential argument, as

$$\lim_{N\to\infty} \phi(\omega) = \ln(A_{tot})\frac{b\omega(\omega^2 - \omega_0^2)}{(\omega^2 - \omega_0^2)^2 + b^2\omega^2}, \tag{8.42c}$$

which has a $1/\omega$ dependency at large frequencies, which will be referenced to in Section 8.4. The group delay characteristic can be obtained by a negative derivative of (8.42c), and has a $1/\omega^2$ dependency at large frequencies. Finally, using the group delay value at the resonance frequency, the same NGD-bandwidth product vs. total out-of-band expression as given by the asymptotic limit (8.38) can be obtained, as expected.

8.3.4 Examples of 1 and 2-stage NGD circuits

The particular active gain-compensated resonator-based circuit chosen, exhibiting NGD of a band-limited signal centered at its resonance frequency is shown in Figure 8.24. NGD is achieved through the parallel RLC resonator in the input branch, while the attenuation is compensated through the proper choice of the feedback resistor value $R_F = 2(R + R_s)$. The input impedance is matched to the system impedance Z_0 at the resonance frequency through the resistor R_m, while the output impedance is matched through the resistor R_{OUT}, which makes the circuit cascadable. The design details and component values can be found in [6].

The transfer function of the circuit shown in Figure 8.24 is just a special case of the general formula (8.22), assuming an ideal amplifier, $G = -1$, and neglecting

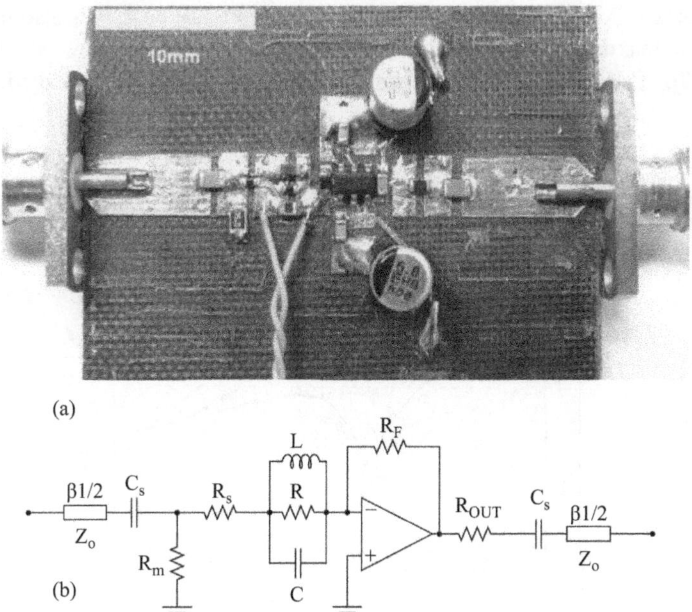

Figure 8.24 A fabricated single-stage, gain-compensated NGD circuit (a) and equivalent circuit model (b) (©2011 IEEE. Reprinted, with permission, from [6])

the DC block capacitors C_S. The circuit parameters for this topology, assuming an ideal op-amp are given by [6]

$$A = \frac{2(R + R_s)/Z_0}{1 + 2R_s/Z_0 - R_s/(R + R_s)}, \tag{8.43a}$$

$$\omega_0 = 1/\sqrt{LC} \qquad \text{and} \qquad Q = R\sqrt{C/L}. \tag{8.43b}$$

The transmission coefficient for a single stage is shown in Figure 8.25 [6]. Also shown are circuit simulation results, obtained using measured *S*-parameters of the op-amp with feedback resistor R_F, and simulating the remaining components of Figure 8.24. Figure 8.25 also shows analytical results obtained using (8.22) with out-of-band gain *A* obtained from the ideal op-amp expression (8.43a). Analytical results are indicated as "ideal" in Figure 8.25. Simulated and measured results are in good agreement over the entire frequency domain, while analytical ideal op-amp results deviate at lower (due to DC blocking capacitors) and higher end frequencies (due to low-pass roll-off). The overall circuit exhibits a measured gain close to the desired 0 dB active gain compensation at the resonance frequency.

The measured maximum NGD of 1.52 ns, a resonance frequency of 454 MHz, and a bandwidth of 103 MHz are close to the simulated values of 1.5 ns, 459 MHz, and 104 MHz, respectively. The theoretical single-stage predicted NGD limit of $1/(\pi\Delta f)$, given by (8.24), is validated in this case.

A cascaded circuit, consisting of two stages of Figure 8.24, was also measured and had a maximum NGD of 2.7 ns at the resonance frequency, as shown in Figure 8.26. This was slightly less than double the value of a single stage measured

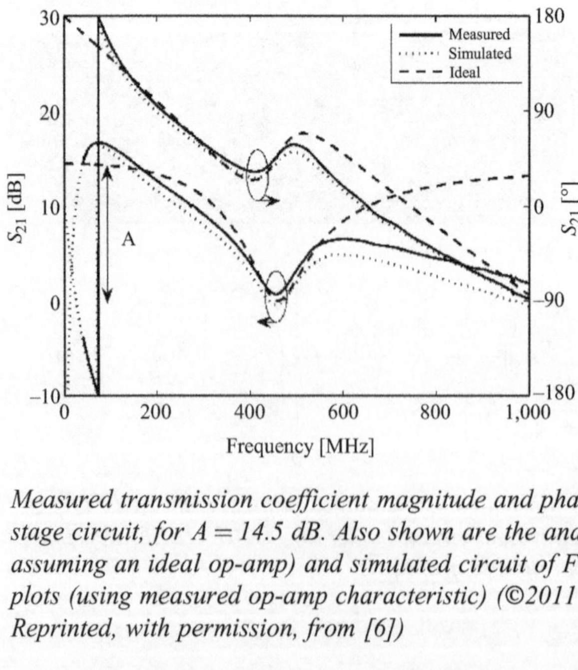

Figure 8.25 Measured transmission coefficient magnitude and phase of a single-stage circuit, for A = 14.5 dB. Also shown are the analytical ((8.22) assuming an ideal op-amp) and simulated circuit of Figure 3.24(b) plots (using measured op-amp characteristic) (©2011 IEEE. Reprinted, with permission, from [6])

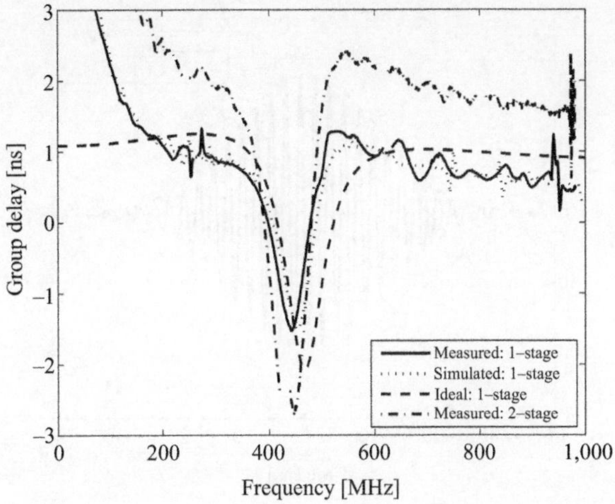

Figure 8.26 Measured group delay of single- and two-stage circuits. Also shown are analytical ((8.22) assuming an ideal op-amp) and simulated (using measured op-amp characteristic) group delay plots (©2011 IEEE. Reprinted, with permission, from [6])

NGD of 1.52 ns, due to non-ideal matching. The overall bandwidth of the 2-stage circuit is also narrower, since the individual stage Q values were not adjusted to maintain the same bandwidth as the single stage.

To validate expression (8.23) in the time domain, a Gaussian pulse modulated carrier was used in [6], with its frequency spectrum centered at ω_0, and a standard deviation $\sigma_\omega = \Delta\omega/6$ is chosen. Most of the waveform power spectrum (99.998%) fits inside the resonator bandwidth ensuring low distortion. The time-standard deviation of the Gaussian pulse is $\sigma_t = 1/\sigma_\omega = 3/(\pi\Delta f) \approx 10$ ns. From (8.24), the maximum single-resonator NGD, expressed as a fraction of the pulse standard deviation is $NGD_{MAX} = 1/(\pi\Delta f) = \sigma_t/3$. The measured time-domain response of the single-stage NGD circuit of Figure 8.24, to this applied Gaussian pulse modulated waveform is shown in Figure 8.27, and test apparatus details are given in [6].

The input waveform envelope is scaled to match the slightly amplified output response (due to the small nonzero gain at ω_0, and the 3 dB in-band amplitude characteristic variation). The pulse has a smooth turn-on point relatively far away from the peak, so there are no visible transients. A 1.5 ns NGD was observed which is in close agreement with the value predicted by frequency domain results from Figure 8.26 and expression (8.23).

8.3.5 Multistage baseband NGD circuits

The transmission coefficient of an N-stage baseband NGD circuit with ideally matched cascaded stages can be derived as

$$S_{21}(\omega) = A_{\text{tot}} \left[\frac{\omega - j\dfrac{\Delta\omega'}{2}}{\omega - j\dfrac{\Delta\omega'}{2} A_{\text{tot}}^{1/N}} \right]^N , \tag{8.44}$$

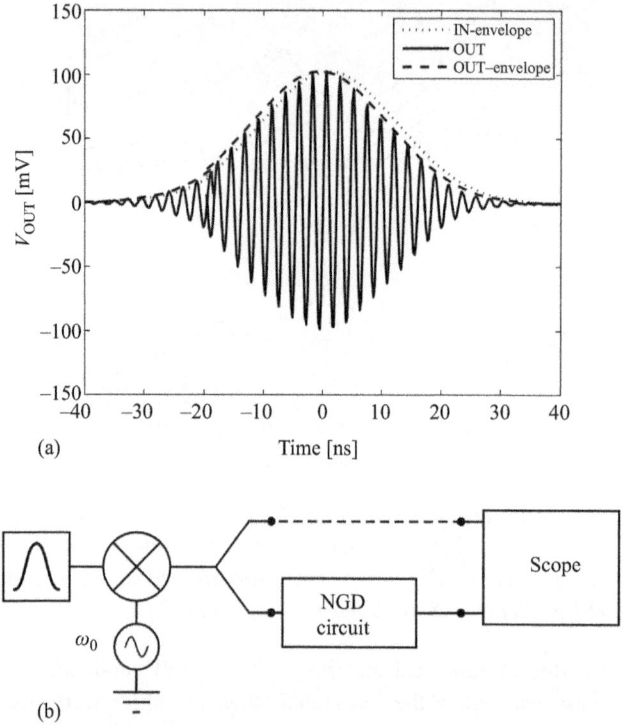

Figure 8.27 (a) Measured single-stage NGD circuit input and output signals for a smooth turn-on Gaussian pulse modulated waveform with a standard deviation of $\sigma_t = 10$ ns and with $f_0 = 458$ MHz. Only the envelope of the input is shown. The experiment setup diagram is shown in (b) (©2011 IEEE. Reprinted, with permission, from [6])

where $\Delta\omega'$ is the bandwidth of each individual stage, which needs to be adjusted to yield an overall desired bandwidth $\Delta\omega$ as given by

$$\Delta\omega' \approx \frac{\Delta\omega\sqrt{1 - 2^{1/N}/A_{\mathrm{tot}}^{2/N}}}{\sqrt{2^{1/N} - 1}}. \tag{8.45}$$

Expression (8.45) is derived by employing a similar procedure as the one used in the derivation of (8.33) for multistage resonator-based circuits. The difference here is that only bandwidths are dealt with, since the Q-factor is not defined for baseband circuits (using $\omega_0 = 0$ in the expression for Q would always yield $Q = 0$, regardless of the $\Delta\omega$ value).

By applying the standard procedure of extracting the phase characteristic to (8.44), then finding its derivative to obtain the group delay and evaluating it at the center frequency, the maximum NGD-bandwidth product in this case can be written as

$$\mathrm{NGD_{MAX}}\Delta f = \frac{1}{\pi} N \sqrt{2^{1/N} - 1} \frac{1 - 1/A_{\mathrm{tot}}^{1/N}}{\sqrt{1 - 2^{1/N}/A_{\mathrm{tot}}^{2/N}}}, \tag{8.46}$$

which is exactly the same as expression (8.36) for resonator-based circuits. Obviously, the asymptotic NGD limits for baseband circuits will be identical to those derived for resonator-based circuits as given by (8.38) and (8.39).

Finally, following a similar derivation as the one in Section 8.3.3, the transmission coefficient of an infinitely distributed baseband NGD circuit can be written as

$$\lim_{N\to\infty} S_{21}(\omega) = A_{\mathrm{tot}} \exp\left[\ln(A_{\mathrm{tot}}) \frac{jb/2}{\omega - jb/2}\right], \tag{8.47a}$$

where

$$b = \Delta\omega \sqrt{\frac{2 \ln(A_{\mathrm{tot}})}{\ln 2} - 1}. \tag{8.47b}$$

It is worth noting that expression (8.47b) for the baseband case is identical to the expression for b in (8.42b) for resonator-based circuits.

Therefore, baseband NGD circuits have a transmission characteristic comprised of a ratio of lower-order frequency functions, as compared to resonator-based circuits (first-order vs. second-order, respectively). However, the same $1/\omega$ dependency for the phase characteristic and a $1/\omega^2$ dependency for the group delay characteristic are observed at large frequencies. The NGD-bandwidth product as a function of the out-of-band gain is identical for both types of circuits as well as the corresponding asymptotic limits.

8.4 Kramers–Kronig engineered NGD medium

In Section 8.3, NGD circuits based on RLC resonators (operating around a non-zero center frequency), and NGD circuits based on high-pass filters (operating at baseband frequencies) were analyzed. For single-stage circuits, the maximum achievable NGD value at the center frequency was shown to be a $1/\pi$ fraction of $1/\Delta f$, which is a measure of the pulse width. For multi-stage circuits, NGD values at the center frequency as a function of the number of stages and the total out-of-band gain were derived, and given by (8.36). Distributed circuits (having a theoretically infinite number of stages) were shown to exhibit an upper asymptotic limit for NGD, given as a square root function of the logarithm of the out-of-band gain magnitude, as given by (8.37).

However, a legitimate question to ask is: can some different circuit topologies perform better than those presented in Section 8.3, in terms of the achieved NGD-bandwidth product (desired quantity) as a function of the out-of-band gain (trade-off quantity)? Rather than experimenting with different circuit topologies, in this section the propagation characteristic will be engineered for a linear and causal medium with a perfectly flat in-band group delay response, which exhibits an optimal asymptotic NGD limit as a function of the out-of-band gain.

The complex transfer function of a causal, physically realizable medium does not have independent real and imaginary parts [3,4]. If the medium was initially at rest, there cannot be any response at the output of a causal medium before an input signal is applied. This requirement translates into causal medium's response to a Dirac delta function input being equal to zero, before the point in time when the input signal is applied. If the Dirac delta function is applied at $t=0$ and the considered medium is linear and causal, the following will apply [12]

$$h(t<0)=\int_{-\infty}^{\infty}H(\omega)e^{j\omega\cdot t}d\omega=0, \tag{8.48a}$$

where

$$H(\omega)=P(\omega)+jQ(\omega)=A(\omega)e^{j\varphi(\omega)}. \tag{8.48b}$$

Since the medium's response to a Dirac delta function in this case takes non-zero values only for $t\geq 0$, it can be decomposed into an odd and an even function of time as [12]

$$h(t)=h_{\text{eve}}(t)+h_{\text{odd}}(t), \tag{8.49a}$$

where

$$h_{\text{eve}}(t)=\begin{cases}\frac{1}{2}h(-t), t<0\\ \frac{1}{2}h(t), t\geq 0\end{cases} \quad \text{and} \quad h_{\text{odd}}(t)=\begin{cases}-\frac{1}{2}h(-t), t<0\\ \frac{1}{2}h(t), t\geq 0\end{cases}. \tag{8.49b}$$

It is clear from (8.49b) that the even and odd components of a causal medium's response are not independent of each other, but can rather be expressed as $h_{\text{eve}}(t)=\text{sgn}(t)\cdot h_{\text{odd}}(t)$, or alternatively $h_{\text{odd}}(t)=\text{sgn}(t)\cdot h_{\text{eve}}(t)$, where sgn(·) denotes a signum function. As known from the Fourier analysis, an even function in time yields a purely real frequency response, while an odd function in time yields a purely imaginary frequency response. Therefore, by comparing (8.48b) and (8.49a) it can be noted that the real part of a causal medium's transfer function, $P(\omega)$, is correlated with $h_{\text{eve}}(t)$ only, while the imaginary part of the transfer function, $Q(\omega)$, is correlated with $h_{\text{odd}}(t)$ only. Since $h_{\text{eve}}(t)$ and $h_{\text{odd}}(t)$ are dependent in this case, then $P(\omega)$ and $Q(\omega)$ will be dependent of each other as well. Furthermore, since the exact functional relationship between $h_{\text{eve}}(t)$ and $h_{\text{odd}}(t)$ is known in this case (a signum function), it can be shown [4] that the real and imaginary parts of a linear causal medium's transfer function are related through Kramers–Kronig relations, given by Cauchy principal value integrals as

$$Q(\omega)=V.P.\left\{-\frac{1}{\pi}\int_{-\infty}^{\infty}\frac{P(\nu)}{\nu-\omega}d\nu\right\} \text{ and } P(\omega)=V.P.\left\{\frac{1}{\pi}\int_{-\infty}^{\infty}\frac{Q(\nu)}{\nu-\omega}d\nu\right\}. \tag{8.50}$$

Since NGD phenomenon stems from the phase response characteristic, it is more useful to relate the amplitude and phase characteristic of a causal medium's transfer function in this case, rather than its real and imaginary parts. Further manipulation of expressions (8.50) yields [12,13]

$$A_{dB}(\omega) = -\frac{20}{\pi \ln 10} V.P.\left\{\int_{-\infty}^{\infty} \frac{\varphi(\nu)}{\nu - \omega} d\nu\right\} \text{ and } \varphi(\omega) = \frac{\ln 10}{20\pi} V.P.\left\{\int_{-\infty}^{\infty} \frac{A_{dB}(\nu)}{\nu - \omega} d\nu\right\}. \tag{8.51}$$

Therefore, if an arbitrary phase characteristic is set for a linear and causal medium's transfer function, the corresponding amplitude characteristic is not arbitrary; it is rather uniquely defined (up to an integration constant). This fact can be taken advantage of and a suitable phase characteristic, and therefore the NGD, can be assumed and the corresponding magnitude characteristic can then be evaluated from (8.51). Then, an attempt can be made to extract the maximum out-of-band gain in that case, and to find its relationship to the achieved NGD.

8.4.1 Frequency-reciprocal decay of the phase response

Group delay as a function of frequency of the media presented in Section 8.3 has a reciprocal square decay ($1/\omega^2$), at high frequencies. Since the group delay is essentially a derivative of the phase response, this means that at high frequencies the phase response of these media has a reciprocal decay ($1/\omega$). For example, distributed media phase characteristic (8.42c) illustrates this functional dependency.

In this section, a phase characteristic with a linear phase response around the center frequency will be examined (extending up to a frequency ω_M), but with different order reciprocal phase decays, ($1/\omega^k$), past the ω_M frequency. The decay functions parameters are chosen such that the phase characteristics are continuous at the reversal point, ω_M, and the decay slopes at the same point, $\tau_g(\omega_M)$, are related to the center frequency phase slope, t_g, such that the NGD/A_{MAX} ratio is maximized in each case.

Figure 8.28 shows phase and corresponding group delay responses for the reciprocal phase decay cases of orders $k = 1/2$, $k = 1$, $k = 2$, and $k \rightarrow \infty$ (equivalent to a linear decay) [12].

For comparison purposes, each case has the same NGD-bandwidth product value at the center frequency, $t_g \cdot \Delta f = 0.6$. The amplitude responses corresponding to Figure 8.28 cases are derived by evaluating (8.51), and they are shown in Figure 8.29. Since all cases have the same NGD-bandwidth product, $t_g \cdot \Delta f = 0.6$, the case with the smallest A_{MAX} value will have the largest NGD/A_{MAX} ratio (optimum case) [12].

As evident from Figure 8.29, the reciprocal decay case ($1/\omega$) has the smallest out-of-band gain, and it is therefore optimal. This is confirmed in Figure 8.30 [12], where curves relating NGD-bandwidth versus maximum out-of-band gain are produced. The asymptotic limit for the optimum case is determined numerically as

$$t_g \Delta f \leq 0.089 \sqrt{A_{\text{MAX-dB}}}. \tag{8.52}$$

The asymptotic limit given by (8.52) has the same functional dependency as the limit for distributed resonator-based media (8.38), but the proportional constant

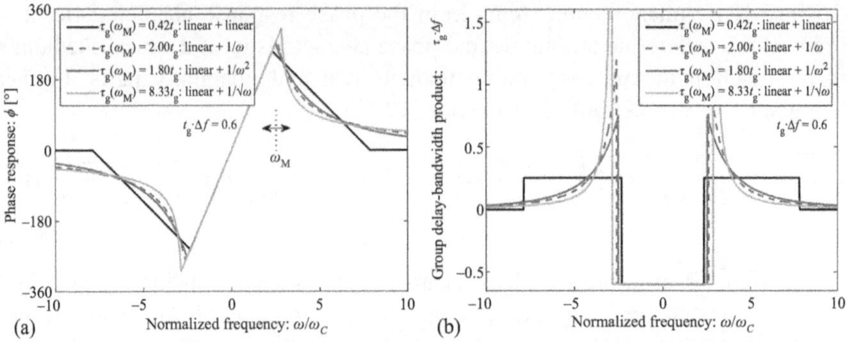

Figure 8.28 Piece-wise linear and reciprocal ($1/\omega^k$) decay (a) phase responses and corresponding (b) group-delay-bandwidth products for $k = 1/2$, $k = 1$, $k = 2$, and $k \to \infty$ (linear decay). Each case shown corresponds to the maximized NGD/A_{MAX} ratio, and it has $t_g \cdot \Delta f = 0.6$ ([12], reproduced, courtesy of The Electromagnetics Academy)

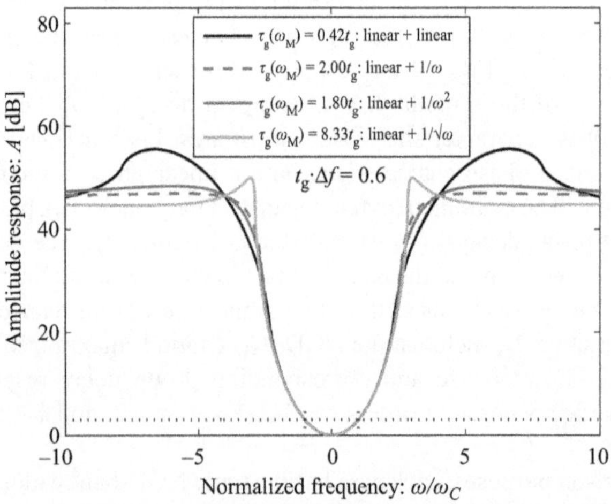

Figure 8.29 Magnitude responses for piece-wise linear and reciprocal ($1/\omega^k$) phase decays with $k = 1/2$, $k = 1$, $k = 2$, and $k \to \infty$ (linear decay). Each case shown corresponds to the maximized NGD/A_{MAX} ratio, and it has $t_g \cdot \Delta f = 0.6$ ([12], reproduced, courtesy of The Electromagnetics Academy)

is larger by about 40%. Changing the order of phase decay to a higher ($1/\omega^2$), or a lower ($1/\omega^{1/2}$) value from the optimum case ($1/\omega$) produces asymptotic limits smaller than (8.52).

All cases shown in Figure 8.29 have their maximum out-of-band values, A_{MAX}, larger than their corresponding out-of-band values at high frequencies, $A_{\mathrm{dB}}(\infty)$.

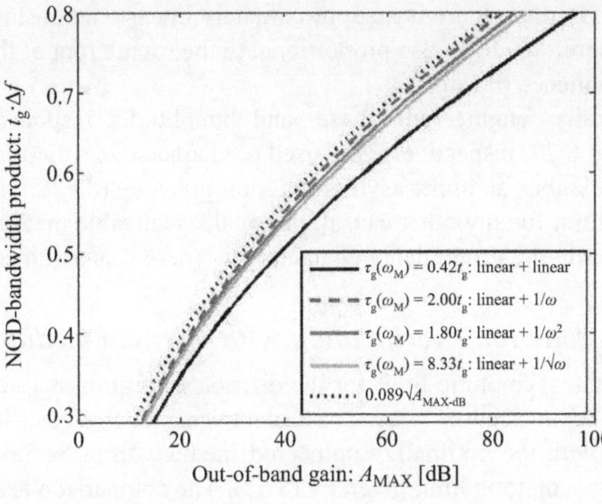

Figure 8.30 Center frequency NGD-bandwidth product as a function of out-of-band gain A_{MAX}, for piece-wise linear and reciprocal ($1/\omega^k$) phase decays with $k = 1/2$, $k = 1$, $k = 2$, and $k \to \infty$ (linear decay) ([12], reproduced, courtesy of The Electromagnetics Academy)

Therefore, amplitude characteristics are not monotonic functions of frequency, unlike the media presented in Section 8.3. Decay slopes at the phase reversal points, $\tau_g(\omega_M)$, can be adjusted for the $k = 1$ and $k = 1/2$ cases to make their amplitude characteristics monotonic functions in frequency half-planes (A_{MAX} is at $A_{dB}(\infty)$). However, changing the decay slopes from the optimum values shown in Figure 8.28 would result in lowering the NGD/A_{MAX} ratio for each case. For example, the limit for the optimal case ($k = 1$) given by (8.52) would decrease by 2.25%, if the phase slope at the reversal point, $\tau_g(\omega_M)$, was changed to make the corresponding amplitude characteristic monotonic. Since the primary objective of this section is to derive the upper asymptotic limit, the limit given by (8.52) will be retained.

The full amplitude characteristic expression corresponding to the optimum linear and reciprocal ($1/\omega$) phase response is obtained from (8.51), and given by [12]

$$A_{dB}(\omega) = \frac{20(t_g\Delta f)}{\ln 10}\left[\frac{2\omega_M}{\omega_C}\ln(2) + \frac{|\omega|}{\omega_C}\ln\left(\frac{\omega_M + |\omega|}{|\omega_M - |\omega||}\right)\right. \\ \left. + \frac{\omega_M^2}{\omega_C}\frac{\ln(2|\omega_M - |\omega||/\omega_M)}{2|\omega| - \omega_M} - \frac{\omega_M^2}{\omega_C}\frac{\ln(2(\omega_M + |\omega|)/\omega_M)}{2|\omega| + \omega_M}\right] \tag{8.53}$$

where the values chosen for the maximum NGD-bandwidth product, $t_g \cdot \Delta f$, and the linear to reciprocal phase transition frequency, ω_M, are related such that $A_{dB}(\omega_C) = 10\log(2) \approx 3$ dB condition is satisfied. In the example depicted in Figures 8.28 and 8.29, the chosen value of $t_g \cdot \Delta f = 0.6$ requires a value of $\omega_M \approx 2.606\omega_C$ for the linear and reciprocal phase response medium, to satisfy the 3-dB bandwidth condition at ω_C. Further, there is an approximately linear relationship between ω_M and $t_g{\cdot}\Delta f$, and therefore ω_M is also proportional to the square root of the out-of-band gain, as a consequence of (8.52).

The optimally engineered phase and amplitude responses shown in Figures 8.28 and 8.29, respectively, are used to characterize a fictitious but causal medium which exhibits an upper asymptotic limit given by (8.52). It attempts to set an upper NGD limit for any other causal, physically realizable medium, with a low in-band variation in the group delay characteristic (here it is assumed ideally flat).

8.4.2 Asymptotic limit verification with physical media

In this section, the asymptotic limit for the distributed high-pass filter-based NGD medium presented in Section 8.3, and a Lorentzian model dielectric medium are both compared with the optimally engineered medium from Section 8.4.1 corresponding to the asymptotic limit given by (8.52). The comparison is carried out for both amplitude and phase frequency responses, for a selected case.

In the first comparison, an infinitely distributed high-pass filter-based medium exhibiting an NGD-bandwidth product of $t_g{\cdot}\Delta f = 0.6$ is chosen, corresponding to a maximum out-of-band gain $A_{\mathrm{MAX}} = 86$ dB. This is compared to a medium with an in-band linear phase approximation with a reciprocal ($1/\omega$) decay at high frequencies, which is chosen to have the same NGD-bandwidth product and with $\tau_g(\omega_M) = 2t_g$, yielding a maximum NGD/A_{MAX} ratio as in Section 8.4.1. The amplitude characteristic for this case is obtained by Kramers–Kronig relations as before. The phase and amplitude response plots for this case are shown in Figure 8.31(a) and (b), respectively [12].

It can be noted that for the same NGD-bandwidth product chosen for both cases, the optimum medium from Section 8.4.1 (linear and reciprocal phase decay combination) has a much smaller maximum out-of-band gain of $A_{\mathrm{MAX}} = 47$ dB. If the parameters of the same optimum medium type from Section 8.4.1 are slightly adjusted to yield a monotonic amplitude characteristic, a slightly larger out-of-band gain value of $A_{\mathrm{MAX}} = 49$ dB results. This case is also shown in Figure 8.31(a) and (b).

In another comparison, a passive Lorentzian dielectric model medium is chosen for $\phi(\omega) = \mathrm{Re}\{-\omega{\cdot}n(\omega)/c\}$. The refractive index of this medium is given by [4]

$$n(\omega) = \sqrt{1 - \frac{\omega_p^2}{\omega^2 - 2j\delta\omega - \omega_0^2}}, \tag{8.54}$$

where ω_0 and ω_p are the medium's resonance and plasma frequencies, respectively, and δ is the damping factor, which is related to medium's absorption bandwidth around the resonance. The chosen example has $\omega_p = 0.53\omega_0$, $\delta = 0.05\omega_0$, and a physical length corresponding to a 360° phase shift of a sinusoidal waveform of frequency ω_0, traveling the same distance in vacuum (light line). The NGD bandwidth of this medium is not centered around $\omega = 0$, but around a point in the vicinity of the resonance frequency instead. The phase and magnitude responses are

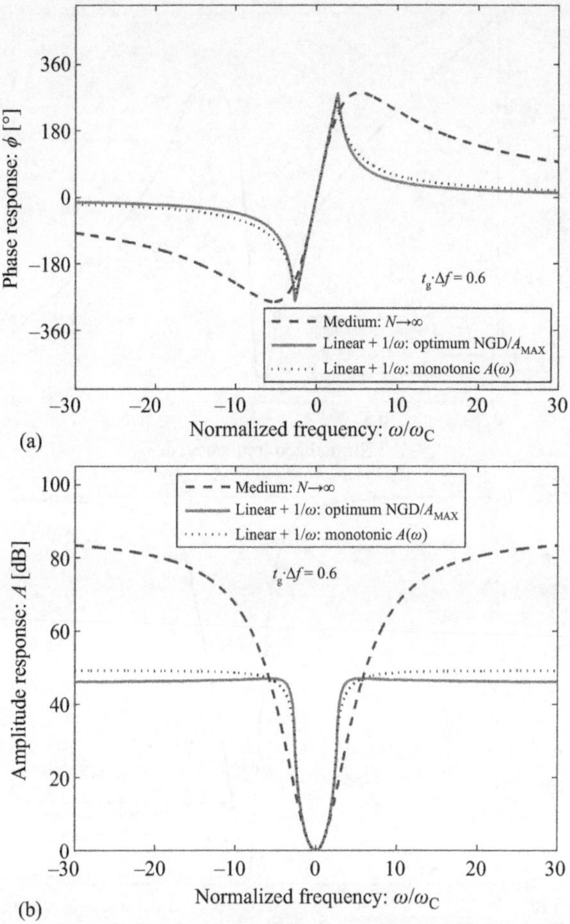

Figure 8.31 (a) Phase and (b) amplitude response characteristics of a distributed high-pass filter-based medium, and a piece-wise linear and reciprocal decay phase approximation medium with NGD-bandwidth product $t_g \cdot \Delta f = 0.6$ ([12], reproduced, courtesy of The Electromagnetics Academy)

not symmetric around the center frequency either. The chosen example exhibits the same NGD-bandwidth product of $t_g{\cdot}\Delta f = 0.6$, corresponding to a maximum out-of-band gain $A_{\text{MAX}} = 99$ dB as shown in Figure 8.32(b) [12].

This medium has an even larger maximum out-of-band gain (or equivalently, center frequency loss, since it is a passive medium) for a given NGD-bandwidth product, as compared with the resonator-based medium from Section 8.3.

The asymptotic limit for this medium will therefore be even smaller, and it is numerically evaluated as shown in Figure 8.33. The same upper limit holds

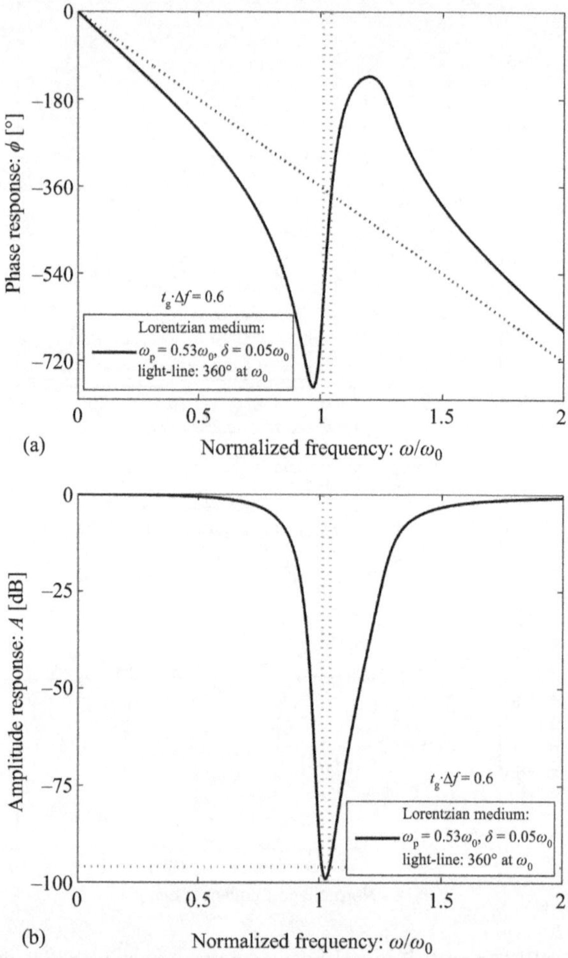

Figure 8.32 *(a) Phase and (b) amplitude response characteristics of a Lorentzian medium with an NGD-bandwidth product $t_g \cdot \Delta f = 0.6$, $\omega_p = 0.53\omega_0$, $\delta = 0.05\omega_0$, and medium length corresponding to a 360° light-line phase at ω_0 ([12] reproduced, courtesy of The Electromagnetics Academy)*

for Lorentzian media with different light-line phase shifts than the one shown. The asymptotic limit is given by

$$t_g\Delta\omega \leq 0.061\sqrt{A_{\text{MAX-dB}}}, \tag{8.55}$$

and it is smaller than the limit given by (8.52). Therefore, the optimally engineered linear and causal medium with a linear phase response around the center frequency and a reciprocal ($1/\omega$) phase decay at high frequencies, indeed has the highest asymptotic NGD limit given by (8.52), compared to other considered media.

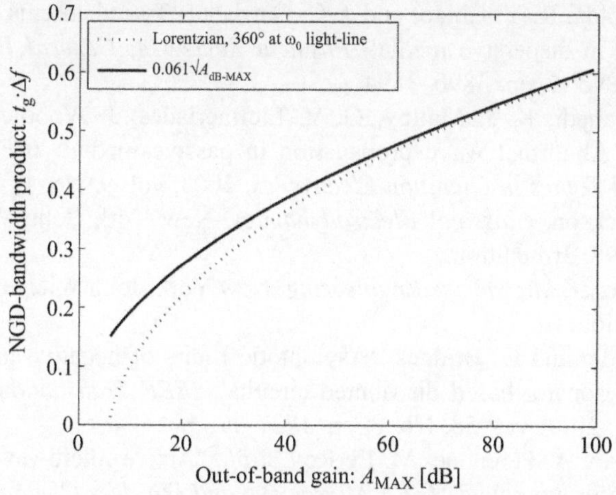

Figure 8.33 Center frequency NGD-bandwidth product as a function of out-of-band gain A_{MAX}, for a Lorentzian medium with $\delta = 0.05\omega_0$, and medium length corresponding to a 360° light-line phase at ω_0 ([12], reproduced, courtesy of The Electromagnetics Academy)

8.5 Conclusion

In this chapter, NGD limits as a function of trade-off quantities were investigated. The desired quantity is relative NGD, which in the frequency domain is characterized by the product of the center-frequency NGD value, and the 3-dB amplitude bandwidth. The NGD-bandwidth product is proportional to time-domain advancement of the peak of a propagated pulse, relative to the temporal width of the pulse. The investigated trade-off quantity is the out-of-band gain in the frequency domain, which for passive devices can also be interpreted as center-frequency attenuation. It was demonstrated that the out-of-band gain is proportional to the amplitude of transients in the time-domain, when signals with defined turn-on/off times are considered. For cascaded resonator-based NGD circuit topologies, the asymptotic limit of NGD-bandwidth product as a function of out-band-gain was derived, and shown to be proportional to a square root function of decibel out-of-band gain value. The final section employs Kramers–Kronig relations to extend the NGD limits analysis to general linear causal media, assuming an ideally flat in-band group delay characteristic. The same square root function for asymptotic NGD-bandwidth limit was observed in this case, only with a higher proportional factor.

References

[1] L. Brillouin, *Wave Propagation and Group Velocity*. New York: Academic Press; 1960.

[2] E. L. Bolda, R. Y. Chiao, and J. C. Garrison, 'Two theorems for the group velocity in dispersive media'. *Physical Review A, General Physics*. 1993, vol. 48, No. 5, pp. 3890–3894.

[3] M. Mojahedi, K. J. Malloy, G. V. Eleftheriades, J. Woodley, and R. Y. Chiao, 'Abnormal wave propagation in passive media'. *IEEE Journal of Selected Topics in Quantum Electronics*. 2003, vol. 9, No. 1, pp. 30–39.

[4] J. D. Jackson, *Classical Electrodynamics*. New York: John Wiley & Sons Inc.; 1998, 3rd edition.

[5] D. M. Pozar, *Microwave Engineering*. New York: John Wiley & Sons; 1998, 2nd edition.

[6] M. Kandic and G. Bridges, 'Asymptotic limits of negative group delay in active resonator-based distributed circuits'. *IEEE Transactions on Circuits Systems*. 2011, vol. 58, No. 8, pp. 1727–1735.

[7] B. Ravelo, A. Perennec, M. Le Roy *et al.*, 'Active microwave circuit with negative group delay'. *IEEE Microwave and Wireless Components Letters*. 2017, vol. 17, No. 12, pp. 861–863.

[8] B. Ravelo, A. Perennec, and M. Le Roy, 'Synthesis of broadband negative group delay active circuits'. 2007. *IEEE/MTT-S International Microwave Symposium*. 2007, pp. 2177–2180.

[9] T. Nakanishi, K. Sugiyama, and M. Kitano, 'Demonstration of negative group delays in a simple electronic circuit'. *American Journal of Physics*. 2002, vol. 70, No. 11, pp. 1117–1121.

[10] M. Kitano, T. Nakanishi, and K. Sugiyama, 'Negative group delay and superluminal propagation: An electronic circuit approach'. *IEEE Journal of selected Topics in Quantum Electronics*. 2003, vol. 9, No. 1, pp. 43–51.

[11] B. Ravelo, A. Perennec, and M. Le Roy, 'Negative group delay active topologies respectively dedicated to microwave frequencies and baseband signals'. *Journal of EuMA*. 2008, vol. 4, No. 2, pp. 124–130.

[12] M. Kandic and G. Bridges, 'Limits of negative group delay phenomenon in linear causal media'. *Progress in Electromagnetics Research*. 2013, vol. 134, pp. 227–246. doi:10.2528/PIER12082915

[13] H. W. Bode, *Network Analysis and Feedback Amplifier Design*. New York: Van Nostrand; 1945.

Chapter 9

Superluminality and detectable information in dispersive channels

Ahmed H. Dorrah[1] *and Mo Mojahedi*[1]

Abstract

Energy exchange between an electromagnetic pulse and dispersive media may result in complicated, yet interesting, phenomena in which the group velocity becomes abnormal (i.e., superluminal or negative). For such cases, signal velocity (velocity of detectable information) remains debatable. In this chapter, we present a systematic study that can be applied to pulse propagation in any dispersive medium in order to quantify the detectable information content and calculate its speed, while accounting for pulse reshaping effects and noise generated in the medium and the detector. Accordingly, we present an operational context within which the constraints of superluminal signaling and its potential applications are shown. We provide scenarios in which the signal velocity is evaluated in microwave circuits with negative group delays and extend the method to include optical pulses in inverted media as well. Such analysis explores the fundamental limitations and capabilities of a broad range of superluminal signaling applications.

Keywords

Negative group delay (NGD), negative group velocity (NGV), superluminal information.

9.1 Introduction

An electromagnetic pulse or wave packet can be regarded as a superposition of many monochromatic waves each propagating with its own phase velocity. In a temporally dispersive medium (channel), these different spectral components of the

[1]Edward S. Rogers Sr. Department of Electrical and Computer Engineering, University of Toronto, Canada

pulse propagate at different phase velocities and can be nonuniformly attenuated (in passive media) or amplified (in active media) depending on the frequency of each component. As these weighted spectral components propagate in the medium, they add up constructively or destructively in such a way that leads to reshaping of the pulse envelope. In general, this would lead to signal distortion; however, under some cleverly designed configurations, it is observed that the peak of the pulse propagating in a dispersive medium may travel faster than the peak of a companion pulse traversing the same distance in free space while maintaining its envelope—an effect called superluminal group velocity (SGV). Under other circumstances, the peak of the transmitted pulse (output peak) in a dispersive medium may even be detected before the peak of the input pulse interacts with the medium—a phenomenon known as negative group velocity (NGV). From a fundamental point-of-view, both SGV and NGV describe the same phenomenon which can be collectively referred as abnormal group velocities (AGVs) or, equivalently, abnormal group delays (AGDs).

The counter-intuitive phenomenon of AGV was first investigated long ago by Sommerfeld [1] and Brillouin [2,3] by studying the transients associated with electromagnetic pulse propagation in absorbing causal dielectrics. Afterwards, Garret and McCumber were able to establish the theoretical possibility of achieving superluminal propagation with minimal pulse distortion and they pointed out to the full compliance of this phenomenon with Einstein's causality [4]. Later on, Chu and Wong successfully realized AGV in an experimental setup using laser pulses [5]. Then afterwards, Segard and Macke extended the experiment to the millimeter waves [6].

More recently, AGV has been observed in passive and active media: using a single microwave pulse [7–9], in under-sized waveguides [10–13], in electronic circuits [14], at optical frequencies [15–19], in the single-photon limit [20–23], in negative index meta-materials [24,25], using ring resonators [26,27], and by encoding information in the spatial degrees of freedom of an optical pulse [28]. Despite the counter-intuitive implication of AGV, it has been considered to be in full agreement with the requirements of special relativity (Einstein causality) which prohibits information to be transmitted faster than the speed of light in vacuum (c) [29–36]. The common line of arguments asserts that any causal pulse used for signaling must have a transitional turn-on in time and space, referred to as the *front*, which can be represented mathematically as nonanalytic point(s) in the complex frequency domain [21,37]. As it has been shown by Sommerfeld and Brillouin long ago, the front and the earliest oscillations associated with it—referred to as precursors or forerunners—always travel luminaly [1–3]. As such, information and the velocity by which it travels must be associated with these points of nonanalyticity and the velocity by which it propagates—a velocity which is never superluminal. Furthermore, even for the cases in which the pulse (typically a smooth Gaussian pulse) is superluminal, it has been shown that the pulse behavior could have been analytically extrapolated from earlier time samples [30–32]. In other words, there is no superluminal transfer of genuine information because information already existed when the analytic continuation was carried out by the medium. Accordingly, information velocity should not be tied to the superluminal effect of the group velocity [30,34,35,37,38].

Although it has been well established, fundamentally, that genuine information is encoded in the points of nonanalyticity at the onset of the pulse (the strictly luminal front); however, in practical situations, this approach is not satisfactory for many reasons: first, the field strength associated with the front and precursor oscillations is extremely small in general and well below the detection level [30]. This makes the detection of the points of nonanalyticity (singularities) if not impossible, certainly very challenging. In fact, when such singularities have been observed to demonstrate the compliance of AGD with Einstein causality [16], special care in generating the singularities and their detection has been employed. Second, information cannot be viewed as a localized quantity (e.g., points of non-analyticity, alone) as its detection requires the act of a detector and a window of time associated with arrival of enough power (sufficient amount of photons) to trigger a detection process. Third, information, its propagation, and its detection are stochastic processes, in which each subsystem of a communication system plays an important role. Each subsystem introduces a certain amount of noise that imposes considerable effects on the stochastic nature of the information and its velocity. In short, any definition for detectable information (signal velocity) should account for the associated noise from the channel and the nonideal detector.

In light of the above considerations, relating the information velocity (signal velocity) to the strictly luminal front is practically non meaningful and, therefore, the possibility of superluminal signaling has always been debatable. In short, there is no universal approach that quantifies the velocity of detectable information (signal velocity) of an electromagnetic pulse propagating in a general dispersive medium.

Although some efforts in this area have been made [16,17,28,39,40], the results are still inconclusive for many reasons: First, the effect of the medium (channel) length—which has a direct impact on the dispersion characteristics of the medium and the reshaping of the pulse—has been widely ignored. In fact, signal velocity has always been studied at a fixed propagation distance instead of studying its evolution at different space points and thus the behavior of the signal velocity at different propagation distances is still unexplored. Second, in the theoretical analysis [39,40], pulse reshaping effects in terms of compression (or broadening), and nonuniform scaling have also been neglected. Given the fact that pulse advancement can be in the order of only 3% of the initial pulse width in some cases [17,39], consequently pulse compression effects can significantly affect the interpretation of the results and cannot be overlooked.

In this chapter, we present a general approach through which the velocity of detectable information (signal velocity) can be calculated in a dispersive medium (channel) while taking into account pulse reshaping effects, in addition to the classical and quantum noise sources associated with the medium and the nonideal detector. Such noise sources play an important role in the stochastic behavior of information and its effect should be included in any definition that describes the evolution of detectable information. We first start by presenting a method for quantifying signal velocity in a general dispersive medium then proceed to apply the methodology into two cases: (i) a negative group delay (NGD) circuit in the microwave domain and (ii) a double-resonance-inverted gain medium at optical frequencies. In this study, we also highlight the capabilities as well the limitations of signaling schemes that utilize superluminal (or negative) group delay.

9.2 Approach

In any noisy communication system, there exists a nonzero probability of encountering errors in the detected signal. Therefore, reliable detection of information can only be described probabilistically. A standard metric that is used to quantify reliable information detection, without loss of generality, is the signal-to-noise (SNR) ratio. In fact, an operational approach for calculating the signal velocity—which depends on the SNR—has been widely utilized in the literature [39–42]. For such cases, the SNR is expressed (as a function of time) in the following general form

$$\mathrm{SNR}(t) = \frac{P_{\mathrm{sig}}(t)}{P_{\mathrm{MedNoise}(t) + P_{\mathrm{DetNoise}}(t)}} \tag{9.1}$$

where $P_{\mathrm{sig}}(t)$ is the signal power, $P_{\mathrm{DetNoise}}(t)$ is the noise associated with the detector circuit, and $P_{\mathrm{MedNoise}}(t)$ is the total classical and quantum noise associated with the medium. For a simple on–off keying communication scheme in which information is carried by either a bit-1 or bit-0, the signal velocity (velocity of detectable information of a pulse) is related to the time instant at which the signal power exceeds the noise floor (typically; $\mathrm{SNR} = 0$ dB). Using this operational definition, the possibility of reducing signal detection latency using AGD media has been demonstrated both theoretically [39–42] and experimentally [28].

In the following, we provide two different case studies in which the above operational approach is applied in more details to evaluate the signal velocity in channels exhibiting abnormal group delay (AGD).

9.3 Microwave NGD channel

Consider the scheme depicted in Figure 9.1. A microstrip transmission line is divided into two equal sections and loaded with an NGD circuit with unity gain. The channel is either excited by a signal represented by a causal, smoothly varying, modulated Gaussian pulse (bit 1) or not excited at all (bit 0). This scheme represents a simple on–off keying (OOK) communication protocol. The generated pulse is then processed at the receiver which consists of a detector, a sampler, and a decision device. The noisy detection apparatus at the receiver end attempts to determine as quickly as possible whether or not a signal has been sent. The same configuration is applied to the case of a conventional positive group delay (PGD) microstrip line for comparison.

Instead of sampling the output of the detector at only a single time instant at which the SNR has a maximum value as in conventional communication systems, the output is sampled continuously as a function of time herein. Accordingly, $\mathrm{SNR}(t)$ can be described and calculated as a function of time. The time instant at which the SNR exceeds a preset threshold implies that the pulse (new information) has been reliably

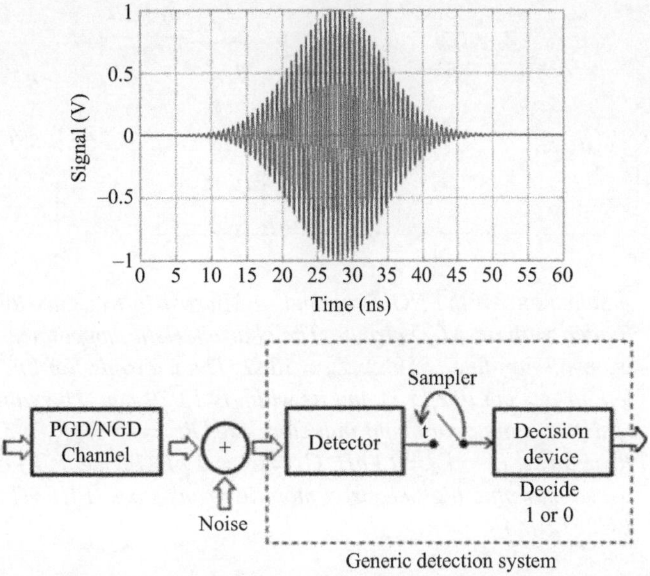

Figure 9.1 A schematic diagram of the proposed thought experiment. (a) A 1.45 GHz modulated Gaussian pulse with FWHM of 15 ns is applied at the input of NGD and PGD channels. The channel noise is added before the detection process [41]

detected within the limit of the preset threshold. The time required to reach the threshold defines the signal delay.

9.3.1 Channel analysis

For the microwave channel, we consider the active NGD circuit proposed by Ravelo *et al.* [8,9]. The circuit is composed of two cascaded cells consisting of a parallel connection of a PHEMT transistor and a series RLC branch. This topology realizes NGD in addition to possible input/output matching. Figure 9.2 shows a schematic of the NGD channel in which a microstrip transmission line is cut into two symmetric sections and loaded with the NGD circuit in the middle.

The circuit considered by Ravelo, however, exhibits gain. This means that the NGD output pulse experience amplification in addition to peak advancement. In the present analysis, the active circuit is tuned to obtain overall unity channel gain so that the output peak is time-advanced with no pulse amplification. This ensures fair comparison with the conventional PGD microstrip line and enforces the SNR to settle at the steady-state value for both the NGD and PGD configurations if the corresponding channel noise contributions are equal. Consequently, any achieved reduction in the detectable information latency here would be attributed to pulse advancement solely and not pulse amplification. Unity gain is attained by reducing the value of R_1 invoked in [9] from 11 to 3.75 Ω herein. This configuration achieves NGD of −6.3 ns at the design frequency of 1.45 GHz as depicted in

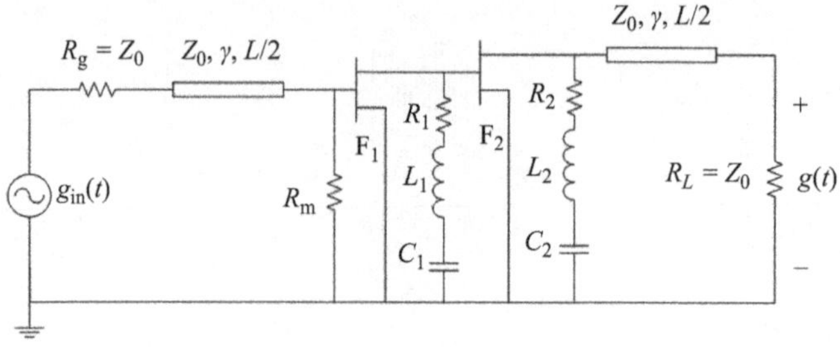

Figure 9.2 A schematic of the NGD channel. A Microstrip transmission line is loaded with an NGD circuit. The characteristic impedance of the transmission line is 50 Ω, $Z_o = 50\ \Omega$. The transmission line total length is 3 cm (0.145 λ) and its width is 1.079 mm. The values for the resistors, capacitors, and inductors are: $R_m = 75\ \Omega$, $R_1 = 3.75\ \Omega$, $R_2 = 36\ \Omega$, $L_1 = L_2 = 12$ nH, $C_1 = C_2 = 1$ pF. Elements F_1 and F_2 are pseudomorphic high-electron mobility transistors (PHEMTs) – model EC2612 [44]

Figure 9.3(a). In Figure 9.3(b), the input pulse spectrum is plotted versus the channel insertion loss. The input bandwidth is adjusted so that the attenuated and amplified frequencies level out yielding a pulse with no appreciable amplification at the peak as shown in the time-domain results in Figure 9.3(c). It is shown that peak advancement (which accounts for 40% of the pulse full-width half-maximum (FWHM)) is achieved with minimal pulse distortion.

In order to account for the noise of the NGD channel, a linear bias-independent noise model for the PHEMT is used. According to [43], the equivalent circuit of the PHEMT and its noise figure at a certain frequency as found in [44], are sufficient for calculating the noise figure and the optimum noise source impedance for the PHEMT at a wide range of frequencies. The advanced design system (ADS) noise simulations of the PHEMT exhibit excellent agreement with [43] in terms of minimum noise figure ($F_{\min}$), noise resistance (R_n), and optimum source reflection coefficient (S_{opt}). The analysis is extended to the two-stage NGD circuit, and the overall noise contribution is calculated and plotted in Figure 9.3(d).

The noise calculations for the passive PGD microstrip line are much more straight-forward. The numerical values for the parameters of the microstrip line are listed in Table 9.1. For such case, the microstrip line is represented in a passive circuit that consists of resistances, inductors, and capacitors. The equivalent Johnson noise is then calculated and the output noise power is verified using ADS simulations. The noise voltage spectral densities at 1.45 (GHz) for the NGD and PGD channels are found to be 1.22 (nV/$\sqrt{\text{Hz}}$) and 0.447 (nV/$\sqrt{\text{Hz}}$), respectively.

9.3.2 Detector noise analysis

It is inadequate to study the problem of information latency in NGD circuits without considering practical scenarios in which the pulse is detected at the output

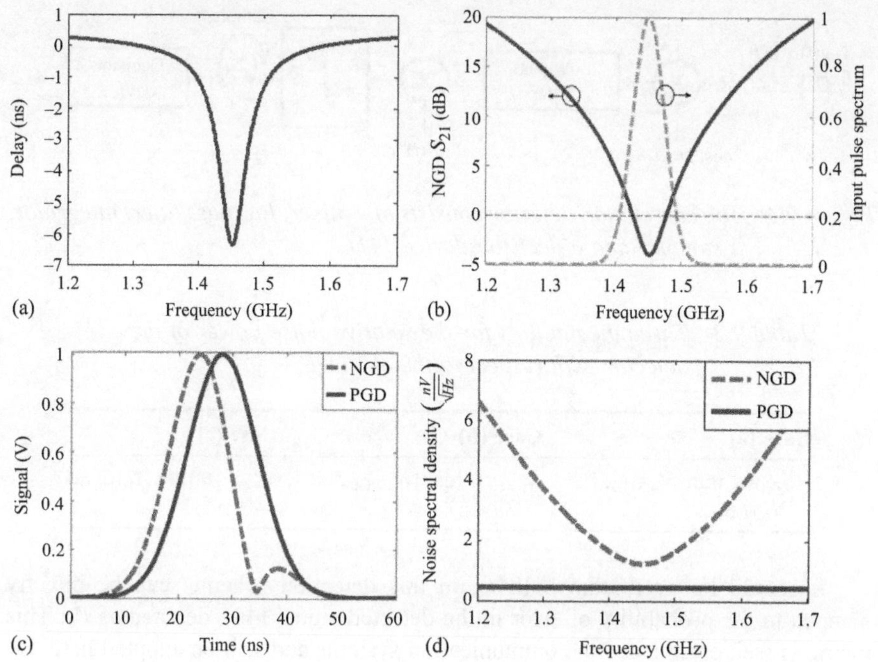

Figure 9.3 *(a) NGD channel delay; (b) normalized input Gaussian pulse spectrum fitted on the NGD S_{21} plot; (c) NGD and PGD channel output response; and (d) noise spectral densities of the NGD and PGD channels in units of $nV/\sqrt{Hz}$ [41]*

Table 9.1 *Numerical values of the microstrip line*

Parameter	**Value**
Relative permittivity, ϵ_r	3.5
Relative permeability, μ_r	1.0
Dielectric thickness, d	0.508 mm
Loss tangent, tan δ	0.0018
Conductivity, σ	5.8×10^7 S/m
Characteristic impedance, z_0	50 Ω
Microstrip line length, L	3 cm
Microstrip line width, W	1.079 mm

of the channel using a nonideal detector. In this subsection, we consider the correlation detector in order to highlight the role played by the detector in defining the information velocity. The correlation detector consists of a mixer, low pass filter (demodulator stage) and a correlator block which multiplies the demodulated baseband signal (with added channel noise) by a locally generated version of itself and integrates the result as depicted in Figure 9.4.

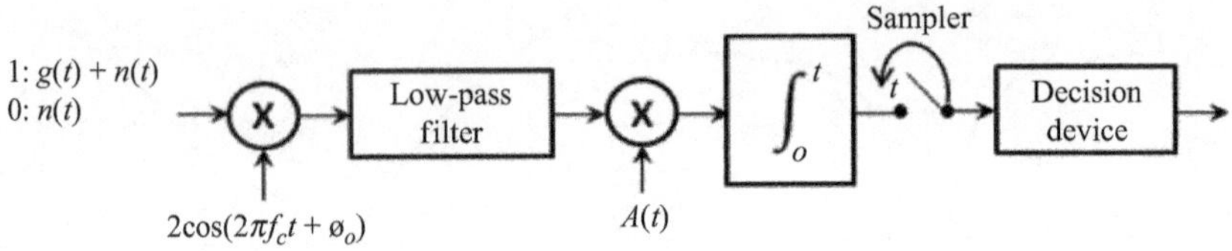

Figure 9.4 The correlation detector consists of a mixer, low pass filter, integrator, a sampler and a decision device [41]

Table 9.2 Three inequalities for the relative noise power of the detector with respect to the channel

Case (a)	Case (b)	Case (c)
$N_{o,\text{det}} \ll \min(N_{o,\text{NGD}}, N_{o,\text{PGD}})$	$N_{\text{o,det}} = \max(N_{\text{o,NGD}}, N_{\text{o,PGD}})$	$N_{\text{o,det}} \gg \max(N_{\text{o,NGD}}, N_{\text{o,PGD}})$

It should be noted that $\text{SNR}(t)$, in this detection scheme, can be directly mapped to the probability of error in the detected signal [45], denoted as P_e. This metric is well established in communication systems and will be adopted here. By extending this metric to be time-dependent function, i.e., $(P_e(t))$, the following expression can be derived [41]:

$$P_e(t) = Q\left(\frac{E_A(t)}{4N_0}\right) \tag{9.2}$$

$$E_{A(t)} = \int_0^t E_{A(u)} du \tag{9.3}$$

where $Q(x)$ is the well-known tail function, N_0 is the noise spectral density (in units of V^2/Hz) and it consists of the noise contributions of the channel ($N_{o,\text{ch}}$) and the detection system ($N_{o,\text{det}}$), i.e., $N_o = N_{o,\text{ch}} + N_{o,\text{det}}$. The term $E_A(t)$ is the time-integrated signal power (instantaneous energy) at the output of the correlator block (in units of V^2/Hz). The correlator block directly precedes the sampler stage as shown in Figure 9.4.

In order to calculate $P_e(t)$, the noise contribution of the correlation detector should be specified. We consider three different profiles for the relative power strength of the detector noise as compared to the channel noise. These profiles are summarized in Table 9.2.

9.3.3 Calculating signal velocity in the NGD channel

We note that reliable detection of information can be defined to take place when $P_e(t)$ falls below a certain threshold. We set this threshold here to be 10^{-5} for a reliable communication scheme. First, we consider case (a) in which the detector has negligible noise contribution. In this case, the channel noise is dominant. It is

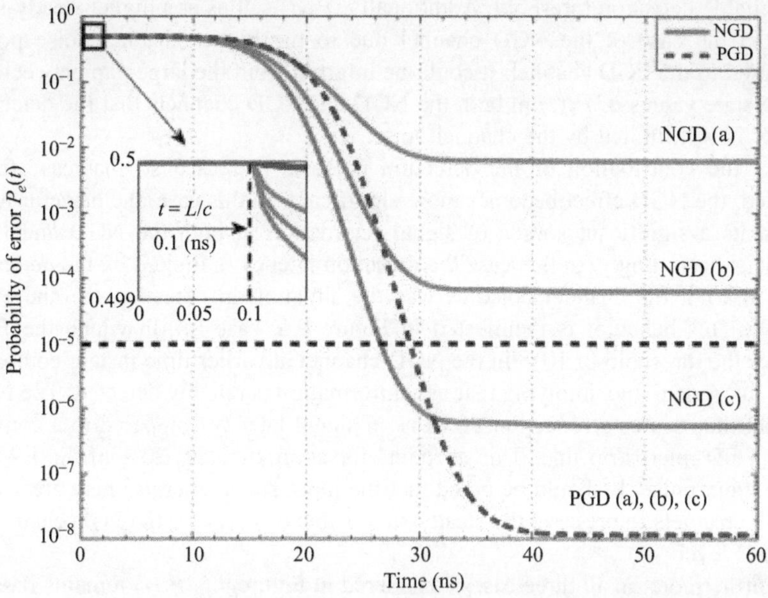

Figure 9.5 $P_e(t)$ for different scenarios of the correlation detector noise strength (a) $N_{o,det} \ll min\ (N_{o,ch,NGD}, N_{o,ch,PGD})$. (b) $N_{o,det} = max\ (N_{o,ch,NGD}, N_{o,ch,PGD})$. (c) $N_{o,det} \gg max\ (N_{o,NGD}, N_{o,PGD})$. Here, L is the channel's length and c is the speed of light in vacuum. The input energies are adjusted accordingly to preserve the same steady-state SNR for the PGD channel (denoted by the dashed curve) [41]

observed that $P_e(t)$ in the NGD channel is significantly time delayed regardless of the 6.3 (ns) pulse peak advancement exhibited in the NGD channel (as compared to the PGD channel). This is depicted in Figure 9.5, case (a). Such observation can be attributed to the large noise contribution of the NGD channel as compared to the noise generated in the normal PGD microstrip line. The other source of retardation for the $P_e(t)$ in the PGD microstrip line is the detector noise. However, in case (a), this source of retardation is not significant compared to the retardation caused by the NGD circuit noise. Therefore, despite the pulse peak advancement, the NGD channel fails to achieve reduction in information latency compared to the micro-strip line. It is worth noting that this observation is in agreement with [16,39] in which the detection latency was limited by the channel noise as well.

In case (b), the noise contribution of the detection system is comparable to the NGD channel. The input pulse energy in both channels is scaled (increased) accordingly to compensate the increase in the noise level in order to preserve the same steady-state $P_e(t)$ for the PGD channel at the 10^{-8} level (as a fixed reference). Although $P_e(t)$ in the NGD channel acquires lower values at earlier time instants, as illustrated in Figure 9.5, case (b), this is not accompanied by reliable signal detection; there is no new information detected according to our definition of 10^{-5}

as a reliable detection threshold. Additionally, $P_e(t)$ settles at a higher steady-state value in the case of the NGD channel due to the higher channel noise power compared to the PGD channel. It could be inferred from the large gap between the steady-state values of $P_e(t)$ in both the NGD and PGD channels that the detection latency is still limited by the channel noise.

As the contribution of the detection noise is increased so that case (c) is satisfied, the NGD effect becomes more significant. In this case, the detector noise represents a significant source of signal retardation in both the NGD and PGD channels; accordingly, in this case the detection latency is limited by the *detection noise* and not the channel noise as was the situation in cases (a)–(b) and Refs. [16,39]. This behavior is manifested in Figure 9.5, case (c), in which the $P_e(t)$ reaches the threshold of 10^{-5} in the NGD channel at earlier time instant compared to the microstrip line, implying that new information is reliably detected. The NGD channel thus scores a reduction of ~3 ns in signal latency compared to a conventional PGD microstrip line. This accounts for approximately 20% of the FWHM of the input pulse. It should be noted that the input signal energies here are scaled in both channels to preserve the steady-state value of $P_e(t)$ for the PGD channel at the 10^{-8} level.

Furthermore, in all three cases considered in Figure 9.5, $P_e(t)$ remains fixed at a value of 0.5 for all time instants $t < \frac{L}{c}$, where L is the channel length, and c is the speed of light in vacuum. This implies void reliable detection at earlier time instants before the arrival of the pulse front (precursor field) and validates Einstein's causality. In effect, this affirms that such reduction in signal latency does not imply superluminal information transfer. In other words, information has been detected in the NGD channel after the arrival of the strictly luminal front and before the arrival of detectable information in a PGD channel (with noisy detector).

Thus far, it has been shown that while information is delocalized and spread out over the entire spectral components of an electromagnetic pulse (from a fundamental point-of-view); however, from a practical consideration, it is possible to define a finite time window within which information can be reliably detected. This has been illustrated by applying a statistical approach to a noisy NGD microwave circuit terminated with a nonideal detector. Moreover, the possibility of reducing the detection latency of information (under certain conditions on the detector and the channel) has been demonstrated. It should be noted that information transfer is a matter of interaction between the input pulse energy level, temporal support, induced reshaping, medium noise and detector noise. This is depicted in Figure 9.6 where cases (a)–(c) correspond to the cases listed inTable 9.2. In all three cases, the noise in the NGD channel (marked in the black dotted line) is the same. In order to achieve reduction in signal detection latency [which corresponds to case (c)], the detector noise should be much larger than the NGD channel noise (~10 times larger in our analysis). In order to establish a fair comparison, the input pulse energy (for both the PGD and NGD channels) should be scaled accordingly (by the same amount) to preserve the same steady-state SNR for all the scenarios of the detector noise (a)–(c). It is worth mentioning that cases (a), (b), and (c) correspond to realistic scenarios that are fully valid in practice. For instance, as for the

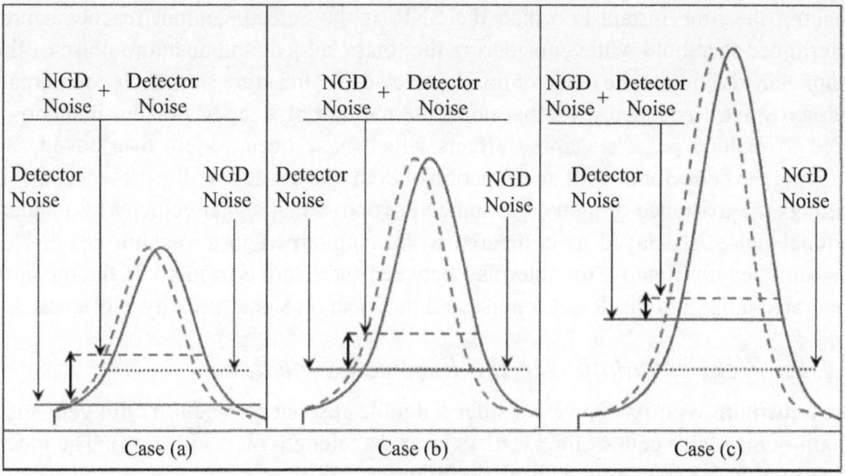

Figure 9.6 Schematic diagram illustrating the dynamics of the three inequalities (a), (b), and (c) discussed in Figure 9.5. The interplay between the medium noise, detector noise and input energy determines the information detection latency. Energy in case (c) > energy in (b) > energy in (a) to obtain a constant steady-state SNR for the PGD medium. In each case, the input energy for the NGD and the PGD medium is the same to ensure fair comparison [41]

particular NGD circuit considered in this analysis, the correlation detectors that satisfy the three noise profiles (a)–(c) are obtained from Refs. [46,47].

It should be added that the above results have been obtained while considering three different noise profiles in the same correlator detector at the receiver side. Indeed, the type of detector deployed at the receiver side also plays an important role in the signal detection latency. The interested reader is advised to check Ref. [41] for other detector types such as the envelope detector and the matched filter detector. Moreover, we note that the above conclusions have been conducted for the case of an NGD circuit in the microwave domain. In the remaining part of this chapter, we extend the calculations of signal velocity to the case of a double-resonance inverted medium with optical gain. At optical frequencies, the generated noise within the medium (channel) and the detector differ in nature from their counterpart at microwave frequencies, which can lead to new interpretations. However, the ability to reduce signal detection latency is still feasible, as will be shown.

9.4 Double-resonance Lorentzian medium with gain

In this section, we continue with our calculations of signal velocity in superluminal media but this time at optical frequencies. That is to say the velocity of detectable information (signal velocity) in a medium capable of supporting abnormal (superluminal or negative) group velocities is calculated. Again, this is carried out by tracking the time instant at which the SNR at the detector output reaches a predetermined threshold while considering the total classical and quantum noise of the channel in addition to the detector noise, which differ in nature from their counterpart at microwave frequencies. Furthermore, the method of steepest descent is incorporated to include pulse reshaping effects which have been widely overlooked. By studying the behavior of SNR as a function of both space and time, the present analysis predicts the existence of a cutoff distance beyond which signal velocity of a superluminal pulse is delayed as compared to its counterpart in a vacuum channel of the same length. Finally, the interplay between the relative strength of the medium generated noise and the detector noise and its effect on signal velocity is discussed.

9.4.1 Field calculation in the Lorentzian medium

For a medium with AGD, we consider a double-resonance medium with gain such as ammonia vapor cells at the GaAlAs laser wavelength ($\lambda = 780$ (nm)). The index of refraction for such a medium is described as [48]:

$$n(\omega) = \sqrt{1 + \frac{\omega_{p,1}^2}{\omega^2 - \omega_{0,1}^2 + 2i\delta\omega} + \frac{\omega_{p,2}^2}{\omega^2 - \omega_{0,2}^2 + 2i\delta\omega}} \qquad (9.4)$$

where $\omega_{p,j}$ and $\omega_{0,j}$ ($j = 1$ or 2) are the plasma and resonance frequencies associated with the first and second resonances, and δ is the phenomenological line-width ($\sim$400 MHz). For each of the resonances, the typical condition $\delta < \omega_{p,j} < \omega_{0,j}$ is satisfied. Such medium has a causal response and, consequently, the Kramers–Kronig relations are valid. The frequency-dependent index of refraction and the group delay of the gain medium, as a function of detuning, are depicted in Figure 9.7.

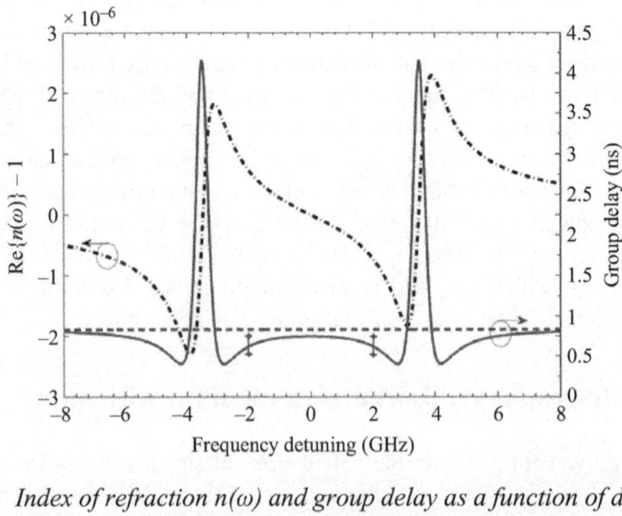

Figure 9.7 Index of refraction n(ω) and group delay as a function of detuning (the horizontal dashed line denotes the luminal group delay in vacuum). The vertical arrows mark the spectral width of the input pulse [42]

The ammonia medium exhibits abnormal group delay (with ~0.085 ns advancement) with a flat response within a detuning of 4 GHz around $\omega_c = 2.4166 \times 10^{15}$ rad/s ($\lambda = 780$ nm). We chose an input modulated Gaussian pulse centered at the carrier frequency $\omega_c = 2.4166 \times 10^{15}$ rad/s and an initial pulse width $2T = 0.8$ ns centered at $t_0 > 0$ (~3T). The pulse is excited at the plane $z = 0$ and propagates inside a medium with an index of refraction that is given in (9.4). The input Gaussian pulse is written in the form:

$$f(t) = e^{-\left(\frac{t-t_0}{2T}\right)^2 sin(\omega_c t)} \tag{9.5}$$

For such signal, the initial spectrum is expressed as:

$$F(\omega) = \sqrt{\pi}T\left[e^{-\frac{T^2(\omega-\omega_c)^2}{4}}\right]e^{-i\omega_c t_0} \tag{9.6}$$

Accordingly, the general description of the field $E(z,t)$ is obtained by solving the integral:

$$E(z,t) = \frac{1}{2\pi}Re\left\{i\oint \tilde{F}e^{z\phi(\omega,\theta')/c}d\omega\right\} \tag{9.7}$$

where $i = \sqrt{-1}$ and $\tilde{F}$ is the spectral amplitude, which is given by:

$$\tilde{F} = \sqrt{\pi}Te^{-i\omega_c t_0} \tag{9.8}$$

and $\phi(\omega, \theta')$ is the unified phase function for the Gaussian pulse, expressed as:

$$\phi(\omega,\theta') = \left\{i\omega\left[n(\omega) - \theta'\right] - \frac{cT^2}{4z}(\omega - \omega_c)^2\right\} \tag{9.9}$$

The term $\theta' = c(t - t_0)/z$ represents the dimensionless space-time parameter [49]. At a given propagation distance z, different values of θ' correspond to different instants of time.

The total field is calculated by deforming the integration contour in (9.6) over the path of steepest descent of the phase function $[\phi(\omega, \theta')]$. Such deformation is achieved by adding the contributions of the saddle points (ω_{SP}) that satisfy $\frac{\partial\phi(\omega=\omega_{SP},\theta')}{\partial\omega} = 0$. This approach provides useful insights to the time-frequency dynamics of the field and accounts for the precursor oscillations and pulse reshaping.

Figure 9.8 depicts the paths of the different saddle points in the complex frequency domain as a function of θ'. Each saddle point contribution ($A_{\omega_{SP}}$) to the total field can be expressed in the following closed form expression [49]:

$$A_{\omega_{SP}}(\theta') = \sqrt{\frac{c}{2\pi z}}Re\left\{\frac{i\tilde{F}e^{\left\{\frac{z}{c}\phi(\omega_{SP},\theta')\right\}}}{\sqrt{\frac{-\partial^2}{\partial\omega^2}\phi(\omega_{SP},\theta')}}\right\} \tag{9.10}$$

Additionally, the method of saddle points has another advantage: it highlights the saddle points that contribute significantly to the construction of the pulse as opposed to saddle points with minimal contributions. In Figure 9.8, the dominant saddle point, denoted as $\omega_{SP_{Middle}}$, follows the vertical path moving downwards at the carrier frequency $\omega_c = 2.4166 \times 10^{15}$ rad/s in the middle of the superluminal region. The other saddle points, denoted as $\omega_{SP_{Left}}$ and $\omega_{SP_{Right}}$, follow a circular path toward the branch cuts $(\omega^+(0), \omega^+(1))$ and $(\omega^+(2), \omega^+(3))$, respectively. These latter saddle points are located within the subluminal region and their contributions to the total generated pulse energy are negligible. As seen later, the information content of the pulse is recoverable without the need to detect these spectral components at the receiver side.

Note that for an input pulse with a broader spectral width (or at longer distances), the contributions of the saddle points in the subluminal region $(\omega_{SP_{Left}})$ and $(\omega_{SP_{Right}})$ become more pronounced and may even lead to a scenario in which the pulse becomes subluminal. In this analysis, we consider propagation lengths at which the pulse remains in the superluminal regime. This superluminality is realized as long as the following condition is satisfied [50]:

$$z \leq \frac{c\delta(\omega_{0,2} - \omega_{0,1})^2 T^2}{4\omega_p^2} \tag{9.11}$$

All saddle points contributions are added and the medium response $A(L, t)$ is calculated as depicted in Figure 9.9. The result is verified using the inverse Laplace method [51] and shows a perfect match as demonstrated in the same figure.

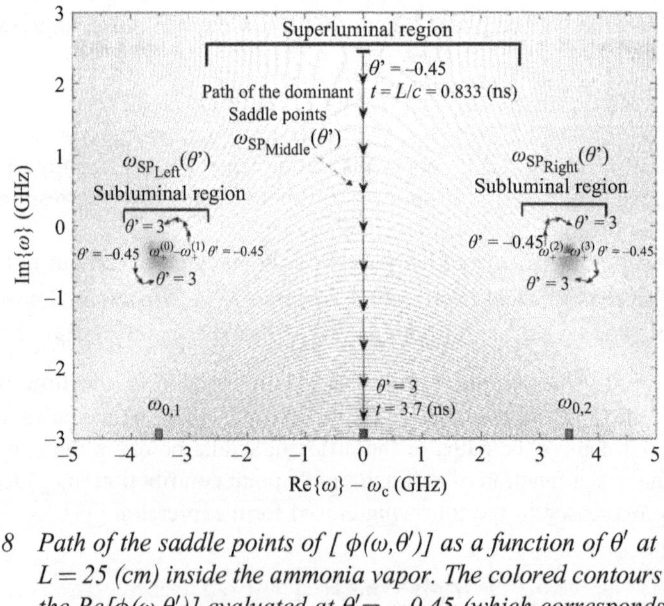

Figure 9.8 Path of the saddle points of $[\phi(\omega,\theta')]$ as a function of θ' at a distance $L = 25$ (cm) inside the ammonia vapor. The colored contours illustrates the $Re[\phi(\omega,\theta')]$ evaluated at $\theta' = -0.45$ (which corresponds to time $t = L/c$). The first and second resonances are denoted by $(\omega_0, 1)$ and $(\omega_0, 2)$, respectively. $(\omega^+(0), \omega^+(1))$, and $(\omega^+(2), \omega^+(3))$ signify the branch cuts for the resonances $\omega_{0,1}$ and $\omega_{0,2}$ [42]

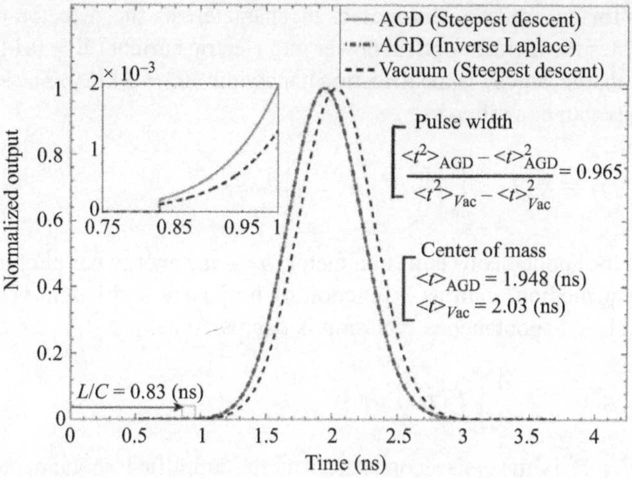

Figure 9.9 Normalized medium response at propagation distance L = 25 (cm). The insets depict the arrival of the pulse front after duration L/c in compliance with Einstein's causality [42]

The normalized Gaussian pulse, after traveling 25 (cm) inside the ammonia vapor, experiences slight compression (3.5%) and exhibits peak advancement (10.25% of the FWHM). The calculations of the expectation of arrival time, following Ref. [52], demonstrate 0.0842 ns advancement for the AGD medium in agreement with our results. The inset in Figure 9.9 depicts the arrival of the front in both ammonia and vacuum (of the same length L) with a luminal delay ($t = L/c$) in compliance with Einstein's causality and Sommerfeld's predictions.

Thus far the output response of the channel (dispersive medium) has been evaluated. At this point, the detection process needs to be included in our formulation. To do so, we must account for the noise generated in both the channel and the detector and calculate its impact on the SNR, as discussed next.

9.4.2 Noise analysis for the inverted Lorentzian medium

In order to calculate the signal velocity in an inverted medium, the noise generated in the medium and the detection process are taken into account. Since the ammonia vapor is acting as a gain medium, we include both the quantum and classical sources of noise in our analysis. We consider the following sources: shot noise ($P_{\text{Shot}}(t)$), spontaneous emission noise ($P_{\text{ASE}}(t)$), mutual beat noise between the signal and amplified spontaneous emission ($P_{\text{Sig-ASE}}(t)$), and self-beat noise of the amplified spontaneous emission ($P_{\text{ASE-ASE}}(t)$). The expressions for these noise sources are derived from [39,40] and extended here to be a function of both space and time. As such, the shot noise is given as:

$$P_{\text{shot}}(z,t) = 2eP_{\text{sig}}(z,t)\Re B_e \tag{9.12}$$

The term $P_{\text{sig}}(z,t)$ is the instantaneous power incident on the detector surface and can be expressed as $P_{\text{sig}}(z,t) = \frac{1}{t}\int_0^t E(z,\tau)^2 d\tau$, where $E(z,\tau)$ is the optical field after propagating for an arbitrary distance z. $\Re$ characterizes the detector responsivity which translates the incident optical power into electric current ($\Re = 0.1$ (A/W)), e is the electron charge, and B_e is the detection bandwidth ($B_e = 5$ kHz). The spontaneous emission noise can be written as:

$$P_{\mathrm{ASE}}(z,t) = N_{sp}\hbar\nu\frac{1}{t}\int_0^t\int\left[g(z,\omega)^2 - 1\right]d\omega d\tau \tag{9.13}$$

where N_{sp} is the spontaneous emission factor, $\hbar\nu$ is the energy per photon, $g(z,\omega)$ is the equivalent medium gain as a function of frequency and length. The self-beat noise of amplified spontaneous emission is expressed as:

$$P_{\mathrm{ASE-ASE}}(z,t) = \frac{1}{t}\int_0^t F(z,\tau)^2 d\tau \tag{9.14}$$

Here, $F(z,t)$ is the noise correlation of the amplified spontaneous emission expressed as $F(z,t) = N_{\mathrm{sp}}\hbar\nu\int[g(z,\omega)^2-1]e^{-i\omega t}d\omega$. Finally, the beat noise between the signal and the amplified spontaneous emission is expressed as:

$$P_{\mathrm{Sig-ASE}}(z,t) = \frac{2}{t}\int_0^t\left[F(z,\tau)P_{\mathrm{sig}}(z,\tau)\right]d\tau \tag{9.15}$$

Additionally, we account for the detector noise in terms of its thermal noise and dark current; where dark current arises from the mobility of electron charges in the detector circuit even in the absence of detected signal. For a photomultiplier detector, the $\mathrm{SNR}(z,t)$ at the detector output can be expressed as a function of time and space as follows [39,40,42]:

$$\mathrm{SNR}(z,t) = \frac{2G^2\left[P_{\{\mathrm{sig}\}(z,t)\Re}\right]^2}{G_e^2\left[2e\left(i_d + P_{\mathrm{sig}}(z,t)\Re + P_{\mathrm{ASE}}(z,t)\Re\right) + \frac{4P^2_{\mathrm{ASE-ASE}}(z,t)\Re}{B_0} + \frac{2P^2_{\mathrm{Sig-ASE}}(z,t)\Re}{B_0}\right]B_e + \frac{4KTB_e}{R}} \tag{9.16}$$

Here, G is the detector photo-multiplication gain ($G = 80$), i_d is the dark current ($i_d = 10^{-14}$ A), K is Boltzmann constant, and T is the temperature in Kelvins ($T = 300$ K). B_0 denotes the optical bandwidth ($B_0 = 10$ GHz) and $4KTB_eR$ represents the thermal noise associated with the equivalent detector resistance R ($R = 500\ \Omega$). For the vacuum channel, the noise sources only include the shot noise, thermal noise, and dark current.

9.4.3 Calculating signal velocity for the inverted medium

In this section, we proceed with our calculations for the signal velocity. At each space point z, the $\mathrm{SNR}(z,t)$ in (9.16) is solved for the time instant that corresponds to

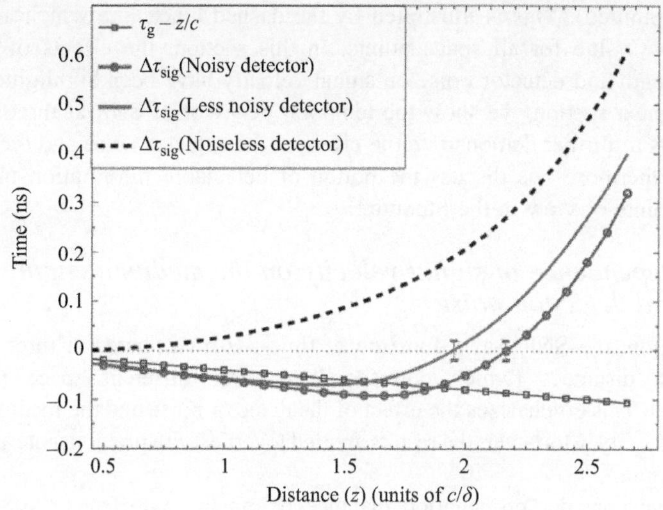

Figure 9.10 Relative signal delay in ammonia vapor cells with respect to a vacuum channel as a function of propagation distance in units of (c/δ). The relative signal delay is compared with the corresponding relative group delay (denoted by the line with square markers). This comparison is carried out for three different noise levels at the detector side [42]

0 dB (arrival of detectable information) for both the case of ammonia vapor and a vacuum channel. The time difference between the arrival of detectable information in the ammonia vapor medium and its arrival in vacuum is denoted as $\Delta\tau_{\text{Sig}}$ and is depicted in Figure 9.10.

A negative value for $\Delta\tau_{\text{Sig}}$ implies faster signal detection in the ammonia vapor medium. Figure 9.10 implies that there exists a range of space points within which the signal velocity is faster in the AGD medium than its counterpart in vacuum (as depicted by the negative values of the solid line and the line with circle markers).

Additionally, it is shown that there exists a cutoff distance beyond which the signal velocity is delayed as compared to vacuum despite the superluminal group delay (as denoted by the vertical markers). Accordingly, the medium length plays an important role in the proper interpretation of signal velocity and should not be overlooked in any experimental setup. This will be further discussed in the next section.

Moreover, the noise level of the detector (in terms of its thermal noise) determines whether signal velocity can be time advanced with respect to a vacuum channel. This is also depicted in Figure 9.10 in which the noisy and less noisy detectors correspond to detector equivalent resistance $R = 0.5$ and 5 kΩ, respectively. It is shown that signal latency can be reduced when the detector noise is much more pronounced. For instance, when the detector noise level is reduced, the cut off distance (at which signal velocity is delayed) shifts to a shorter length. Whereas in the limit of noiseless detector, it is shown that signal velocity is always delayed with respect to vacuum in agreement with Ref. [39] (in which the detector noise was omitted). This is illustrated by the dashed black line which acquires a non-negative value for all space points. In this section, the effects of both the medium length and detector noise on signal velocity have been highlighted.

In the next section, we show the temporal behavior of SNR at three different space points to further demonstrate the effect of the detector noise and the medium length. Furthermore, we discuss the notion of detectable information in light of previous points-of-view in the literature.

9.4.4 Dependence of signal velocity on the medium length and detection noise

In this section, the SNR—as a function of time—is investigated at three different propagation distances (which correspond to three different space points in Figure 9.10). This emphasizes the effect of the detector noise and the medium length on signal velocity. Moreover, the results are tied together with previous interpretations in the literature.

First we consider propagation distance below the cutoff ($L = 2.083\ (c/\delta)$), which corresponds to 25 (cm) inside the ammonia vapor. The SNR(t) associated with the input Gaussian pulse of (9.5), after propagation for distance L is calculated at the detector output and is depicted in Figure 9.11 (dot-dashed curve). The $SNR(t)$ of the same input pulse at the output of the same detector, but this time traveling 2.083 (c/δ) (25 (cm)) in vacuum is also calculated and shown in Figure 9.11 (solid black curve). Figure 9.11 shows that the front and high frequency oscillations associated with it (Sommerfeld precursors) maintain a strictly luminal speed in both

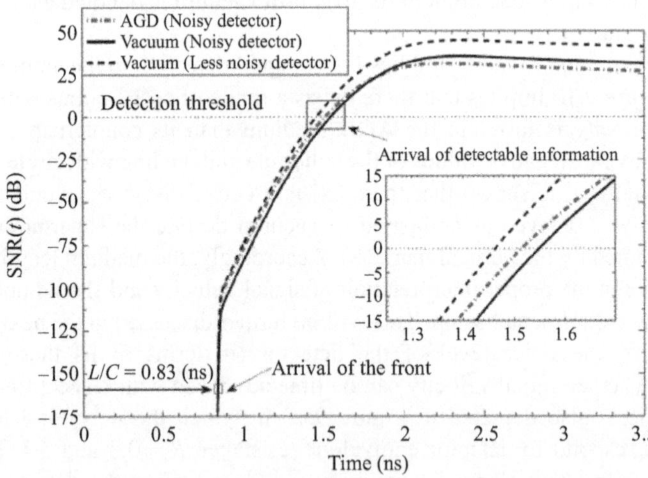

Figure 9.11 SNR as a function of time at a propagation depth of 25 (cm) in the ammonia vapor. The inset shows the information window at vicinity of the detection threshold as compared to a vacuum channel of the same length

media in agreement with Sommerfeld's and Brillouin's prediction [1–3]. This is demonstrated by the fact that the earliest response of both media occurs simultaneously after $t = L/c$ (0.83 (ns)) in compliance with Einstein causality.

From a communication system point-of-view, although the front or points of nonanalyticity, in which genuine information of a wave packet can be encoded, arrive luminaly; however, the association of information with these points is troublesome. For instance, in Figure 9.11, the SNR at onset of the pulse arrival (the front) is as low as −175 dB, which makes its detection extremely difficult if not impossible. Therefore, associating information and its velocity with the front (or singularities) when such field quantities cannot be measured makes little practical sense. For most communication systems, the decision threshold is set at 0 dB, when signal power is equal to the noise power. In light of Figure 9.11, it seems, apparently, that the information velocity associated with the AGD channel is superluminal at the 0 dB level—i.e., $\mathrm{SNR}(t)$ of the AGD channel is advanced in time as compared to the $\mathrm{SNR}(t)$ of vacuum. However, these results must be interpreted properly as discussed below.

The vacuum channel in the above discussion should no longer be regarded as "pure vacuum" but rather a vacuum loaded with a noisy detector. As such, the SNR of the AGD medium is compared to this perturbed vacuum channel. If the noise power associated with the detector is reduced—for example, by increasing the detector equivalent resistance (R, in (9.16) from 0.5 kΩ to 5 kΩ)—the $\mathrm{SNR}(t)$ of the loaded vacuum channel will be shifted in Figure 9.11 from the solid black line to the dashed black line on the same figure, while the $\mathrm{SNR}(t)$ for AGD channel remains almost unchanged (dominated by its inherent noise). In other words, any advantage that the AGD channel would have provided in terms of advanced detection of information washes away in comparison to the vacuum channel with less noisy detector. This effect is in agreement with the prediction in Figure 9.10. Consequently, compliance to Einstein's causality is imposed without the need to refer to the immeasurable front. It should be noted that the aforementioned finite increase in the detector equivalent resistance is always a possibility, although it may not be pursued in all cases due to the cost considerations. In such cases, this analysis clearly shows that there is an advantage to be gained in reducing the signal detection latency by utilizing AGD channels in particular applications in agreement with other studies which have considered electronic circuits as presented in Refs. [14,41].

The presented results bring together three points-of-view: First, it indicates that the recent experimental observations by Vogl *et al.* [28] in early detection of information encoded in an optical pulse traveling in an AGD medium as compared to free space, is a consequence of operating in a situation similar to the one described by the pair of solid black and dot-dashed curves in Figure 9.11. On the other hand, the observation by Chiao *et al.* [39], on the limitation imposed by quantum fluctuation on the advancement of the information in AGD medium, is a consequence of operating in a situation similar to the one described by the pair of dot-dashed and dashed black curves in Figure 9.11. Third, previous assertions by Sommerfeld and Brillouin [1–3], in associating information velocity with the velocity of front and singularities, correspond to the theoretical case of having an infinitely sensitive detector which in Figure 9.11 pushes the detection threshold to

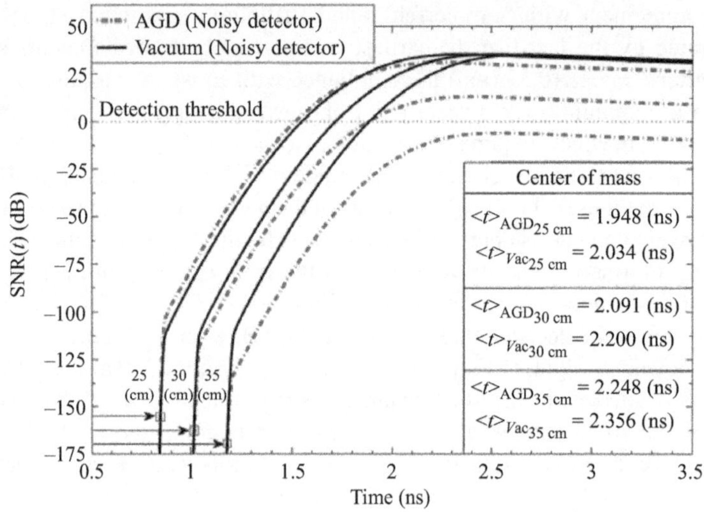

Figure 9.12 SNR(L,t) at different propagation depths L = 2.083 (c/δ), 2.5 (c/δ) and 2.917 (c/δ) (which corresponds to 25, 30, and 35 cm, respectively) in an AGD medium and compared to a vacuum channel. Evidently, information velocity is retarded in an AGD medium despite the exhibited superluminal behavior indicated by the center of mass calculations

the bottom of the figure at −175 dB. Finally, the deciding factor on which scenario applies to which experimental observation depends on the relative strength of the detector noise as compared to the total noise of the AGD channel in addition to the length of the medium.

In addition, the distance that an optical pulse, encoded with information, travels through an AGD medium greatly affects the velocity by which information is transmitted. This has been shown in Figure 9.10. To further investigate this dependency, Figure 9.12 shows the temporal behavior of SNR(L, t) for the AGD and vacuum channels at three different lengths where the detector used is the same as the noisy detector depicted by the dot-dashed line in Figure 9.11 ($R = 0.5$ kΩ). As the propagation depth increases, any advantage of the AGD channel (in early detection of information) over the vacuum channel of the same length washes away, which is manifested at $z = 25$ and 30 cm. This diminished advantage can be discerned as a result of the increased quantum noise generated in the longer gain medium as compared to the perturbed vacuum.

9.5 Conclusion

In this chapter, we have studied the notions of group, front, and information velocities in media exhibiting AGD from a practical communications system point-of-view, our analysis included the case of an NGD microwave circuit, as well as a double-resonance inverted gain medium at optical frequencies. We have considered propagation of electromagnetic pulses in dispersive media with AGD while including pulse reshaping effects, noise generated in the medium, and detection noise. It is observed that, depending on the relative strength of the noise power in the AGD channel as compared to the detector noise, the AGD medium can be used to reduce the information detection latency as compared to a vacuum channel of equal length. For the case the inverted medium, it has been shown that, for a superluminal pulse, there exists a cutoff distance beyond which the signal velocity is delayed with respect to vacuum. We envision these observations to be very useful guidelines for the design of any signaling scheme that utilizes AGD to reduce signal detection latency.

References

[1] A. Sommerfeld, "About the propagation of light in dispersive media," Annals of Physics, Vol. 44, pp. 177–202, 1914.

[2] L. Brillouin, "About the propagation of light in dispersive media," Annals of Physics, Vol. 44, p. 203, 1914.

[3] L. Brillouin, "Wave Propagation and Group Velocity," Academic, New York, 1960.

[4] G. B. Garrett and D. E. McCumber, "Propagation of a Gaussian light pulse through an anomalous dispersion medium," Physical Review A, Vol. 1, pp. 305–313, 1970.

[5] S. Chu and S. Wong, "Linear pulse-propagation in an absorbing medium," Physical Review Letters, Vol. 48, pp. 738–741, 1982.

[6] B. Segard and B. Macke, "Observation of negative velocity pulse propagation," Physics Letters, Vol. 109A, pp. 213–216, 1985.

[7] M. Mojahedi, E. Schamiloglu, F. Hegeler, and K. J. Malloy, "Time domain detection of superluminal group velocity for single microwave pulses," Physical Review E, Vol. 62, pp. 5758–5766, 2000.

[8] B. Ravelo, A. Prennec, and M. Le Roy, "Synthesis of broadband negative group delay active circuits," IEEE International Microwave Symposium, Honolulu, HI, pp. 2177–2180, 2007.

[9] B. Ravelo, A. Perennec, M. Le Roy, and Y.G. Boucher, "Active microwave circuit with negative group delay," IEEE Microwave and Wireless Components Letters, Vol. 17, pp. 861–863, 2007.

[10] P. Barbero, H. E. Hernandez-Figueroa, and E. Recami, "Propagation speed of evanescent modes," Physical Review E, Vol. 62, p. 8628, 2000.

[11] D. Mugnai, A. Ranfagni, and L. Ronchi, "The question of tunneling time duration: A new experimental test at microwave scale," Physics Letters A, Vol. 247, pp. 281–286, 1998.

[12] A. Enders and G. Nimtz, "Evanescent mode propagation and quantum tunneling," Physical Review E, Vol. 48, pp. 632–634, 1993.

[13] A. Ranfagni, D. Mugnai, P. Fabeni, and G. P. Pazzi, "Delay-time measurements in narrowed wave-guides as a test of tunneling," Applied Physics Letters, Vol. 58, pp. 774–776, 1991.

[14] D. Solli, R. Y. Chiao, and J. M. Hickmann, "Superluminal effects and negative group delays in electronics, and their applications," Physical Review E, Vol. 66, p. 056601, 2002.
[15] J. Keaveney, I. G. Hughes, A. Sargsyan, D. Sarkisyan, and C. S. Adams, "Maximal refraction and superluminal propagation in a gaseous nanolayer," Physical Review Letters, Vol. 109, p. 233001, 2012.
[16] M. D. Stenner, D. J. Gauthier, and M. A. Neifeld, "The speed of information in a fast-light optical medium," Nature, Vol. 425, pp. 695–698, 2003.
[17] L. J. Wang, A. Kuzmich, and A. Dogariu, "Gain-assisted superluminal light propagation," Nature, Vol. 406, pp. 277–279, 2000.
[18] K. Kim, H. Seb Moon, C. Lee, S. Kyoung Kim, and J. Bog Kim, "Observation of arbitrary group velocities of light from superluminal to subluminal on a single atomic transition line," Physical Review A, Vol. 68, p. 013810, 2003.
[19] C. Spielmann, R. Szipocs, A. Stingl, and F. Krausz, "Tunneling of optical pulses through photonic band-gaps," Physical Review Letters, Vol. 73, pp. 2308–2311, 1994.
[20] A. M. Steinberg, P. G. Kwiat, and R. Y. Chiao, "Measurement of the single-photon tunneling time," Physical Review Letters, Vol. 71, p. 708, 1993.
[21] R. Y. Chiao and A. M. Steinberg, Progress in Optics, Vol. 37, edited by: E. Wolf, Elsevier, pp. 347–406, 1997.
[22] I. I. Smolyaninov, A. V. Zayats, A. Gungor, and C. C. Davis, "Single-photon tunneling via localized surface plasmons," Physical Review Letters, Vol. 88, p. 187402, 2002.
[23] S. Zhang, J. F. Chen, C. Liu, M. M. T. Loy, G. K. L. Wong, and S. Du, "Optical precursor of a single photon," Physical Review Letters, Vol. 106, p. 243602, 2011.
[24] J. F. Woodley and M. Mojahedi, "Negative group velocity and group delay in left-handed media," Physical Review E, Vol. 70, p. 046603, 2004.
[25] O. F. Siddiqui, S. J. Erickson, G. V. Eleftheriades, and M. Mojahedi, "Time-domain measurement of negative group delay in negative-refractive-index transmission-line metamaterials," IEEE Transactions on Microwave Theory and Techniques, Vol. 52, pp. 1449–1454, 2004.
[26] Q. Li, Z. Zhang, J. Wang, M. Qiu, and Y. Su, "Fast light in silicon ring resonator with resonance-splitting," Optics Express, Vol. 17, pp. 933–940, 2009.
[27] C. Ciminelli, C. E. Campanella, F. Dell Olio, and M. N. Armenise, "Fast light generation through velocity manipulation in two vertically-stacked ring resonators," Optics Express, Vol. 18, pp. 2973–2986, 2010.
[28] U. Vogl, R. T. Glasser, and P. D. Lett, "Advanced detection of information in optical pulses with negative group velocity," Physical Review A, Vol. 86, p. 031806, 2012.
[29] E. L. Bolda, R. Y. Chiao, and J. C. Garrison, "Two theorems for the group velocity in dispersive media," Physical Review A, Vol. 48, pp. 3890–3894, 1993.

[30] K. Wynne, "Causality and the Nature of Information," Elsevier, Optics Communications, Vol. 209, pp. 85–100, 2002.

[31] M. W. Mitchell and R. Y. Chiao, "Causality and negative group delays in a simple bandpass amplifier," American Journal of Physics, Vol. 66, p. 14, 1998.

[32] M. Tomita, H. Amano, S. Masegi, and A. I. Talukder, "Direct observation of a pulse peak using a peak-removed Gaussian optical pulse in a superluminal medium," Physical Review Letters, Vol. 112, p. 093903, 2014.

[33] T. Martin and R. Landauer, "Time delay of evanescent electromagnetic waves and the analogy to particle tunneling," Physical Review A, Vol. 45, p. 2611, 1992.

[34] G. Diener, "Superluminal group velocities and information transfer," Physics Letters A, Vol. 223, pp. 327–331, 1996.

[35] Z. Yang, "Physical mechanism and information velocity of fast light: A time domain analysis," Physical Review A, Vol. 87, p. 023801, 2013.

[36] M. Pawlowski, T. Paterek, D. Kaszlikowski, V. Scarani, A. Winter, and M. Zukowski, "Information causality as a physical principle," Nature Letters, Vol. 461, pp. 1101–1104, 2009.

[37] M. Mojahedi, K. J. Malloy, G. Eleftheriades, J. Woodley, and R. Y. Chiao, "Abnormal wave propagation in passive media," IEEE Journal of Select Topics in Quantum Electronics, Vol. 9, pp. 30–39, 2003.

[38] A. H. Dorrah and M. Mojahedi, "Nonanalytic pulse discontinuities as carriers of information," Physical Review A, Vol. 93, p. 013823, 2016.

[39] A. Kuzmich, A. Dogariu, L. J. Wang, P. W. Milonni, and R. Y. Chiao, "Signal velocity, causality, and quantum noise in superluminal light pulse propagation," Physical Review Letters, Vol. 86, p. 3925, 2001.

[40] M. Z. Feng, W. Sorin, and R. S. Tucker, "Fast light and the speed of information transfer in the presence of detector noise," IEEE Photonics Journal, Vol. 1, pp. 213–224, 2009.

[41] A. H. Dorrah, L. Kayili, and M. Mojahedi, "Superluminal propagation and information transfer: A statistical approach in the microwave domain," Physics Letters A, Vol. 378, pp. 3218–3224, 2014.

[42] A. H. Dorrah and M. Mojahedi, "Velocity of detectable information in faster-than-light pulses," Physical Review A, Vol. 90, p. 033822, 2014.

[43] A. Podell, "A functional GaAs FET noise model," IEEE Transactions on Electronic Devices ED, Vol. 28, pp. 511–517, 1981.

[44] EC2612 pseudomorphic high-electron-mobility transistor, United Monolithic Semiconductors, (2000). Available: http://www.datasheetcatalog.org/datasheet/UnitedMonolithicSemiconductors/EC2612-99F_00.pdf

[45] R. E. Ziemer and W. H. Tranter, "Principles of Communication Systems, Modulation, and Noise," 3rd ed. Boston, MA: Houghton Mifflin, 1990, pp. 451–474.

[46] J. Ryckaert, M. Badaroglu, V. De Heyn, *et al.*, "A 16 mA UWB 3-to-5 GHz 20 Mpulses/s quadrature analog correlation receiver in 0.18 μm CMOS" IEEE International Solid-State Circuits Conference, San Francisco, CA, 2006, p. 368.

[47] T. Kan, D. Ma, R. Mak, and H. C. Luong, "Design theory and performance of 1-GHz CMOS downconversion and upconversion mixers," Analog Integrated Circuits and Signal Processing, Vol. 24, pp. 101–111, 2000.

[48] A. M. Steinberg and R. Y. Chiao, "Dispersionless, highly superluminal propagation in a medium with a gain doublet," Physical Review A, Vol. 49, p. 2071, 1991.

[49] C. M. Balictsis and K. E. Oughstun, "Generalized asymptotic description of the propagated Field dynamics in Gaussian pulse propagation in a linear, causally dispersive medium," Physical Review E, Vol. 55, p. 1910, 1997.

[50] A. H. Dorrah and M. Mojahedi, "Time-frequency dynamics of superluminal pulse transition to the subluminal regime," Physical Review E, Vol. 91, p. 033206, 2015.

[51] P. Wyns, D Foty, and K. E Oughstun, "Numerical analysis of the precursor fields in linear dispersive pulse propagation," Journal of the Optical Society of America A, Vol. 6, pp. 1421–1429, 1989.

[52] J. Peatross, S. A. Glasgow, and M. Ware, "Average energy flow of optical pulses in dispersive media," Physical Review Letters, Vol. 84, p. 2370, 2000.

Chapter 10

Split-ring resonator-loaded X-band waveguide supporting negative group velocity

Joaquim J. Barroso[1], Joaquim P. Leite Neto[2], Pedro J. Castro[2], Ugur C. Hasar[3], and José Edimar B. Oliveira[1]

Abstract

Analytical discussions and experimental results are presented for an X-band (8.2–12.4 GHz) waveguide loaded with an array of split-ring resonators (SRR) designed to exhibit negative group delay and negative refractive index in the 2.8–3.6 GHz frequency range, thus below the cutoff frequency (6.557 GHz) of the empty waveguide. Negative-group-velocity effects arise from resonant absorption losses of the split-rings, whereby such effects are demonstrated by time-domain simulations of propagating modulated Gaussian pulses. It is shown that an injected pulse of 80-ns time width and central frequency of 3.26 GHz experiences a negative delay of −14 ns after traversing the waveguide while retaining its Gaussian shape although with strong attenuation. A retrieval procedure is used to determine the complex propagation factor for which the role of the attenuation constant in reshaping the output spectrum amplitude is discussed. The SRR-loaded waveguide examined here has attractive features, as the system is easily scalable to higher frequencies and proves to show potential applications for the delay compensations of signal in microwave waveguide systems by keeping a constant group delay to avoid signal distortion.

Keywords

Metamaterials, split-ring resonator, negative group velocity, retrieval of constitutive parameters.

[1]Divisão de Engenharia Eletrônica, Instituto Tecnológico de Aeronáutica, Brazil
[2]Laboratório Associado de Plasmas, Instituto Nacional de Pesquisas Nacionais, Brazil
[3]Department of Electrical and Electronics Engineering, Faculty of Engineering, University of Gaziantep, Turkey

10.1 Introduction

Fundamentally related to causality [1,2], negative group delay (NGD) on account of its time advancement property finds practical applications in communication systems by introducing a compensatory delay in electronics devices to improve their performance, constrained by group delay caused by frequency selective components [3]. The realization of NGD effects has been achieved in reflection-type planar NGD circuits [4] and in the gigahertz range by using active circuit topologies [5,6]. In addition, NGD has been demonstrated in a variety of physical systems, namely, in gain-assisted anomalous dispersive atomic medium [7], Bragg grating [8], metamaterial [9], erbium-doped optical fiber [10], Mach–Zehnder interferometer [11], birefringent crystal [12], and recently in a non-Foster loaded waveguide [13].

While superluminal dispersive effects can arise in a medium with neither absorption nor reflection, e.g., a photonic band gap structure [14], NGD propagation occurs in metamaterials, which are intrinsically characterized by anomalous dispersion such as a lossy left-handed transmission line [15]. In fact, NGD phenomena have been observed in engineered materials to provide anomalous dispersion with metamaterial media implemented with transmission lines periodically loaded by lumped resonant RLC circuits [16,17] and with microstrip line coupled with split-ring resonators (SRRs) [18].

Experiments on the propagation and transmission of electromagnetic waves in waveguides loaded with anisotropic magnetic metamaterials (e.g., periodic arrangements SRRs) have enabled the development of subwavelength waveguides, resonators, filters, and dephasers, for instance. Early experiments in this area [19–21] showed that in a waveguide loaded with an SRR resonator array a backward wave travels in a narrow frequency range and below the cutoff frequency of the fundamental TE_{10} mode of the empty waveguide. The counterintuitive phenomenon of below-cutoff wave propagation was explained by means of the negative permeability introduced by the split rings inside the waveguide in conjunction with the negative electrical permittivity provided by the waveguide at frequencies below the cutoff frequency. Metamaterial-loaded waveguides attract interest as metamaterials can be properly engineered to modify the propagation characteristics of the waveguide according to user-defined dispersion relations, in addition to enabling miniaturization of guiding systems.

In the present work, analytical discussions and experimental results are presented for an X-band waveguide loaded with an array of SRRs composed of two, four, and six, unit cells designed to exhibit negative group velocity in a frequency band of anomalous dispersion region. Measured and simulated spectra show the transmission characteristics of the composite medium and how the array length relates to anomalously dispersive behavior of the effective propagation medium. This is done by retrieving the complex propagation factor $\gamma = \alpha + j\beta$ from *S*-parameters, with α and β denoting, respectively, the attenuation and phase constants, for which dispersion curves of α and β as function of angular frequency, ω, are obtained. NGD phenomenon is evidenced by a branch of negative slope in the ω vs. β curve, in a way such that the negative-slope branch is associated with high attenuation in the corresponding ω vs. α curve. These features are further examined through electromagnetic simulations by injecting modulated Gaussian pulses into the SRR-loaded waveguide whereby NGD is demonstrated in the time domain. Namely, an injected pulse of 80-ns time width experiences a negative delay of -14 ns after traversing the waveguide while retaining its Gaussian shape although with strong attenuation. The remainder of the paper is organized as follows. The design and constructional features of the SRR along with the experimental setup are given in Section 10.2. The retrieval procedure to determine the propagation factor is given in Section 10.3, followed by discussion of experimental results in Section 10.4. Electromagnetic simulations are presented in Section 10.5, where the role of attenuation in reshaping the spectrum amplitude is pointed out. The paper is concluded in Section 10.6.

10.2 Waveguide loaded with SRRs and experimental setup

The SRR used in our experiments and simulations is depicted in Figure 10.1 along with its geometric parameters [22,23]. The metallic inclusion (copper with conductivity 5.7×10^7 S/m) is 37 μm thick and patterned on a kapton substrate (dielectric constant 3.2 and loss tangent 18.0×10^{-4}). The gap is introduced in the rings to create a resonant response, such that the gaps acting as capacitors make the circulating induced current oscillate back and forth giving rise to a local magnetic field. Thus, having a capacitance and a distributed self-inductance provided by the ring arms, the resulting split ring behaves like an LC resonant circuit. Two concentric rings are used to increase the capacitance which reduces the resonance frequency.

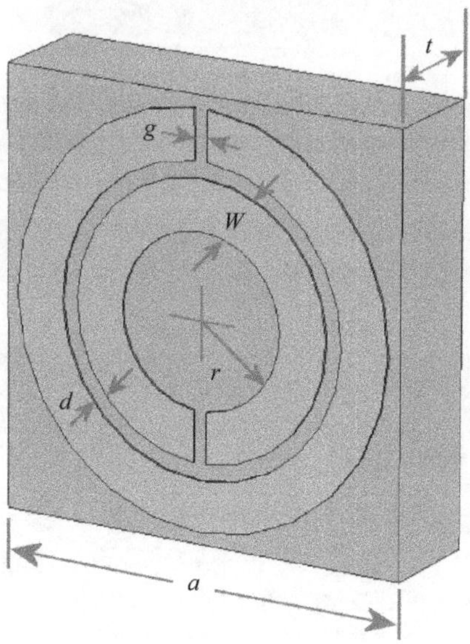

Figure 10.1 Schematics of the SRR used in the experiments and simulations. Parameters: $a = 10.2$ mm, $d = 0.75$ mm, $w = 0.75$ mm, $t = 1.6$ mm, $r = 2.25$ mm, $g = 1.13$ mm, and substrate electrical permittivity $\varepsilon = 3.2$

The SRR dimensions were determined from design formulas [22] so that the square unit cell size $a = 1.02$ cm is much smaller than the resonant wavelength 8.70 cm, corresponding to the magnetic resonance frequency of 3.45 GHz. Closing the gaps suppresses the magnetic response, but without substantially affecting the electric response since the metallic rings act as wires. In this situation, the induced current is still allowed to flow but no more can the current oscillate independently of the external electromagnetic field.

To load the X-band waveguide, we placed the SRR periodic array in the longitudinal plane of symmetry of a WR90 waveguide (cross-section area of 22.86×10.16 mm^2 and cutoff frequency of 6.557 GHz). Three arrays of two, four, and six SRRs were fabricated to examine the effect of the array length on the transmission band [23]. The experimental arrangement is shown in Figures 10.2 and 10.3, where both ends of the loaded waveguide were symmetrically attached to

Figure 10.2 WR90 waveguide loaded with six SRRs. The lattice constant (distance between the centers of two adjacent rings is $a = 10.2$ mm)

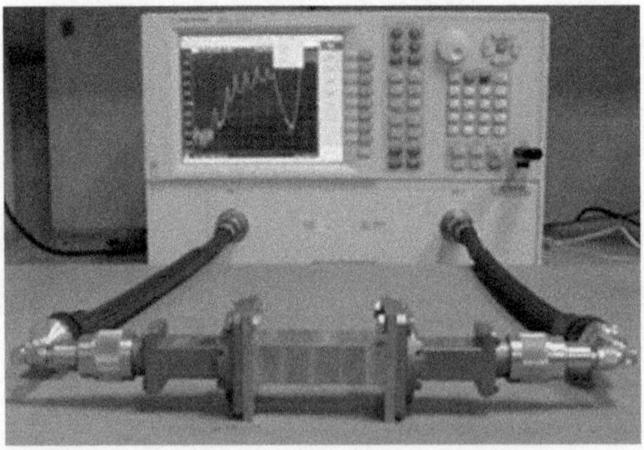

Figure 10.3 Setup to measure the transmission band of the X-band waveguide loaded with the SRR array using waveguide coaxial adapters

waveguide-to-coaxial adapters, through which the input and output signals are injected and detected by a vector network analyzer. Also, we note from Figure 10.2 that each array is made a bit longer than the corresponding waveguide section to facilitate the coupling of the input signal to the array of SRRs.

10.3 Retrieval of transmission parameters

The properties of the metamaterial inserted in the rectangular waveguide operating in the fundamental TE_{10} mode are determined by measuring the scattering (*S*-) parameters through a network analyzer. In terms of the semi-infinite reflection and transmission coefficients, the backward and forward complex *S*-parameters can be written as [24,25]:

$$S_{11} = \frac{\Gamma(1 - \mathrm{T}^2)}{1 - \Gamma^2\mathrm{T}^2}, \quad S_{21} = \frac{\mathrm{T}(1 - \Gamma^2)}{1 - \Gamma^2\mathrm{T}^2} \tag{10.1}$$

where Γ and T are the semi-infinite intrinsic reflection and transmission coefficients at the air-SRR array interfaces. Assuming the time harmonic variation $\exp(j\omega t)$, these coefficients are expressed as:

$$\Gamma = \frac{\mathrm{z} - 1}{\mathrm{z} + 1}, \quad \mathrm{T} = e^{-\gamma L} \tag{10.2}$$

where L is the length of the SRR array and $\gamma = \alpha + i\beta$ is the complex propagation factor, including the phase propagation constant β and the attenuation coefficient α. From (10.1) and (10.2), the normalized wave impedance z, the transmission coefficient T, and the propagation factor γ are retrieved from the *S*-parameters as:

$$\mathrm{z} = \pm\sqrt{\frac{(1 + S_{11})^2 - S_{21}^2}{(1 - S_{11})^2 - S_{21}^2}}, \quad \mathrm{T} = \frac{S_{21}}{1 - S_{11}\left(\dfrac{\mathrm{z} - 1}{\mathrm{z} + 1}\right)} \tag{10.3}$$

$$\gamma = \frac{1}{L}[-\ln T + j2\pi m], m = 0, \pm 1, \pm 2 \ldots \tag{10.4}$$

Since the material under consideration is a passive medium, the sign in (10.3) is determined by the requirement that $\mathrm{Re}[z] \geq 0$. In (10.4), m is an integer denoting the branch of the logarithmic function $\ln T$ where the phase $2\pi m$ is added to make the complex function $\ln T$ single-valued and continuous [26,27]. The refractive index is determined as $n = -j\gamma/\beta_0$, where β_0 is the free-space wavenumber. Normalized to β_0, the index n quantifies the phase delay per unit length in the loaded waveguide, relative to the phase delay in vacuum. Also, the passivity condition requires $|T| \leq 1$ implying that the imaginary part of the refractive index is $\mathrm{Im}\{n\} \leq 0$.

10.4 Experimental measurements

First, the waveguide loaded with the SRR array was symmetrically connected at both ends to identical X-band waveguide-to-coaxial adapters, by means of which the input and output signals were injected and detected by an Agilent N5230C PNA-L (Keysight Technologies) microwave network analyzer, as can be seen in Figure 10.3. The transmission-type technique (S_{21} parameter) was used for calibration of the measurement setup (vector network analyzer Agilent N5230C). Once the setup was calibrated, the unloaded waveguide was attached directly to the VNA via a coaxial cable and then the waveguide response was measured. We inserted the metamaterial array into the waveguide and measured the transmission spectra for samples of different lengths. Calibration was performed in the 1–6 GHz range using a coaxial-to-waveguide probe.

The magnitudes and phases of the reflection and transmission coefficients, S_{11} and S_{21}, are shown in Figures 10.4 and 10.5, respectively. The magnitude parameters displayed in Figure 10.4 indicate that maximum transmission intensity ($S_{21} \sim 0.22$) occurs at around 3.32 GHz, well below the cutoff frequency (~6.56 GHz) of the empty X-band waveguide. The positive slope in the 3.1–3.3 GHz range of the S_{21} phase curve in Figure 10.5 indicates a negative group delay, defined as $\tau = -d\varphi/d\omega$, where φ is the phase of S_{21}.

A dispersion diagram obtained from S-parameters is presented in Figure 10.6. From 3.24 to 3.34 GHz, the propagation is backward since in this frequency range the phase velocity is negative and the group velocity is positive. For higher frequencies ($f > 3.34$ GHz), both the phase velocity and the group velocity are positive and then the propagation is forward. There is no gap between the forward and backward modes, which indicates that this structure is similar to a balanced composite right left-handed (CRLH) transmission line as described in [28] and [29].

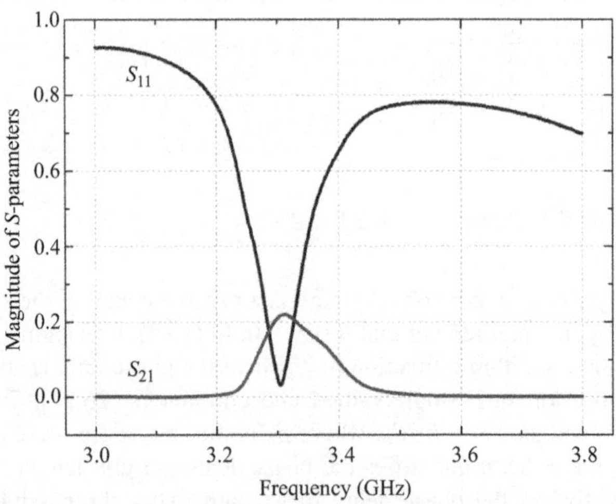

Figure 10.4 Measured magnitudes of reflection (S_{11}) and transmission (S_{21}) scattering parameters

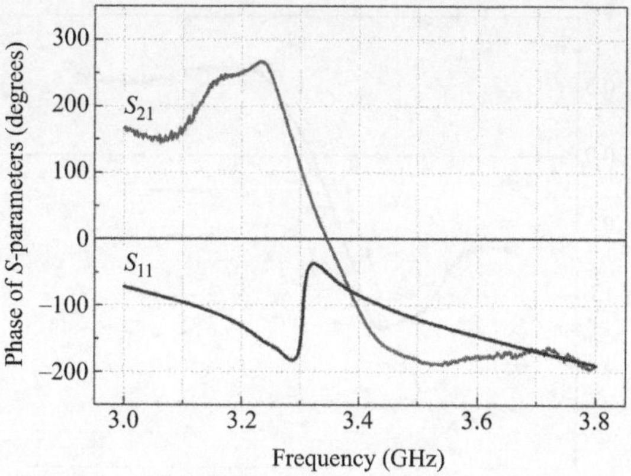

Figure 10.5 Measured unwrapped phases of reflection and transmission coefficients for the SRR-loaded X-band waveguide

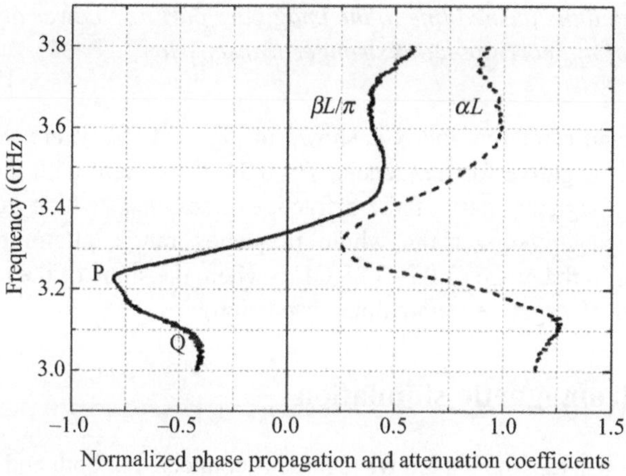

Figure 10.6 Dispersion diagram for the SRR array inserted in the WR90 waveguide

The dashed curve indicates that the highest attenuation corresponds to point *Q*, at the end of branch PQ which shows negative slope. More importantly, the dispersion curve exhibits a knee (point P) at 3.24 GHz below of which the curve bends downward with a negative slope, such that the group velocity is negative, thus characterizing anomalous dispersion. Such a feature was observed in a coplanar waveguide loaded with kinetic inductances [30], leading to negative index refraction as large as −700 and examined analytically in a left-handed lossy transmission line [15]. These experimental results and the anomalous-dispersion curve (Figure 10.6) arising from the SRR-loaded waveguide will be discussed in the next section through full-wave electromagnetic simulations.

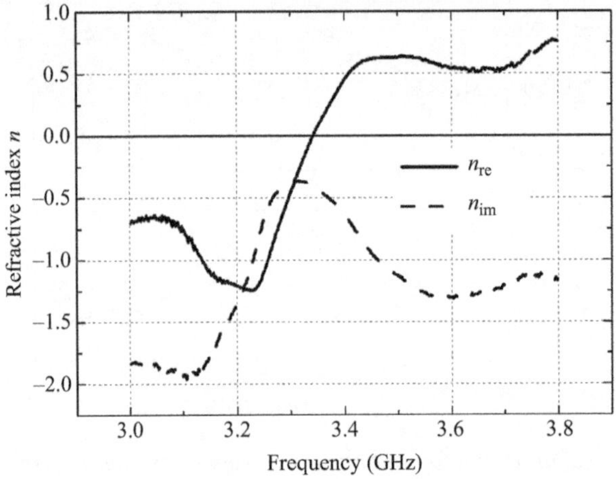

Figure 10.7 Refractive index extracted from scattering parameters measured from the loaded waveguide. The solid line refers to the real part n_{re} and the dashed line to the imaginary part n_{im}. Lower absolute values of n_{im} correspond to stronger transmission

The retrieved refractive index is shown in Figure 10.7, where the real refractive index n_{re} is negative for frequencies $f \lesssim 3.34$. Consistent with the principle of passivity, the imaginary part of the refractive index, n_{im}, which corresponds to absorption, is negative over the whole frequency range, attaining the highest absolute values in the range (3.05–3.25 GHz) where the slope of the real part of the refractive index is negative (anomalous dispersion).

10.5 Electromagnetic simulation

In the simulation setup, the loaded waveguide is connected at both ends to an S-band waveguide (WR-284) with the cross-sectional area of 72.13×34.04 mm^2 and the cutoff frequency of 2.078 GHz; the guide on the left is used for injecting waves to an X-band waveguide (WR-90) operated at below its cutoff frequency 6.557 GHz, while the WR-284 guide on the right detects the transmitted signal. Full-wave electromagnetic simulations were performed in the frequency range from 3.0 to 3.7 GHz.

To simulate the loaded waveguide system, excitation and detection ports are placed perpendicular to the propagation z-direction of the guide with PEC (perfect electrical conductor) boundaries in the simulation 3-D structure. As displayed in Figures 10.8 and 10.9, the ports are located at distances $L_1 = L_2 = 10$ mm from the SRR array. The system is excited by injecting a TE_{10}-mode wave for the S-band waveguide with the electric field polarized along the y-direction, i.e., parallel to the line of slits and with the propagation vector along the z-direction (Figure 10.8).

A full-wave 3-D electromagnetic simulator (CST Microwave Studio) was employed [31] by using hexahedral mesh with ten cells per maximum wavelength

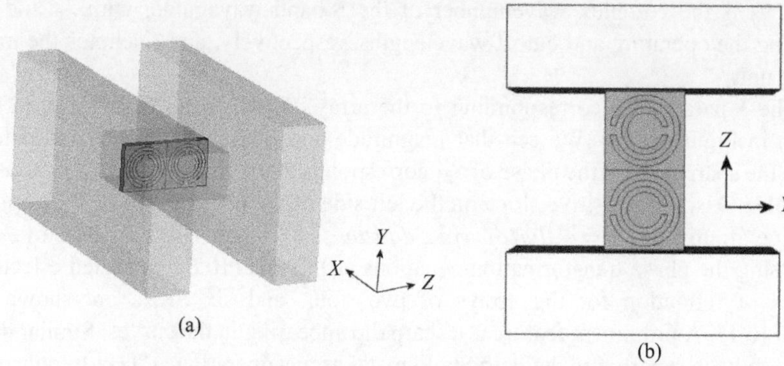

Figure 10.8 *(a) Perspective and (b) side views of the SRR loaded X-band waveguide with input (excitation) and output (detection) ports at the openings of the S-band waveguide*

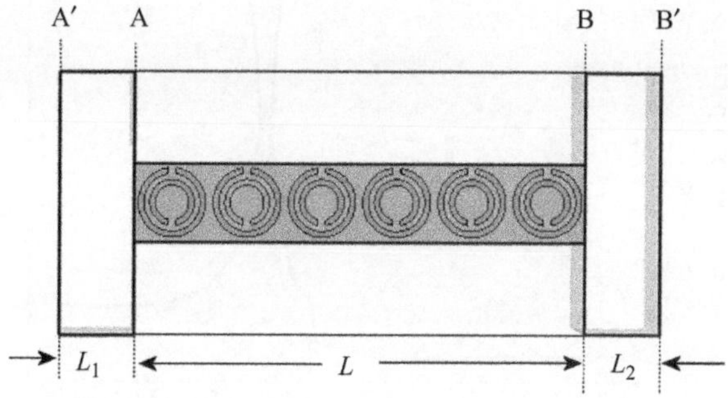

Figure 10.9 *Side view of the waveguide loaded with an array of six SRRs. Simulated S-parameters are referred to the input and output ports at planes A′ and B′, respectively, and then referenced back to the sending and receiving terminations of the SRR array at planes A and B array to proceed with the retrieval procedure described in Section 10.3*

with a total of 11,500 cells. Simulated *S*-parameters are calculated at planes A' and B' as pictured in Figure 10.9 and then referenced to planes A and B by shifting phase references planes from the ports to the SRR-array terminations according to

$$S_{11,A} = S_{11,A'} \exp(2\gamma_0 L_1) \tag{10.5a}$$

$$S_{21,B} = S_{21,B'} \exp[\gamma_0 (L_1 + L_2)] \tag{10.5b}$$

where

$$\gamma_0 = j2\pi\sqrt{(1/\lambda_0^2) - (1/\lambda_c^2)} \tag{10.6}$$

where γ_0 is the complex wavenumber of the S-band waveguide, with λ_0 and λ_c denoting the operating and cutoff wavelengths, respectively, and j denotes the imaginary unit.

The *S*-parameters corresponding to the array of two SRRs (Figure 10.8) are shown in Figure 10.10. We see that magnitude and phase are closely correlated, where the sharp peak in the phase of S_{21} corresponds to the first dip of $|S_{11}|$ at around 3.43 GHz. Also, the positive slope on the left side of the phase curve of S_{21} implies negative group delay $\tau = -d\varphi/d\omega$ (i.e., $d\varphi/d\omega > 0$), where φ is the phase of S_{21}.

Using the phase transformation equations (10.5) and (10.6), extracted effective indices of refraction for the arrays of two, four, and six SRRs are shown in Figure 10.11. A prominent feature is a sharp dip appearing in the curves. Similar dips have also been reported in dielectric/hole metal arrays operating at THz frequencies [32] and in multilayered fishnet metamaterials [33]. The refractive-index curves show

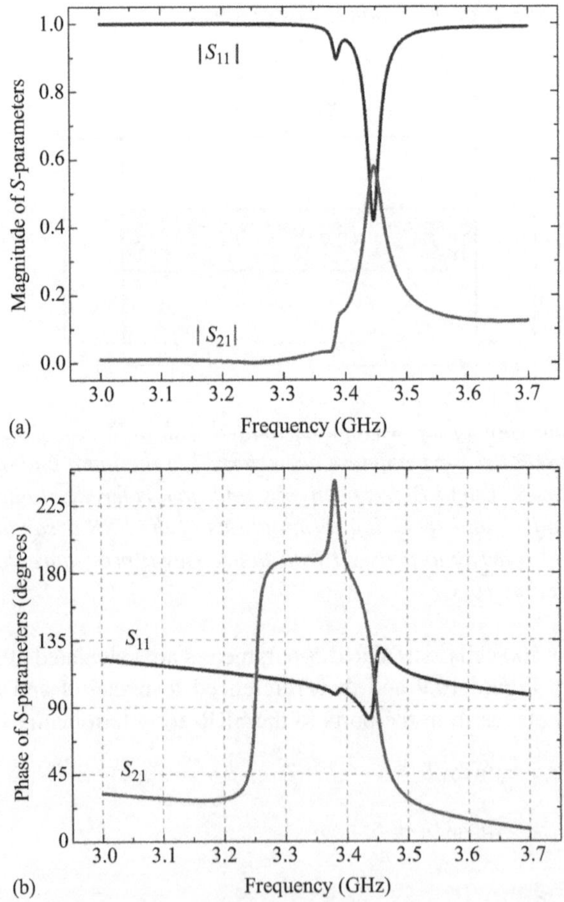

Figure 10.10 (a) Magnitude and (b) phase of S-parameters corresponding to the array with two SRRs

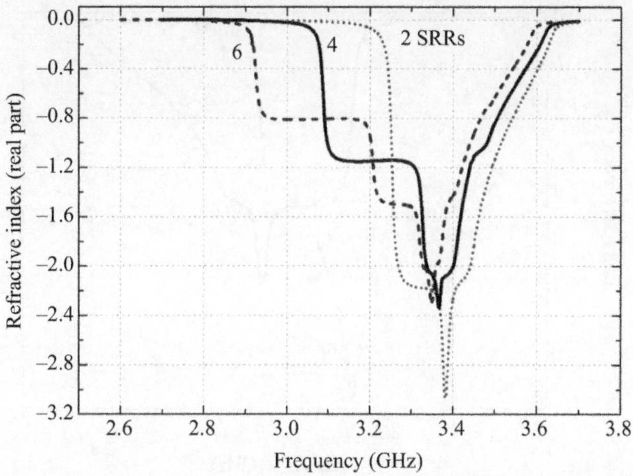

Figure 10.11 Real parts of the effective refractive indices for the waveguide loaded with two, four, and six SRRs

a peculiar characteristic in that the left and right branches differ in shape and have different properties. In fact, the branches on the right tend to converge to a common bulk value as the number of SRRs increases, while the left branch shows a staircase pattern, where the number of steps in each curve scales as half the number of SRRs. In addition, we note that the real part of the refractive indices for the two, four, and six SRRs in the normal dispersion region ($f > 3.4$ GHz) decreases in magnitude (becomes less negative) with an increase in the number of SRR cells in the propagation direction due to the finite conductivity of metallic inclusions and the dielectric loss of the substrate of each SRR cell.

To examine this point further, the real and imaginary parts of the refractive index corresponding to the array with two SRRs are shown in Figure 10.12. In the 3.2–3.3 GHz range, n_{re} is highly dispersive, rapidly varying with frequency, whereas above 3.4 GHz the SRR array shows resonant behavior, with $|S_{21}|$ peaking at around 3.44 GHz (Figure 10.10). In the low-frequency limit ($<$ 3.25 GHz) the real part n_{re} goes to zero, with $|n_{im}| > |n_{re}|$ and $|S_{21}| \approx 0$ (Figure 10.10), such that the SRR array behaves as a highly lossy dielectric. Indeed, $|n_{im}|$ increases with frequency from 3.00 to 3.25 GHz due to skin effect showing a typical metal-like behavior.

Differently from the low-frequency regime, $|n_{im}|$ decreases quickly with frequency and reaches $n_{im} \approx 0$ in the $3.45 - 3.60$ GHz range, where $|S_{21}|$ is relatively large (Figure 10.10). Accordingly, $|n_{re}|$ decreases with frequency and levels off to zero above 3.65 GHz.

To demonstrate NGD propagation in the time domain a sinusoidally modulated Gaussian pulse of unit voltage

$$V_{in}(t) = \sin[2\pi f(t - t_0)] \exp\left[-\left(\frac{t - t_0}{2T_{env}}\right)^2\right] \tag{10.7}$$

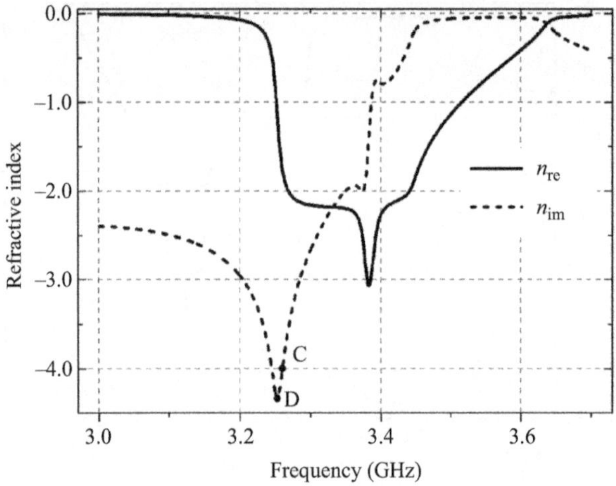

Figure 10.12 Retrieved effective index of refraction for the waveguide loaded with the array of two SRRs marked. Marked point D (3.253 GHz, −4.33) corresponds to the highest attenuation (dip on the n_{im} curve) and point C to (3.260 GHz, −4.00)

where t_0 is the time center of the wave packet, f the carrier frequency, T_{env} the width of the pulse envelope, is injected into the waveguide system from the input port. First, we consider the array of six SRRs driven by a pulse with a center frequency of 3.525 GHz, located on the right branch (normal dispersion) of the refractive-index curve in Figure 10.11. The corresponding input and output waveforms are displayed in Figure 10.13, in which the time center, t_c, is calculated by [15]:

$$t_c = \frac{\int_0^{\infty} t[V(t)]^2 dt}{\int_0^{\infty} [V(t)]^2 dt} \tag{10.8}$$

In this case, the pulse experiences a positive delay of 2.73 ns and suffers from a fivefold attenuation. On the other hand, for a center frequency of 2.92 GHz on the leftmost stair step in Figure 10.11, the pulse after traversing the 6-SRR array experiences a negative time delay of −13.68 ns and is strongly attenuated (by the factor of 1.0×10^5) as displayed in Figure 10.14. Nevertheless, losses can be compensated by introducing negative differential resistance through active transistors incorporated into each SRR unit cell, while retaining the anomalous dispersion of the SRR medium [34,35].

By considering the array of two SRRs and an input pulse with center frequency of 3.46 GHz, on the right branch (normal dispersion) of the refractive index in Figure 10.11, the propagating pulse after reaching the output port is delayed by 7.46 ns and its amplitude is reduced by half as given in Figure 10.15. But for a center frequency of 3.26 GHz, on the anomalous-dispersive branch in Figure 10.11,

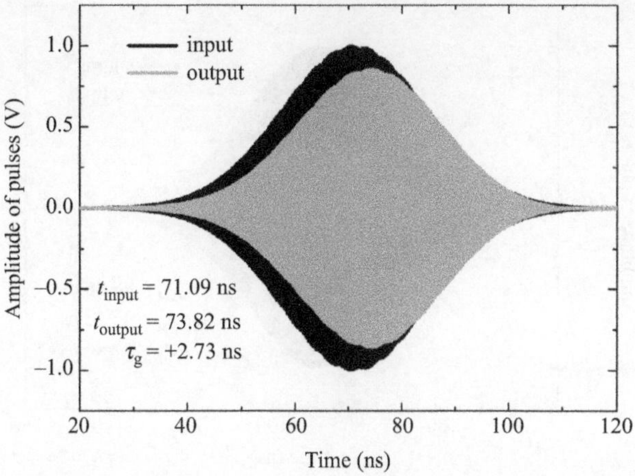

Figure 10.13 Output and input (center frequency of 3.525 GHz) pulses for the waveguide loaded with six SRRs. For visualization, the amplitude of the output pulse is multiplied by 5.0

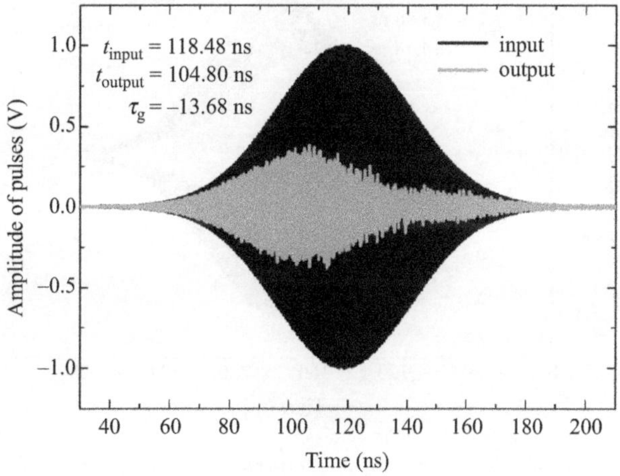

Figure 10.14 Output and input (center frequency of 2.92 GHz) pulses for the waveguide loaded with six SRRs. For visualization, the amplitude of the output pulse is multiplied by 1.0 × 10^5

the pulse experiences a negative delay of −14.14 ns, with its amplitude reduced by the factor of 500 (Figure 10.16). The associated group delay (Figure 10.17) for the whole frequency range of 3.20–3.32 GHz is negative and peaks at −26.0 ns, giving an average value of −13.0 ns in consistency with τ_g indicated in Figure 10.16.

Now, we discuss a kind of attenuation dispersion induced by the imaginary of the refractive index in Figure 10.12 (array of two SRRs) with regard to the

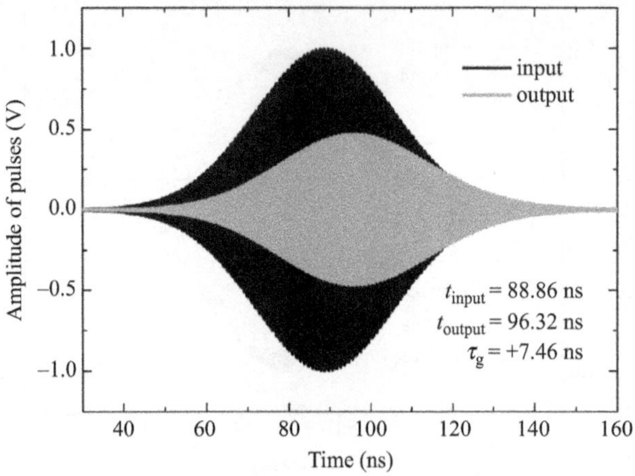

Figure 10.15 Output and input (center frequency of 3.46 GHz) pulses for the waveguide loaded with two SRRs

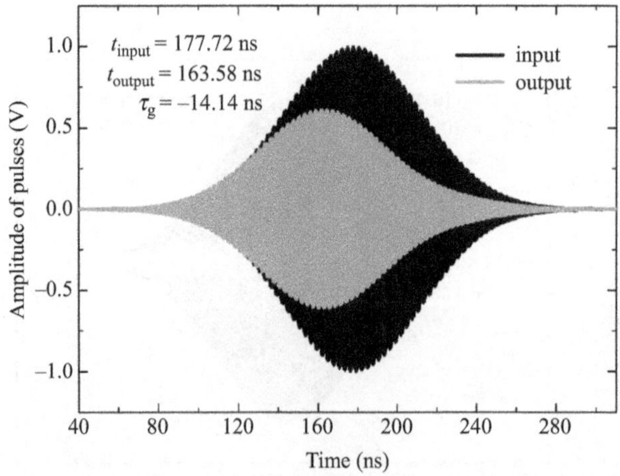

Figure 10.16 Output and input (center frequency of 3.26 GHz) pulses for the waveguide loaded with two SRRs. For visualization, the amplitude of the output pulse is multiplied by 500

frequency range of 3.25–3.27 GHz, which starts from the dip in the n_{im} curve. Quantitatively, the ratios of the input and output amplitudes at points D and C in Figure 10.18 are, respectively, 2.18 and 1.84. Thus the highest attenuation occurs on the left of the central frequency (3.26 GHz), showing that the lower-frequency components are more attenuated than the higher ones. This effect translates into a reshaping of the output frequency spectrum as demonstrated in Figure 10.18, where the amplitude peak appears to be shifted to the right.

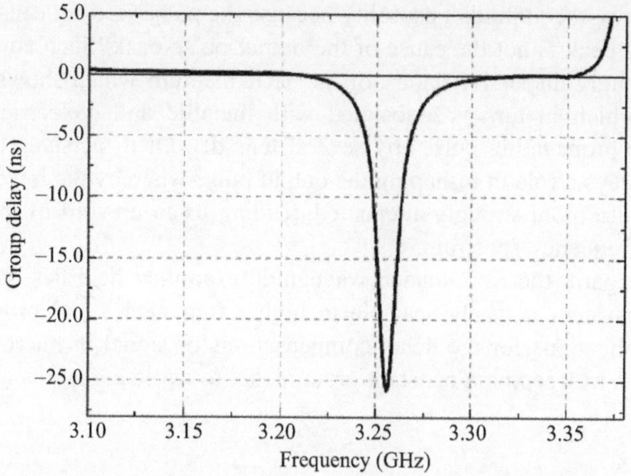

Figure 10.17 Group delay corresponding to the array of two SRRs driven by a Gaussian modulated pulse with center frequency of 3.26 GHz

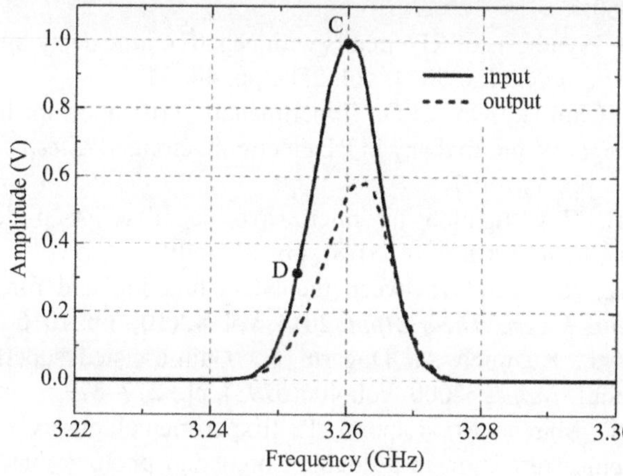

Figure 10.18 Frequency spectra relative to the array of two SRRs driven by a modulated pulse with center frequency of 3.26 GHz. Marked points are in correspondence with points D and C in Figure 10.12

10.6 Conclusions

NGD in the time domain has been demonstrated for Gaussian pulses propagating through an array of SRRs loading a below-cutoff X-band waveguide. Driving the SRR-loaded waveguide with a Gaussian modulated pulse with half-maximum width of 80.0 ns and central frequency of 3.26 GHz, NGD as large as −14.0 ns has been observed in the output pulse. The fact that the output pulse peak precedes the input peak does not contradict causality because the pulses are not causally related, i.e., the input peak is not the cause of the output pulse peak. Such an effect arises from the strongly dispersive nature of the SRR medium which shows anomalous dispersion, which in turn is associated with metallic and dielectric losses that attenuate the propagating pulses by several tens dB. Of dispersive character, the attenuation plays a role in reshaping the output pulse whereby the lower-frequency components are more strongly attenuated, leading to an up-shift of the amplitude peak of the frequency spectrum.

In this regard, the SRR-loaded waveguide examined here has attractive features as the system is easily scalable to higher frequencies and proves to show potential applications for the delay compensations of signal in microwave waveguide systems by keeping a constant group delay to avoid signal distortion.

References

[1] Toll J. S. 'Causality and the dispersion relation: logical foundations'. *Phy. Rev.* 1956, vol. 104(6), pp. 1760–1770.

[2] Brillouin L. *Wave Propagation and Group Velocity*. New York: Academic Press; 1960.

[3] Ravelo B. 'Synthesis of RF circuits with negative time delay by using LNA'. *Adv. Electromagnetics*. 2013, vol. 2(1), pp. 44–54.

[4] Jeong Y., Choi H., Kim C. D. 'Experimental verification for time advancement of negative group delay in RF electronic circuits'. *Electron. Lett.* 2010, vol. 46(4), pp. 306–307.

[5] Ravelo B. 'Investigation on microwave negative group delay circuit'. *Electromagnetics*. 2011, vol. 31(8), pp. 537–549.

[6] Ravelo B.: 'Similitude between the NGD function and filter gain behaviours'. *Int. J. Circ. Theor. Appl.* 2014, vol. 42(10), pp. 1016–1032.

[7] Wang L. J., Kuzmich A., Dogariu A. 'Gain-assisted superluminal light propagation'. *Nature.* 2000, vol. 406(6793), pp. 277–279.

[8] Longhi S., Marano M., Laporta, P. 'Experimental observation of superluminal pulse reflection in a double-Lorentzian photonic band gap'. *Phys. Rev E.* 2002, vol. 65(045602), pp. 0456021–4.

[9] Dolling G., Enkrich C., Wegener M., Soukoulis C. M., Linden S. 'Simultaneous negative phase and group velocity of light in a metamaterial'. *Science.* 2006, vol. 312(5775), pp. 892–894.

[10] Gehring G. M., Schweinsberg A., Barsi C., Kostinski N., Boyd R. W. 'Observation of backward pulse propagation through a medium with a negative group velocity'. *Science.* 2006, vol. 312(5775), pp. 895–897.

[11] Sánchez-López M. M., Sánchez-Meroño A., Aria J., Davis J. A., Moreno I. 'Observation of superluminal and negative group velocities in a Mach–Zehnder interferometer'. *App. Phys. Lett.* 2008, vol. 93(074102), pp. 0741021–3.

[12] Halvorsen T. G., Leinaas J. M. 'Superluminal group velocity in a birefringent crystal'. *Phys. Rev. A.* 2008, vol. 77(023808), pp. 1–7.

[13] Long J., Sievenpiper D. F. 'The observation of dispersionless superluminal propagation in a non-Foster loaded waveguide and its fundamental limitations'. *IEEE Trans. Microw. Theory Techn.* 2018, vol. 66(2), pp. 762–773.

[14] Solli D. R., McCormick C. F., Ropers C., Morehead J. J., Chiao R. Y., Hickman J. M. 'Demonstration of superluminal effects in an absorptionless, nonreflective system'. *Phys. Rev. Lett.* 2003, vol. 91(14), pp. 1439061–4.

[15] Barroso J. J., Oliveira J. E. B., Coutinho O. L., Hasar U. C. 'Negative group velocity in resistive lossy left-handed transmission lines'. *IET Microw. Antennas Propag.* 2017, vol. 15(15), pp. 2235–2240.

[16] Siddiqui O. F., Erickson S. J., Eleftheriades G. V., Mojahedi M. 'Time-domain measurement of negative group delay in negative-refractive-index transmission-line metamaterials'. *IEEE Trans. Microw. Theory Tech.* 2004, vol. 52(5), pp. 1449–1454.

[17] Siddiqui O. F., Mojahedi M., Eleftheriades G. 'Periodically loaded transmission line with effective negative group delay index and negative group velocity'. *IEEE Trans. Antenna Propag.* 2003, vol. 51(10), pp. 2919–2625.

[18] Monti G., Tarricone L. 'Negative group velocity in a split ring resonator-coupled microstrip line'. *Prog. Electromagn. Res.* 2009, vol. PIER 94, pp. 33–37.

[19] Marqués R., Martel. J., Mesa F., Medina F. 'Left-handed-media simulation and transmission of EM waves in subwavelength split-ring-resonator-loaded metallic waveguides'. *Phys. Rev. Lett.* 2002, vol. 89(18), pp. 1839011–4.

[20] Hrabar S., Bartolic J., Sipus Z. 'Waveguide miniaturization using uniaxial negative permeability metamaterial'. *IEEE Trans. Antennas Propag.* 2005, vol. 53(1), pp. 110–119.

[21] Lubkowski G., Damm C., Bandlow B., Schuman R., Schuessler M., Weiland T. 'Broadband transmission below the cutoff frequency of a waveguide loaded with resonant scatterer arrays'. *IET Microw. Antennas Propag.* 2007, vol. 1(1), pp. 165–169.

[22] Barroso J. J., Castro P. J., Leite Neto J. P. 'Experiments on wave propagation at 6.0 GHz in a left-handed waveguide'. *Microw. Opt. Technol. Lett.* 2010, vol. 52(10), pp. 2175–2178.

[23] Castro P. J., Barroso, J. J., Neto J. P. L. 'Microwave experiments on a waveguide loaded by na array of split-ring resonators'. *Proceedings of the 2011 SBMO/IEEE MTT-S International Microwave and Optoelectronics Conference*; Natal, RN, Brazil, Oct. 2011, pp. 308–312.

[24] Chen X., Grzegorczyk T. M., Wu B. I., Pacheco, J., Kong J. A. 'Robust method to retrieve the constitutive effective parameters of metamaterials'. *Phys. Rev. E.* 2004, vol. 70(1), pp. 0166081–7.

[25] Hasar U. C., Muratoglu A., Bute M., Barroso J. J., Ertugrul M. 'Retrieval method for effective constitutive parameters of bi-anisotropic metamaterials using waveguide measurements'. *IEEE Trans. Microw. Theory Techn.* 2017, vol. 65(5), no. 5, pp. 1488–1497.

[26] Barroso J. J., Hasar U. C. 'Constitutive parameters of a metamaterial slab retrieved by the phase unwrapping method'. *J. Infrared Milli Terahz Waves.* 2012, vol. 33(2), pp. 237–244.

[27] Hasar U. C., Barroso, J. J., Sabah, C., Ertugrul, M. 'Stepwise technique for accurate and unique retrieval of electromagnetic properties of bianisotropic metamaterials'. *J. Opt. Soc. Amer. B., Opt. Phys.* 2013, vol. 30(4), pp. 1058–1068.

[28] Lai A., Caloz C., Itoh T. 'Composite right/left-handed transmission line metamaterials'. *IEEE Microw. Mag.* 2004, vol. 5(3), pp. 34–50.

[29] Tang Q., Meng F.-Y., Wu Q., Lee J. C. 'A balanced composite backward and forward waveguide based on resonant metamaterials'. *J. Appl. Phys.* 2011, vol. 109(7), pp. 07A3191–3.

[30] Yoon H., Yeung, K. Y. M., Umansky, V., Ham, D. 'A Newtonian approach to extraordinary strong negative refraction'. *Nature.* 2012, vol. 488(7409), pp. 65–69.

[31] CST Microwave Studio. Version 2013. Darmstadt: CST GmbH.

[32] Wang S., Garet F., Blary K., *et al.* 'Composite left/right-handed stacked hole arrays at submillimeter wavelengths'. *J. Appl. Phys.* 2010, vol. 107(7), pp. 0745101–6.

[33] Szabó Z. 'Closed form Kramers–Kronig relations to extract the refractive index of metamaterials'. *IEEE Trans. Microw. Theory. Techn.* 2017, vol. 65(4), pp. 1150–1159.

[34] Ye D., Chang K., Ran L., Xin H. 'Microwave gain medium with negative refractive index'. *Nature Comm.* 2014, vol. 5, pp. 58411–7.

[35] Xu X., Padilla W. J., Sonkusale S. 'Loss compensation in metamaterials through embedding of active transistor based negative differential resistance circuits'. *Opt. Exp.* 2012, vol. 20(20), pp. 22406–22411.

Chapter 11

Metamaterial-based common-mode noise filter with NGD effect for multilayer PCB

Richard Xian-Ke Gao[1], Si-Ping Gao[2], Hui Min Lee[1], and En-Xiao Liu[1]

Abstract

The ever increase of data transmission rate that runs to tens Gb/s or beyond in the high-speed electronic circuits nowadays makes common-mode noise become a critical issue that could introduce electromagnetic interference (EMI) and adversely affect signal integrity (SI) and power integrity (PI). Together with miniaturization requirement of circuit design, it becomes requisite to explore newly effective common-mode filtering technology to tackle the common-mode noise associated EMI problems. In this chapter, the metamaterial-based common-mode filter design is elaborated. Through the study of a set of compact planar common-mode filters using defected ground structure (DGS), the fundamental filtering methodology and equivalent circuit models are elucidated with the presence of reasonable insertion loss performance. One of the intrinsic characteristics of the filters, termed negative group delay (NGD), has yet been paid much attention before, is also evaluated and discussed in this chapter for addressing SI/PI and EMI issues in electronic circuits.

Keywords

Common-mode noise, defected ground structure, insertion loss, metamaterial, negative group delay, transmission lines.

11.1 Introduction

Differential lines are commonly used in high-speed electronic circuits due to low crosstalk, high immunity to noise and low electromagnetic interference (EMI) of

[1]Institute of High Performance Computing, Agency for Science, Technology and Research (A*STAR), Singapore
[2]Department of Electrical and Computer Engineering, National University of Singapore, Singapore

differential signals [1]. However, in practical printed circuit boards (PCBs), common-mode noises on differential lines are inevitably induced by imbalance of lines or timing skew. The common-mode noise compromises PCB's SI and PI and produces unintentional emission which further degrades the performance of device vicinal. When conventional electromagnetic compatibility (EMC) solutions and approaches to mitigating different kinds of EMI problems in high-speed electronics become less effective, a novel and interesting structure which consist of a series of metallic objects (usually symmetric) etched on or in a dielectric slab has attracted increasingly attention, which is defined as metamaterial or left-handed (LH) material.

11.1.1 Overview of metamaterial

Since the past century, the scientific community has been interested in making artificially engineered materials for manipulating electromagnetic waves. In 1967, Veselago first postulated materials with negative permittivity and permeability [2]. His theoretical study proved that the phase velocity of waves in such materials could be made in the opposite direction of Poynting vector, which is contrary to wave propagation in natural materials. The first such materials, consisting of split-ring resonators (SRRs) and continuous wires, was introduced and realized by Pendry *et al.* [3,4]. Such left-handed metamaterial allows an electromagnetic wave to convey energy against its phase velocity. Many researchers have contributed to the theory, design, and application of metamaterials over the years. Smith *et al.* experimentally demonstrated functioning electromagnetic metamaterials by stacking horizontally periodic SRRs and thin wire structures [5]. Shelby studied negative refraction in left-handed materials [6]. Kong provided a general formulation for the electromagnetic wave interaction with stratified metamaterial structures [7]. Ubeid conducted a numerical study of a structure containing left-handed material waveguide [8]. Ziolkowski studied wave propagation in media having negative permittivity and permeability [9]. Oraiz developed a theory for zero reflection of electromagnetic waves from multilayered metamaterial structures [10]. Yang studied the cancellation of reflection and transmission at metamaterial surfaces [11]. Sabah and Uckun obtained theoretically high reflection coating with negative and positive refractive indices [12]. Cory and Zach analyzed the reflection and transmission characteristics of a multilayered structure consisting of left-handed materials and dielectric slabs [13]. All those efforts show that metamaterial is promising and has found applications such as antennas, absorbers, superlens, cloaking, and so on.

Because of its unique electromagnetic properties, metamaterial has also found many applications in the area of EMC, such as noise filtering, electromagnetic shielding, and wave absorption. Among the numerous explorations of metamaterial structures, electromagnetic bandgap (EBG), defected ground structure (DGS), and frequency selective surface (FSS) have been extensively studied to solve different EMI issues. Shahparnia and Ramahi are among the first to use EBG structures for the suppression of electromagnetic radiation from parallel-plate bus structures caused by switching noise [14]. Wang *et al.* studied the split power planes with EBG structures for the wideband suppression of ground bounce noise in PCB [15].

Naqui *et al.* presented the principle of selective common-mode suppression by applying complementary split-ring resonators (CSRRs) on the ground plane of transmission lines [16]. Wu *et al.* designed a common-mode filter based on two U-shaped and one H-shaped coupled patterned ground structure in the frequency range of 3.6–9.1 GHz [17]. Huang elucidated a design of circular-ring miniaturized-element meta-surface for frequency selective shielding application [18]. Gao *et al.* explored a C-H-C shape DGS filter for common-mode noise suppression at the frequency of 8.4 GHz with bandwidth of 6 GHz [19] and the filter has further been optimized to make its parameters robust to uncontrollable variations [20]. Chaudhary demonstrated a planar circuit made of dual-plane defected structures exhibiting an abnormal wave propagation phenomenon [21], i.e., negative group delay (NGD). The phenomenon will be elucidated in Section 11.1.2.

11.1.2 Negative group delay

Group delay (GD) can be extracted from the forward transmitting scattering parameters (*S*-parameter), and generally it varies over frequency. But within a small amount of bandwidth, GD is usually nearly constant. Usually the rate of change of transmission phase angle with respect to frequency is negative. From signal transmission point of view, GD is the time delay of the amplitude envelopes of various sinusoidal components of a signal in transmission and thus frequency dependent. It is in contrast with phase delay, which is the time delay of the signal phase. The delay variation with respect to frequency means that a signal with multiple frequency components will be distorted after passing through a circuit because these components are not delayed by the same amount of time. In other words, the original shape of the signal is changed because of GD. GD and phase delay are directly related to the group velocity and phase velocity. Einstein's theory of special relativity states that no (real) mass, energy, or information is able to travel faster than the speed of light [22,23]. Consequently, the group velocity of a practical signal should always be less than the speed of light, though the phase velocity can be greater than the speed of light.

In certain circumstances, negative group velocity is an abnormal wave propagation phenomena [24], and it includes superluminal [25] (a frame of reference traveling faster than the light), backward wave propagation (negative refractive index [26]), and simultaneous negative phase and group velocity [27]. Media or structures exhibiting NGD behavior cause the output peak of a signal to precede the input peak. This phenomenon is achieved through pulse reshaping [24] and does not violate causality, since the initial transient of the pulse is still positively delayed and propagates at a speed not exceeding the speed of light in vacuum [28]. NGD occurs only at limited frequency bands of media exhibiting anomalous dispersion, usually accompanied with wave absorption. Anomalous dispersion, which does not violate the relativistic causality requirements, must exist within some frequency bands for all dispersive media [29] as dictated by Kramers–Kronig relations, which are applicable to all physically realizable, causal linear systems [24]. Anomalous dispersion of electromagnetic waves through a medium was initially studied by Brillouin and Sommerfeld [23]. They considered a semi-infinite sinusoid waveform propagating through a medium with Lorentzian dispersion and showed that the wave front velocity is always positive and exactly luminal under all circumstances, thus satisfying relativistic causality. Therefore, the group velocity does not equate with the wave front velocity in such media, but merely characterizes the propagation of distinct features of a well-behaved wave packet, such as pulse maximum [23,24]. The difference between the front velocity and group velocity was recently demonstrated in a medium with a slow group velocity [30].

Many works on NGD-based circuit and systems have been reported in the literature. The NGD synthesizer operating at a microwave frequency was first presented in [31] and [32]. Various active/passive NGD circuits have been designed using RLC resonators (see a brief review in [33]) and artificial media [34], such as left-handed metamaterial. The NGD phenomena for DGS-based common mode filter for noise suppression [19,20] will be systematically studied in this chapter.

11.2 Common-mode filter and DGS technology

When data transmission rate increases higher, for example, gigahertz and even more, differential signal transmission mode is recommended to be deployed. That is because such transmission mode design can help reject common-mode noise that is one of the critical problems in high-speed electronic circuit. Miniaturization and compactness of chips to micro- or nano-size make the signals lower, however, differential signal transmission can effectively double the original signal amplitude level thus the signal-to-noise ratio remains the same level as before. Since differential signals have equal magnitude and opposite phase in an ideal differential circuit, there is no return signal travelling back to signal source through the common ground.

A simple differential circuit, consisted of two single-ended signal sources which have equal amplitude but opposite phase and two loads (for simplicity, two resistors are used, for example), is illustrated in Figure 11.1. The two signal

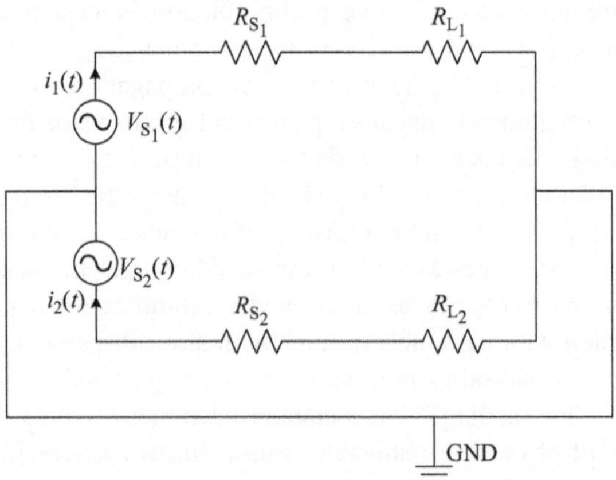

Figure 11.1 A simple differential circuit

sources, $V_{S_1}(t)$ and $V_{S_2}(t)$, use a common reference ground. The two loads R_{L_1} and R_{L_2} together with internal resistors R_{S_1} and R_{S_2} are in series with $V_{S_1}(t)$ and $V_{S_2}(t)$, respectively. To realize the circuitry balance without return current on ground, there should have such electrical relationship for the sources and loads,

$$R_{\mathrm{S1}} = R_{\mathrm{S2}}, \tag{11.1}$$

$$R_{\mathrm{L1}} = R_{\mathrm{L2}}, \tag{11.2}$$

$$V_{\mathrm{S1}}(t) = -V_{\mathrm{S2}}(t), \tag{11.3}$$

$$i_1(t) = \frac{V_{\mathrm{S1}}(t)}{R_{\mathrm{S1}} + R_{\mathrm{L1}}}, \tag{11.4}$$

$$i_2(t) = \frac{V_{\mathrm{S2}}(t)}{R_{\mathrm{S2}} + R_{\mathrm{L2}}}. \tag{11.5}$$

If (11.1) to (11.5) are satisfied, the electric circuit in Figure 11.1 is ideally differential or balanced. All the alternating current (AC) signals between the common ground reference terminals that are seen in the single-ended circuits are eliminated, leading to the minimum electromagnetic leakage to the environment.

11.2.1 Common-mode noise

However, this is an ideal situation. In the practical application, it is almost impossible to realize perfect balance for a differential circuit. For example, the polarizations of signals in the differential circuit are exactly opposite but the amplitudes are not exactly equal to each other. Let one voltage signal be $V_{S_1}(t)$ and the other be $V_{S_2}(t)$ where $V_{S_1}(t)$ and $V_{S_2}(t)$ have the opposite phases, but not exactly equal in magnitude. It results in that the sum of their currents, $\Delta i(t) = (i_1(t) - i_2(t))$ at ground terminal, is not zero. This time-dependent current $\Delta i(t)$ must return to sources through ground reference. This unwanted signal is usually defined as common-mode noise.

Assume that there is difference of the signal line lengths in the differential circuit due to the layout requirement, such as bending. If one signal line (of the differential pair) is longer than the other one, the signals transmitted on the differential circuit are no longer equal since their phases are different at the load side. Hence, this induces certain return signal through ground. In another words, there is common-mode signal on the differential circuit. This can lead to SI and PI issues, which is classified into common-mode EMI problem.

If the signals are not exactly equal and opposite in the differential circuit, we can divide them into differential and common mode components correspondingly. For instance, assume that the one signal ($i_1(t)$ in Figure 11.1) is 100 mA and the return signal ($i_2(t)$) is 92 mA. This is equivalent to the differential mode component being 96 mA and the common mode component being 4 mA, respectively. It is clearly shown that if the transmitted signals are not exactly equal to each other while opposite in the phase, there must be a common mode current component flowing through ground or power system or somewhere else. Such common mode current component is unpredictable and uncontrolled, even though it might be small, it can still generate critical EMI radiations. The clock skew, unbalanced routing, overshoot and undershoot, and discontinuous impedance would shift the differential signals to be unbalanced with different phases. The difference of the phases might generate the common-mode noise too. Even the signal line lengths are perfectly equal and the signals are exactly equal and their phases are opposite, and there is no clock skew and discontinuous impedance. Thereby there is no return current path formed through ground. However, there might be an induced current coupled into the differential circuit from adjacent signal line or plane or external environment due to electromagnetic radiation and flowing back through ground or power system. This phenomenon is defined as crosstalk. The crosstalk also generates common-mode noise and common-mode EMI problems.

11.2.2 Common-mode filter

To suppress common-mode noise is a requisite for EMC design. The most effective approach is filtering technology. There is at least one resonance at certain frequency in a filter, which will cause the circuit impedance change tremendously so that the electric energy could be absorbed or reflected to the source due to high impedance at certain frequency. Therefore, the function of a filter to isolate some frequency components of the input signal is achieved. The transfer function of one filter circuit with an input signal of $V_{in}(\omega)$ and output of $V_{out}(\omega)$, where ω is the signal angular frequency, is defined as

$$H(\omega) = \frac{V_{\text{out}}(\omega)}{V_{\text{in}}(\omega)} = |H(j\omega)|e^{j\varphi(\omega)}, \tag{11.6}$$

where, the transfer phase function, $\varphi(\omega)$, related to the transfer GD through a differential circuit, is described as:

$$\tau_d = -\frac{d\varphi(\omega)}{d\omega}, \tag{11.7}$$

where τ_d is GD time. In most common-mode filters, it is only of interest on the magnitude of the transfer function. However, phase and GD time become important in modern electronic systems recently. Especially, the NGD is a new concern for SI/PI.

In a filter, the frequency band, where the input signal can pass without loss (0 dB insertion loss) up to a few dB insertion loss (conventionally 3 dB), is defined as passband. Inversely, the frequency band where the input signal cannot pass the filter circuit, is defined as the stopband. Based on the interested frequency range where the signal can pass or not, there are four typical kinds of filters: (1) low-pass filter, (2) high-pass filter, (3) band-pass filter, and (4) band-stop filter. Others are the extension of these four types. The response performances of these four type filters are illustrated in Figure 11.2, where ω_c is the cutoff frequency. ω_l and ω_h are the lower and higher bounds of frequency band, respectively.

Low-pass filter (Figure 11.2(a)) is usually realized by using a cascade of series inductors and shunt capacitors. The inductors in series behave as low impedance component while the shunt capacitors produce high impedance at low frequency

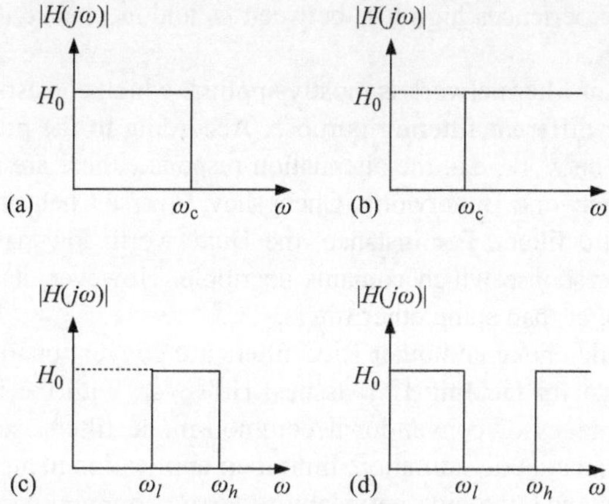

Figure 11.2 The four typical responses of filters: (a) low-pass; (b) high-pass; (c) band-pass; and (d) band-stop

range (below the cutoff frequency), thus the input signal is allowed to pass and transmit to the load side. In turn, at the high frequency range which is above the cutoff frequency, the series inductors produce large impedance and the shunt capacitors behave as low impedance component. Thereby the signal is impeded to be transmitted to the output terminal of filter. The number of these inductors and capacitors determines the response steepness of the filter. More inductors and capacitors result in faster filtering response, but the circuit would be more complex with in-band loss, higher cost, and larger size. The behavior of high-pass filter is the reverse of the low-pass filter. The high-pass filter (Figure 11.2(b)) consists of a series of capacitors with shunt inductors, which allows signals in the high frequency range (above the cutoff frequency) pass through the filter with a minimum loss. The behavior of the filter results in a high attenuation for the low-frequency signals.

The band-pass filter (Figure 11.2(c)) allows the signals pass the circuit in a certain frequency band of which the signal frequency is between the lower cutoff frequency (ω_l) and the upper cutoff frequency (ω_h). Beyond the frequency band, the transmission of signals stops. The center frequency of band, ω_0, is defined by the lower and upper cutoff frequencies,

$$\omega_0 = \frac{(\omega_l + \omega_h)}{2}. \tag{11.8}$$

One more parameter used to identify the performance of band-pass filter is fractional bandwidth (FBW),

$$FBW = \frac{(\omega_h - \omega_l)}{\omega_0}. \tag{11.9}$$

As a complement of the band-pass filter, the band-stop filter (Figure 11.2(d)) impedes the signal to be transmitted if its frequency is within the frequency band.

Since the signal experiences high loss between ω_l and ω_h, hence it is named band stop or band reject.

It is known that filter network is mostly applied, which consists of two or more kind of filters for different filtering purpose. According to the performance characteristic of filter network, e.g. the attenuation response, there are a few of typical designs for filter network: Butterworth, Chebyshev, inverse Chebyshev, Thompson–Bessel, and Elliptic filters. For instance, the Butterworth low pass filter has the flattest passband response which contains no ripple. However, its transition from pass to stop is longer than some other filters.

Common-mode choke and other RLC filters are conventionally applied in the filter circuit design for tackling EMI issues. However, with the increase of data rate, the disadvantages of conventional common-mode filters, such as the large size, material loss, magnetic saturation, limit their application in high-speed circuits nowadays. For instance, the high permeability ferrite common-mode choke is sufficient for filtering in the low frequency range up to megahertz. However, its filtering efficiency drops abruptly when frequency increases further. This is because of the magnetic saturation. Another type of filter uses multilayer low-temperature co-fired ceramic (LTCC) technology. However, the high fabrication cost and process complexity impede its wide application. In the past ten to fifteen years, quarter-wavelength open stub resonator, CSRR, and EBG have been developed for mitigating common-mode noise. However, these designs have the drawback of large footprint, which compete layout space with signal circuits on PCB.

Alternatively, one effective approach evolved from EBG has been proposed and developed quickly. By etching some parts of ground plane under the differential circuit, the return current distribution can be perturbed thereby the filtering effect is achieved. This technology is denominated as DGS, which is rather applicable for high-speed electronic circuit design.

11.2.3 Defected ground structure

In DGSs, slots and apertures are intentionally etched on the solid ground planes to form a transmission zero [17,19]. A thorough introduction of the application of DGSs to single transmission line can be found in Chapter 5 of [35], while this chapter focuses on the effect of DGSs on differential lines. Common-mode noise mitigation using DGSs for differential lines has been extensively studied for its low cost and compact size [17,19,20,36–38]. In these DGS filters, the defects on the ground plane disturb the current distribution on it to achieve a common-mode filtering yet preserve no detriment to differential signals. The working principle of a DGS filter is elucidated with the aid of an equivalent circuit model in Section 11.2.3.1. It is followed by an introduction of the NGD effect caused by DGS.

11.2.3.1 Equivalent circuit of DGS

A typical dumbbell DGS is etched on the ground plane under a pair of coupled microstrip lines as illustrated in Figure 11.3. Such a DGS perturbs the current distribution on the ground plane due to large impedance within the stopband. It works as a parallel RLC resonator for common-mode signals [39,40]. It is then

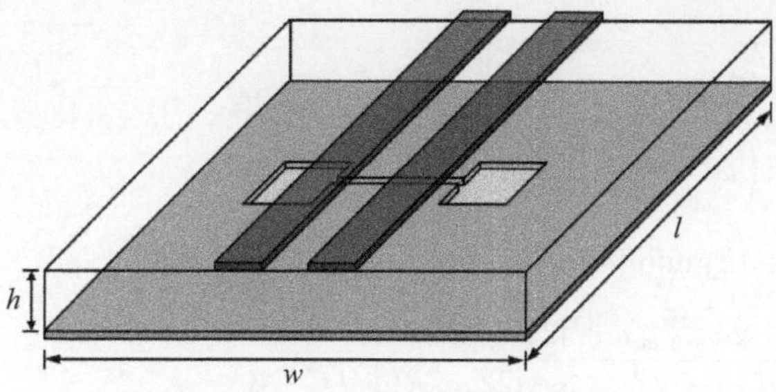

Figure 11.3 A DGS beneath the differential microstrip lines

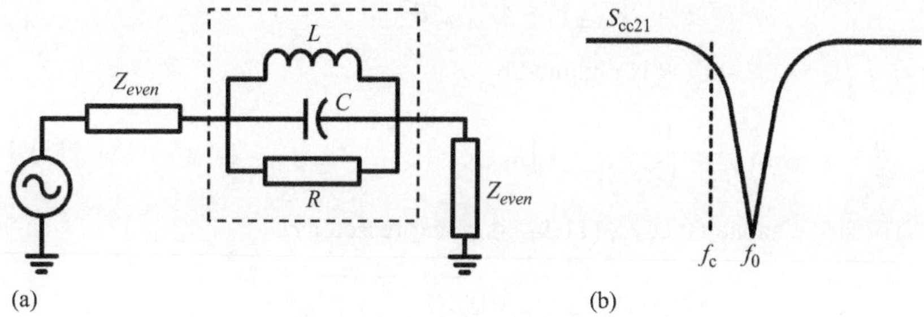

Figure 11.4 (a) Equivalent circuit and (b) the typical frequency response of DGS filter

inserted into the transmission line model with even-mode reference impedance Z_{even} at two terminals due to the common-mode scenario, as shown in Figure 11.4 (a). Figure 11.4(b) shows a typical transmission coefficient $|S_{21}|$, of which f_c and f_0 are the 3 dB cutoff frequency and the resonant frequency, respectively. The equivalent value of the elements in the equivalent circuit model can be extracted from a simulated $|S_{21}|$ using the equations,

$$\begin{cases} R = 2Z_{\text{even}}\left(\dfrac{1}{|S_{\text{cc21}}(2\pi f_0)|} - 1\right) \\ L = \dfrac{1}{(2\pi f_0)^2 C} \\ C = \dfrac{f_c\sqrt{R^2 + 4RZ_{\text{even}} - 4Z_{\text{even}}^2}}{4\pi R Z_{\text{even}}\left(f_0^2 - f_c^2\right)} \end{cases} \tag{11.10}$$

where the $|S_{\text{cc21}}(2\pi f_0)|$ is derived from

$$S_{\text{cc21}} = \frac{2Z_{\text{even}}}{Z + 2Z_{\text{even}}}, \tag{11.11}$$

where

$$Z = \frac{1}{\frac{1}{R} + j\left(\omega C - \frac{1}{\omega L}\right)}. \tag{11.12}$$

Substituting (11.12) into (11.11),

$$S_{cc21} = \frac{2\omega Z_{even} L + j2Z_{even} R(\omega^2 LC - 1)}{\omega L(R + 2Z_{even}) + j2Z_{even} R(\omega^2 LC - 1)}, \tag{11.13}$$

whose magnitude form is

$$|S_{cc21}|^2 = \frac{4\omega^2 {Z_{even}}^2 L^2 + 4{Z_{even}}^2 R^2 (\omega^2 LC - 1)^2}{\omega^2 L^2 (R + 2Z_{even})^2 + 4{Z_{even}}^2 R^2 (\omega^2 LC - 1)^2}. \tag{11.14}$$

Let $\omega = \omega_0 = 1/\sqrt{LC}$, R is obtained by

$$R = 2Z_{even}\left(\frac{1}{|S_{cc21}(2\pi f_0)|} - 1\right). \tag{11.15}$$

To derive capacitance C, (11.14) is re-expressed as

$$|\omega^2 LC - 1| = \frac{\omega L \sqrt{(R + 2Z_{even})^2 |S_{cc21}|^2 - 4{Z_{even}}^2}}{2Z_{even} R \sqrt{1 - |S_{cc21}|^2}}. \tag{11.16}$$

By substituting $S_{cc21}(\omega_c) = \sqrt{2}/2$ into (11.16), it thus has

$$C = \frac{f_c \sqrt{R^2 + 4RZ_{even} - 4Z_{even}^2}}{4\pi R Z_{even}\left(f_0^2 - f_c^2\right)}. \tag{11.17}$$

Lastly, L can be derived from (11.10) after C is obtained.

Figure 11.5 shows the common-mode S-parameters obtained using full-wave simulation [41] and the equivalent circuit model. Good agreement can be observed which proves the equivalent circuit model is effective in modeling the behavior of the DGS common-mode filter.

According to the property of a parallel RLC resonant circuit, the 3 dB FBW of a DGS common-mode filter can be characterized using the equivalent R, L, and C as

$$\text{FBW} = \frac{1}{Q_L} = \frac{1}{Q_e} + \frac{1}{Q} = \left(\frac{1}{2Z_{even}} + \frac{1}{R}\right)\sqrt{\frac{L}{C}}, \tag{11.18}$$

where Q, Q_e, Q_L are the unloaded, external, and loaded quality factors, respectively. Commonly R is much larger than Z_{even}, thus (11.18) can be simplified as

$$\text{FBW} \approx \frac{1}{2Z_{even}}\sqrt{\frac{L}{C}}. \tag{11.19}$$

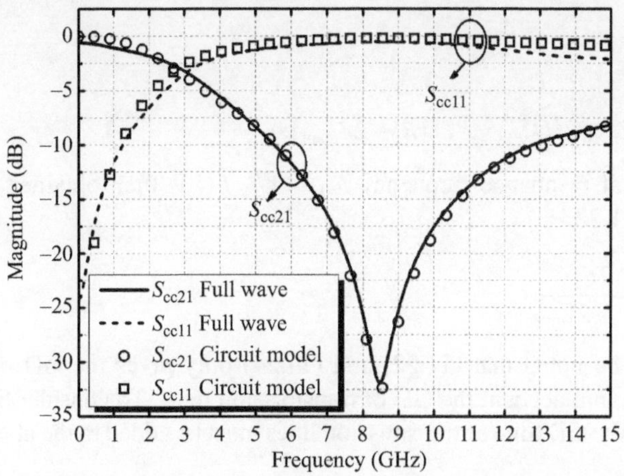

Figure 11.5 Common-mode S-parameters: full-wave vs. equivalent circuit model

From (11.19), it is seen that FBW is proportional to $\sqrt{L/C}$. The actual bandwidth (BW) can be easily obtained by

$$\mathrm{BW} = \mathrm{FBW} \times f_0 = \frac{1}{4\pi Z_{\mathrm{even}} C}. \tag{11.20}$$

Different from FBW, BW is only inversely proportional to C.

11.2.3.2 NGD effect of DGS

NGD, a counterintuitive phenomenon, has been discovered to relate time advance to the wave propagation in certain frequency intervals within artificial medium [42–45]. The inductance–capacitance (LC) resonators were found to exhibit NGD effect around resonant frequencies [46]. Furthermore, DGS that can be considered as an LC resonator had the same effect when applied to single transmission line [47]. In this section, we will explain and demonstrate the NGD of DGS for differential lines.

The GD time of a circuit can be characterized by investigating the phase variation of transmission coefficient. A signal transmits through the whole circuit with additional phase due to the travelling time or delay, which can be defined as the negative rate of phase change of S_{cc21} with respect to frequency,

$$\tau = -\frac{\partial \angle S_{cc21}}{\partial \omega}. \tag{11.21}$$

It can also be obtained from the equivalent circuit model as

$$\begin{aligned}
\tau &= -\frac{\partial}{\partial \omega} \arctan\left(\frac{\mathrm{Im}[S_{cc21}]}{\mathrm{Re}[S_{cc21}]}\right) \\
&= -cos^2(\angle S_{cc21}) \frac{\partial}{\partial \omega}\left(\frac{\omega R^2 L \tilde{U}}{\tilde{V}}\right) \\
&= \frac{2\omega^2 R^2 L^2 \tilde{U}\left[4Z_{\mathrm{even}} R^2 C \tilde{U} + L(R + 2Z_{\mathrm{even}})\right] - \tilde{V}\left(R^2 L \tilde{U} + 2\omega^2 R^2 L^2 C\right)}{\left(\omega^2 L^2 + R^2 \tilde{U}^2\right)\left[\omega^2 L^2 (R + 2Z_{\mathrm{even}})^2 + 4{Z_{\mathrm{even}}}^2 R^2 \tilde{U}^2\right]}
\end{aligned} \tag{11.22}$$

where

$$\tilde{U} = \omega^2 LC - 1,$$
$$\tilde{V} = 2Z_{\text{even}} R^2 \tilde{U}^2 + \omega^2 L^2 (R + 2Z_{\text{even}}). \tag{11.23}$$

GD time at resonance frequency $\omega_0 = 1/\sqrt{LC}$ is then obtained by inserting $\tilde{U} = 0$ into (11.22)

$$\tau = -\frac{2R^2 C}{R + 2Z_{\text{even}}}. \tag{11.24}$$

It should be noted that (11.22) and (11.24) only gives the GD of DGS itself without taking into account the GD of transmission lines. To consider the GD of the whole circuit, the GD time of transmission lines must be added to the above equations

$$\tau_{\text{TL}} = \frac{l}{c}, \tag{11.25}$$

where l and c are the total length of and the light speed in the transmission lines, respectively.

Figure 11.6 illustrates the phase variation of S_{cc21} for the same case in Figure 11.5. It can be seen that the phase of S_{cc21} decreases as frequency increases except for a small frequency range (shaded area) covering the resonant frequency of the DGS. This kind of phase variation will lead to negative GD at the immediate proximity of the resonant frequency while positive GD elsewhere. Figure 11.7 confirms the GD effect against frequency for the whole circuit including a pair of differential lines and a DGS underneath, through full-wave simulation and equivalent circuit model. For most of the frequencies, GD is positive due to the

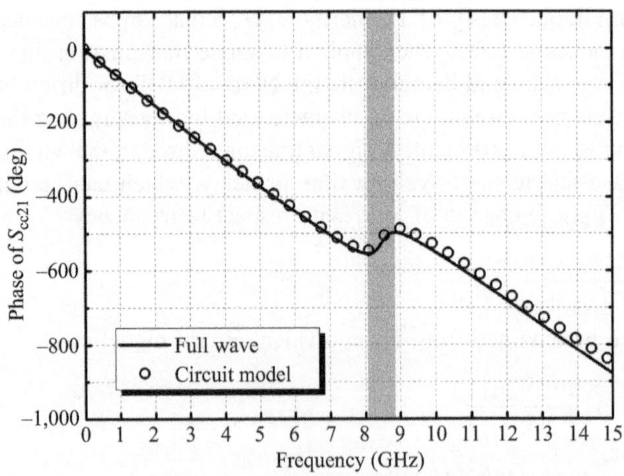

Figure 11.6 Phase variation of S_{cc21} against frequency

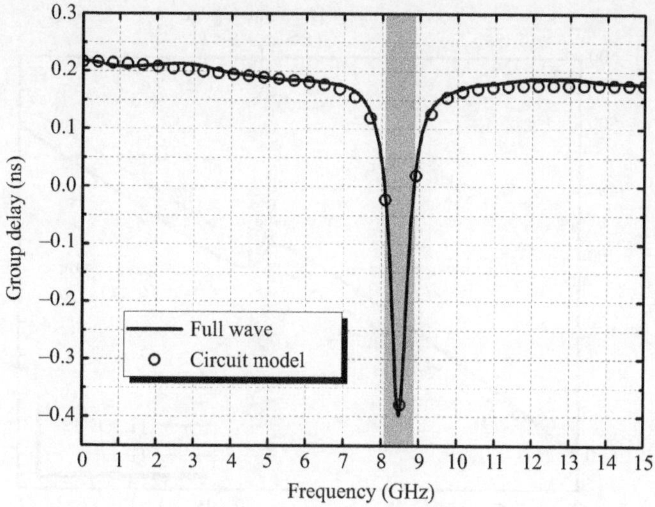

Figure 11.7 GD of the circuit including a pair of differential lines and a DGS underneath

delay of the differential lines, while for the shaded frequency band, GD is negative due to the abnormal increasing of phase caused by the DGS.

It is clear from (11.24) that the NGD time can be tuned by varying the equivalent elements R and C of DGS, i.e., a DGS with larger R or C leads to a longer NGD time at the resonance. However, the bandwidth of S_{cc21} of the DGS will decrease with larger C according to (11.19). Therefore, to increase the NGD time while maintaining the bandwidth of the DGS, it is recommended (1) increasing R solely or (2) increasing L and C proportionally.

In addition, the NGD phenomenon can be explained by the variation of propagation constant caused by adding a DGS to a pair of differential lines. Figure 11.8 shows the dispersion diagram of the differential lines with and without DGS by using the same case in Figure 11.6. For the differential lines only, the propagation constant k is in linear relation with frequency. However, adding a DGS generates additional L and C to the transmission lines, which results in increase of the propagation constant of the whole structure (denoted as β) at lower frequencies and decrease at higher frequencies. This effect makes the phase velocity of the waves propagating in a DGS-loaded transmission line be lower than that of a pure transmission line in the slow-wave region; conversely, the phase velocity in a DGS-loaded transmission line is higher than that of a pure transmission line in the fast-wave region. Most importantly, between the slow- and fast-regions, there is a very narrow transition region where β suddenly drops from larger than k to smaller than k. This drastic plunge of β with respect to frequency leads to an NGD within this region.

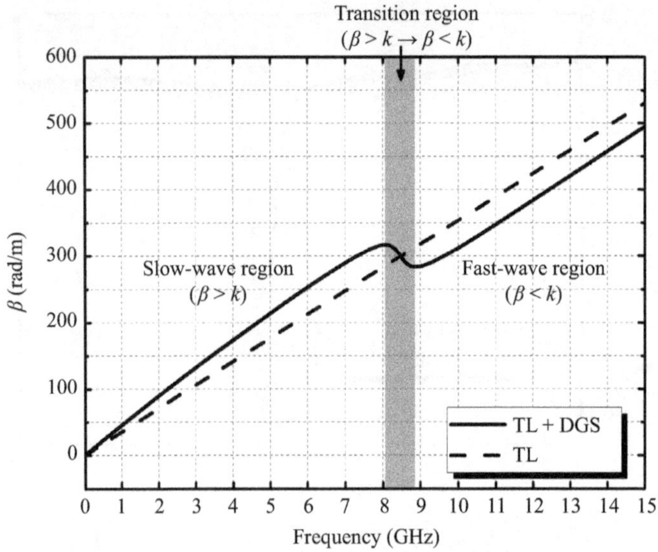

Figure 11.8 Dispersion diagram of the differential lines (TL) with and without DGS

11.3 DGS-based common-mode filter for two-layer PCB

11.3.1 Dumbbell DGS common-mode noise filter design

In this section, the dumbbell-DGS filter (Figure 11.3) is used and re-illustrated in Figure 11.9 for characterizing the common-mode filtering performance and NGD effect. The DGS is to be symmetrical to the central line of the coupled microstrip lines [40]. In this example, the PCB size is 30 mm × 20 mm. The substrate material is Rogers 4350 ($\varepsilon_r = 3.48$ and $\tan\delta = 0.0037$), which is commonly used for high-speed circuits. The parameters for the coupled microstrip lines are given as $w_m = 0.3$ mm, $s_m = 0.32$ mm, $h_s = 0.168$ mm and $t_m = 0.03$ mm. Its length is 30 mm. The dimensions of the dumbbell slot are given as $a = b = c = 3$ mm, $g = 0.2$ mm, and $t_{DGS} = 0.03$ mm, where w_m and s_m are chosen to achieve a 100 Ω differential impedance and even-mode characteristic impedance Z_{even} is thus calculated as 59.132 Ω.

11.3.1.1 Common-mode filtering characteristic

The presence of DGS (Figure 11.9) suppresses the common-mode signal within a certain stopband while maintaining a good signal transmission for the differential signals as shown in Figure 11.10. A stopband of 7.3 GHz is obtained when the attenuation level is 10 dB. It is noted that for the convenience, the DGS common-mode filter represents the coupled differential microstrip lines with DGS beneath in the rest sections.

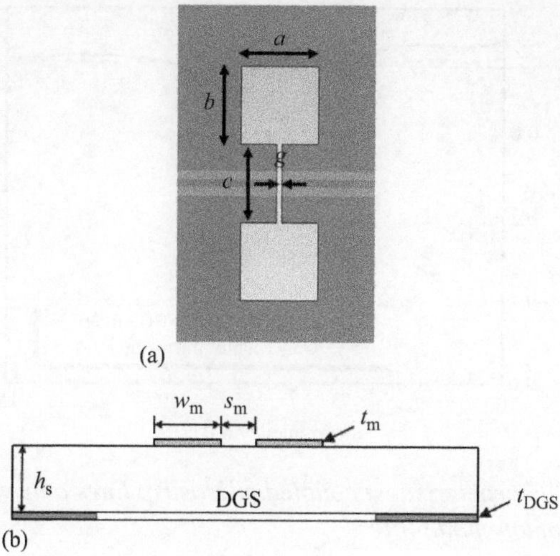

Figure 11.9 (a) Top view and (b) cross-sectional view of a simple dumbbell DGS in a two-layer PCB

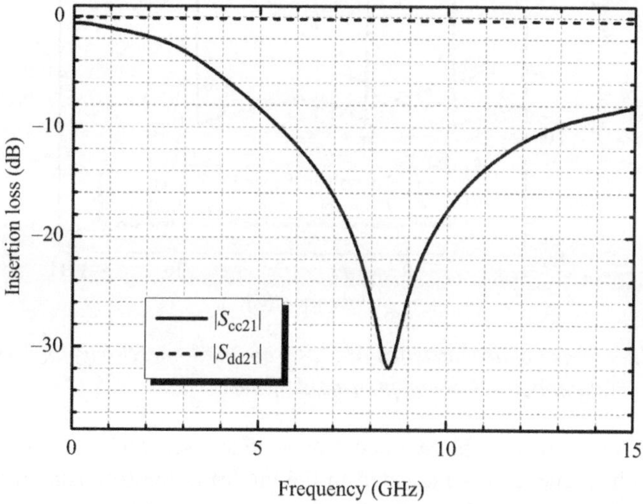

Figure 11.10 The common-mode insertion loss $|S_{cc21}|$ and differential-mode insertion loss $|S_{dd21}|$ of a dumbbell DGS common-mode filter

11.3.1.2 NGD effect of dumbbell DGS common-mode filter

Based on the study in Section 11.2, the NGD effect of the dumbbell DGS common-mode filter is evaluated and clearly seen in Figure 11.11. An NGD time of 0.397 ns is obtained at the resonant frequency of 8.47 GHz with an NGD bandwidth (NGD-BW) of 779 MHz, where NGD-BW is defined as the bandwidth at GD time of 0 ns.

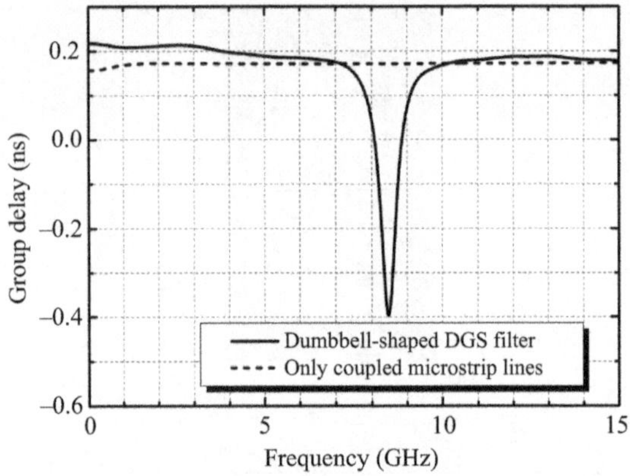

Figure 11.11 GD characteristics: coupled microstrip lines only vs. dumbbell DGS common-mode filter

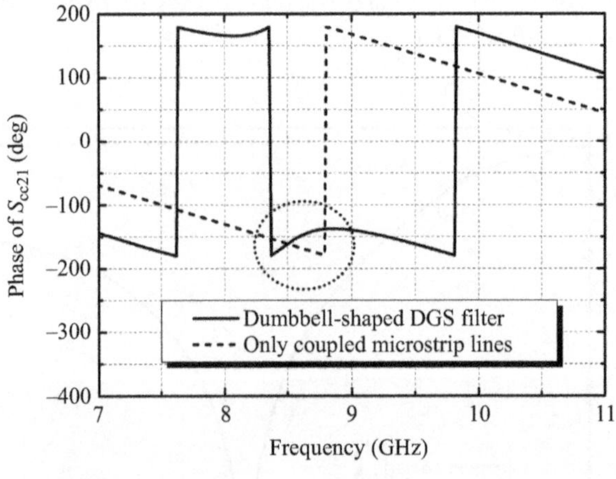

Figure 11.12 Phase characteristics of S_{cc21}: coupled microstrip lines only vs. dumbbell DGS common-mode filter

Figure 11.12 shows that the phase of common-mode insertion loss (S_{cc21}, solid line) increases with frequency (as indicated in the circle) at the resonant frequency for the case with DGS filter, which signifies the presence of NGD.

11.3.2 Parametric study of DGS filter on common-mode noise filtering and NGD effect

In a DGS common-mode noise filter design, the shape and size of the slots etched on the ground plane play a critical role in the determination of the resistance, inductance and capacitance of the equivalent circuit as discussed before. The equivalent resistance, inductance and capacitance define the filtering frequency response as well as the NGD effect.

11.3.2.1 Effect of DGS geometrical parameters

Based on the equivalent circuit model derived in Section 11.2, the effect of DGS geometrical parameters on the frequency response and NGD is studied.

Slot gap width of DGS

Figure 11.13 shows the E-field distribution on the DGS etched ground plane at resonant frequency. It is observed that the electric field concentrates at the thin slot gap of the dumbbell DGS. This implies that this vertical slot gap width determines the equivalent capacitance of the DGS filter. Hence, a smaller slot gap width results in a larger equivalent capacitance. This implication is applicable to any vertical slot gap of DGS beneath the microstrip lines.

In general, a larger gap width g results in higher resonant frequency and wider stopband bandwidth (Figure 11.14). Table 11.1 presents the values of the resonant

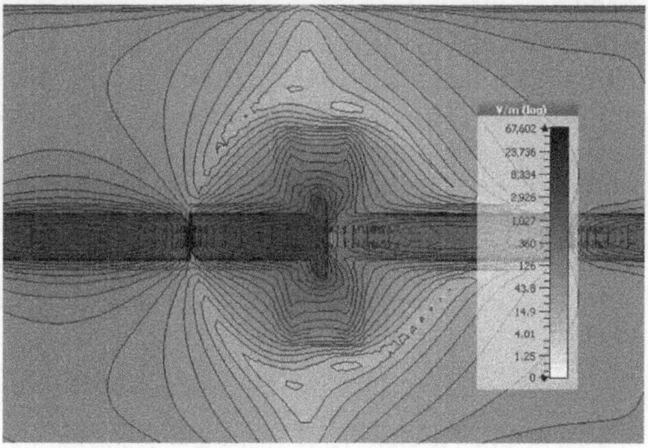

Figure 11.13 E-field distribution of the dumbbell DGS filter at resonant frequency

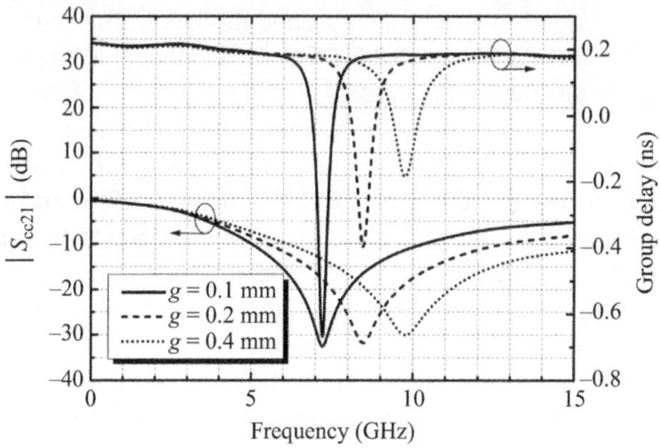

Figure 11.14 The common-mode insertion loss $|S_{cc21}|$ and GD time for different slot gap g of a dumbbell DGS filter

Table 11.1 The resonant frequencies, bandwidths at −10 dB, equivalent resistances, inductances, capacitances, NGD times, and NGD bandwidths obtained for different slot gap width g of a dumbbell DGS filter. The other dimensions are fixed

g/mm	f_0/GHz	BW/GHz	FBW/%	R/kΩ	L/nH	C/pF	NGD/ns	NGD-BW/ MHz
0.1	7.215	5.352	74.18	4.878	5.564	0.0874	0.668	675
0.2	8.475	7.274	85.83	4.529	5.564	0.0634	0.397	779
0.4	9.780	10.318	105.50	3.642	5.423	0.0488	0.186	847

frequency f_0, both actual BW and FBW at −10 dB, extracted equivalent resistance R, inductance L, and capacitance C associated with different slot gap widths. It is observed that the decrease in C results in the increases of the resonant frequency, BW and FBW (refer to (11.20) in Section 11.2). Hence, to achieve a higher (lower) resonant frequency with a wider (narrower) stopband, a larger (smaller) vertical slot gap width is recommended.

However, the NGD time decreases as the gap width increases, which is depicted in Figure 11.14. But bandwidth of NGD increases. The various NGD times and bandwidths are also illustrated in the Table 11.1. This is mainly due to the decrease in R and C (refer to (11.24) in Section 11.2) which results in shorter NGD time.

Hence, to achieve a wider common-mode stopband and NGD-BW, a larger vertical slot gap is recommended. However, this will result in a shorter NGD time. Thus, there is a compromise between achieving NGD effect and common-mode noise filtering in a DGS filter design. Note that there is also a trade-off between longer NGD time and wider NGD-BW.

Horizontal length of DGS

The surface current distribution (Figure 11.15) reveals that instead of a right to left flow of the current, it is forced to travel around the DGS. This current path determines the equivalent inductance of the DGS filter. Hence, a longer current path results in a larger equivalent inductance.

In general, the increase in etched head dimension a decreases the resonant frequency and increases FBW as shown in Figure 11.16. Table 11.2 gives the values of the resonant frequencies f_0, both BW and FBW at −10 dB, extracted equivalent resistance R, inductance L, and capacitance C for the different head length a of the DGS. The decrease in resonant frequency and increase in FBW are due to the increase in current path which increases the equivalent inductance (refer to (11.19)). FBW is directly proportional to L while resonant frequency is inversely proportional to L. The impact on the BW is not obvious as the effect of equivalent capacitance is not significant in this case. To achieve a higher (lower) resonant frequency with a smaller (larger) FBW, a smaller (larger) slot horizontal length is recommended.

Figure 11.16 also shows that the NGD time increases but NGD-BW decreases as the head length a increases. The various NGD times and bandwidths are also

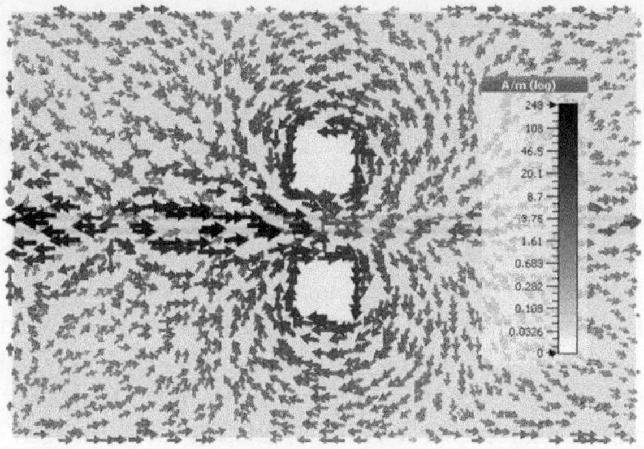

Figure 11.15 Surface current distribution of the dumbbell DGS filter at resonant frequency

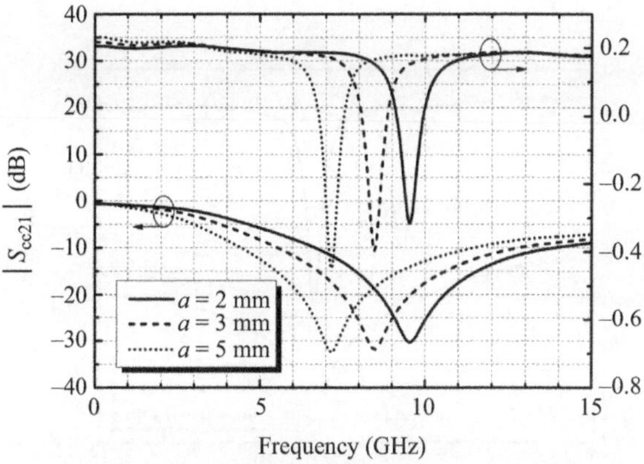

Figure 11.16 The common-mode insertion loss $|S_{cc21}|$ and GD time for different head length a of the dumbbell DGS filter

Table 11.2 The resonant frequencies, bandwidths at −10 dB, equivalent resistances, inductances, capacitances, NGD times, and NGD bandwidths obtained for different head length a of the dumbbell DGS filter. The other dimensions are fixed

a/mm	f_0/GHz	BW/GHz	FBW/%	*R*/kΩ	*L*/nH	*C*/pF	NGD/ns	NGD-BW/ MHz
2	9.54	7.233	75.82	3.755	4.351	0.064	0.315	812
3	8.475	7.274	85.83	4.529	5.564	0.0634	0.397	779
5	7.16	7.058	98.58	4.792	7.735	0.0639	0.449	772

illustrated in the Table 11.2. This is due to the increase in R since NGD time is directly proportional R^2. The effect of C is negligible.

Vertical length of DGS

In general, the increase in vertical length of DGS, i.e. slot length c, decreases both the resonant frequency and the BW/FBW as illustrated in Figure 11.17. The phenomenon for the case of varying DGS etched head dimension b is similar. The values of the resonant frequencies f_0, both BW and FBW at -10 dB, extracted equivalent resistance R, inductance L, and capacitance C for the different slot length c of the DGS are tabulated (Table 11.3). The decrease in resonant frequency and bandwidth are mainly due to the increase in equivalent capacitance at the slot gap. Hence, to achieve a higher (lower) resonant frequency with a wider (narrower) bandwidth, a smaller (larger) slot length c of DGS is recommended.

From Figure 11.17, it is observed that the NGD time decreases and NGD-BW increases even though there is an increase in C as slot length c increases. This is because that the effect of R is more dominant than that of C. A longer slot length results in a decrease in R which leads to a shorter NGD time.

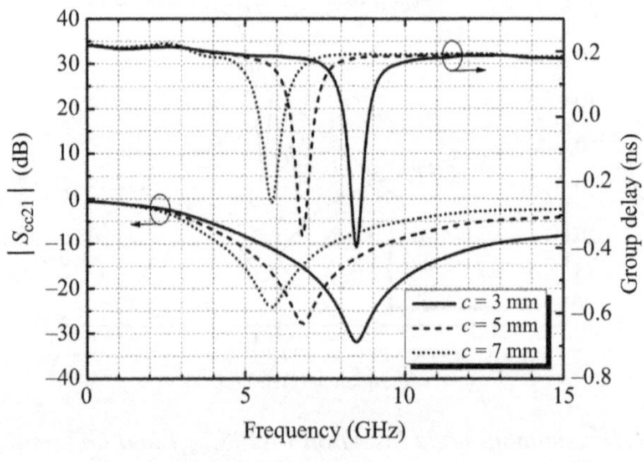

Figure 11.17 The common-mode insertion loss $|S_{cc21}|$ and GD time for different slot length c of a dumbbell DGS filter

Table 11.3 The resonant frequencies, bandwidths at -10 dB, equivalent resistances, inductances, capacitances, NGD times, and NGD bandwidths obtained for different slot length c of a dumbbell DGS filter. The other dimensions are fixed

c/mm	f_0/GHz	BW/GHz	FBW/%	R/kΩ	L/nH	C/pF	NGD/ns	NGD-BW/MHz
3	8.475	7.274	85.83	4.529	5.564	0.0634	0.397	779
5	6.810	4.623	67.89	2.795	5.690	0.0960	0.363	780
7	5.835	3.496	59.91	1.783	5.975	0.125	0.262	784

Hence, to achieve a wider S_{cc21} stopband and longer NGD time, a shorter vertical DGS slot length is recommended. However, the NGD-BW will be narrower.

DGS modification for compact design

As discussed before, the way to achieve a lower resonant frequency is to have larger DGS dimensions. However, due to space limitation in modern circuits, a compact DGS design is preferred. Hence, another DGS shape which can achieve lower resonant frequency without increasing the dimensions is introduced. Instead of increasing the etched dimension (a or b), the DGS shape is modified to a C-shaped dumbbell type to allow more charge accumulation at the slot gap which increases the equivalent capacitance as shown in Figure 11.18. Hence, this decreases the resonant frequency without the need to increase the overall DGS size.

The values of the resonant frequencies f_0, both BW and FBW at −10 dB, extracted equivalent resistance R, inductance L, capacitance C, NGD time, and NGD-BW for the two different DGS designs are illustrated in Table 11.4. It is seen that the stopband bandwidth is narrower when using C-shaped dumbbell DGS (Figure 11.19). This is due to the larger equivalent capacitance for C-shaped

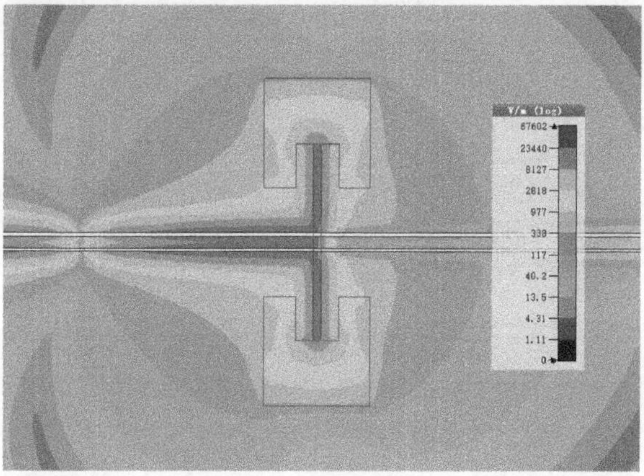

Figure 11.18 E-field distribution of a C-shaped dumbbell DGS filter at resonant frequency

Table 11.4 The resonant frequencies, bandwidths at −10 dB, equivalent resistances, inductances, capacitances, NGD times, and NGD bandwidths obtained for a dumbbell and a C-shaped dumbbell DGS filters. The overall DGS size is fixed

DGS shape	f_0/GHz	BW/GHz	FBW/%	R/kΩ	L/nH	C/pF	NGD/ns	NGD-BW/MHz
Dumbbell	8.475	7.274	85.83	4.529	5.564	0.0634	0.397	779
C-shaped dumbbell	7.125	4.853	68.11	4.690	5.515	0.0905	0.697	677

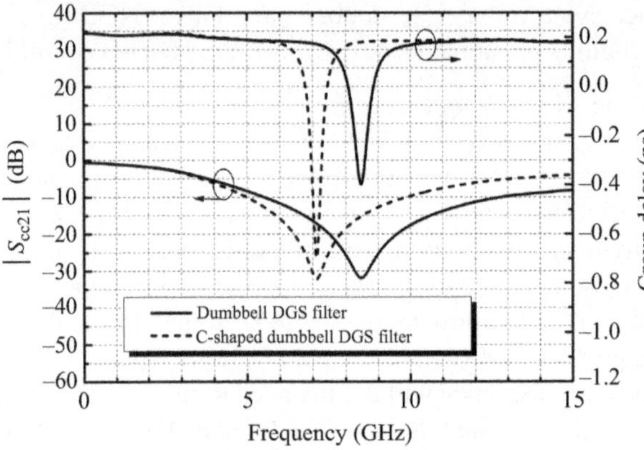

Figure 11.19 The common-mode insertion loss $|S_{cc21}|$ and GD times of a dumbbell DGS filter and a C-shaped dumbbell DGS filter

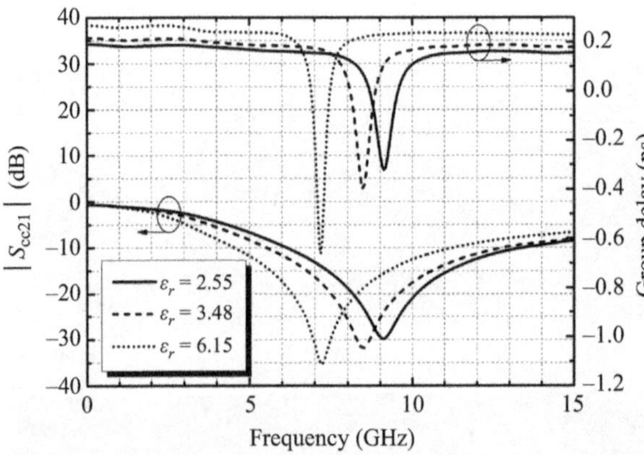

Figure 11.20 The common-mode insertion loss $|S_{cc21}|$ and GD time for different substrate dielectric constant ε_r of a dumbbell DGS filter

dumbbell DGS. The NGD time is longer for C-shaped dumbbell DGS as the R and C are larger as shown in Table 11.4. Therefore, a more compact design and a longer NGD time can be achieved for such C-shaped dumbbell DGS but at the expense of both stopband bandwidth and NGD-BW.

11.3.2.2 Effect of substrate material in PCB

The substrate between the layer with DGS and microstrip lines also has impact on the performance of the filter because substrate material has impact on the equivalent R, L, C, and Z_{even} of the coupled microstrip lines. Note that the width of coupled microstrip lines is adjusted to fix the differential-mode impedance at 100 Ω for all cases.

From Figure 11.20, it is observed that the increase in dielectric constant decreases the resonant frequency and slightly reduces the BW and FBW. This is

Table 11.5 The resonant frequencies, bandwidths at −10 dB, equivalent resistances, inductances, capacitances, NGD times, and NGD bandwidths obtained for different substrate dielectric constant ε_r of a dumbbell DGS filter. The other dimensions are fixed

ε_r	f_0/GHz	BW/GHz	FBW/%	R/kΩ	L/nH	C/pF	NGD/ns	NGD-BW/ MHz
2.55	9.12	8.165	89.53	3.992	5.302	0.0574	0.320	878
3.48	8.475	7.274	85.83	4.529	5.564	0.0634	0.397	779
6.15	7.185	5.495	76.37	5.366	5.920	0.0829	0.661	577

due to the increase in equivalent capacitance. Table 11.5 presents the values of the resonant frequencies f_0, both BW and FBW at −10 dB, extracted equivalent resistance R, inductance L, capacitance C, NGD time, and NGD-BW for the different substrate dielectric constants. The substrate dielectric constant has a dominant effect on resonant frequency. Hence, to achieve a higher (lower) resonant frequency and a wider (narrower) bandwidth, a smaller (larger) dielectric constant for the substrate is recommended.

Figure 11.20 also reveals the NGD time increases but NGD-BW decreases as dielectric constant increases. This can explain by increase in R and C. Hence, a larger dielectric constant is recommended to achieve a longer NGD time but a narrower common-mode stopband and NGD-BW.

11.4 Cavity-backed DGS filter with NGD effect for multilayer PCB

11.4.1 Drawback of conventional DGS filter

The DGS filters work well for common-mode noise suppression in a two-layer PCB environment, however, it has a drawback of back-radiation or back-leakage due to the slot openings on the ground plane. A multilayer PCB that has at least three layers would not have such kind of EMI issue that is because, when DGS is applied on one ground plane in the multilayer PCB, the additional or other ground/power plane could serve the shielding purpose which blocks electromagnetic field from emitting to external environment. However, the filtering performance (common-mode insertion loss) is also significantly degraded in the multilayer PCB [37]. For instance, the same dumbbell DGS used in the previous study is applied in a three-layer PCB. The insertion loss is deteriorated with the ripples or resonances induced due to the ground bounce noises generated between the DGS etched ground plane and the second ground or power plane beneath (Figure 11.21). The ground bounce perturbs the electromagnetic behavior of the DGS which in turn affects the current distribution in the DGS layer and leads to a poor filtering behavior. To reinstate the filtering performance of DGS on multilayer PCB, an effective solution is proposed.

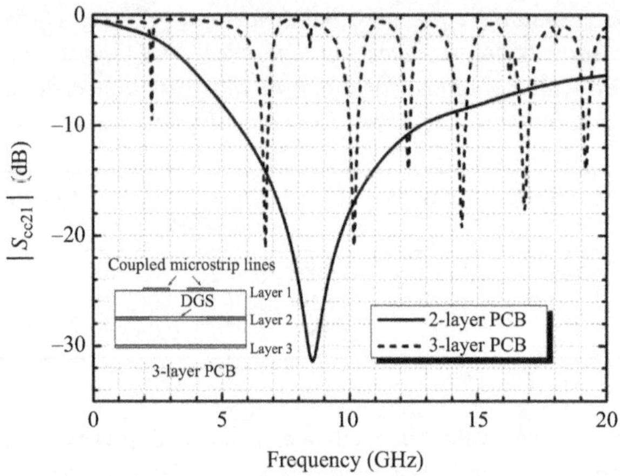

Figure 11.21 Comparison of the common-mode insertion loss |S_{cc21}| of a dumbbell DGS common-mode filter in a two-layer and three-layer PCB

11.4.2 Cavity-backed DGS filter design

A cavity is introduced beneath the conventional DGS filter in an attempt to shield the ground bounce noises generated between the ground plane with DGS and the power/ground plane directly below the DGS. Figure 11.22 shows the cavity-backed dumbbell DGS filter in a three-layer PCB. The PCB size is $w \times l$, while the thicknesses of Substrates 1 and 2 are h_1 and h_2, respectively. The coupled microstrip lines on the top layer (Layer 1) have a width of w_m and a space distance of s_m. The dumbbell DGS in Layer 2 has design parameters a, b, c, and g as shown in the inset of Figure 11.22(a). Beneath the DGS is a cuboid metallic cavity introduced between Layers 2 and 3, which has a depth of h_2. The material used for metallic cavity is the same as the power/ground planes, which is usually copper.

As an example to demonstrate the effectiveness of cavity-backed dumbbell DGS filter design concept in a multilayer PCB, a three-layer PCB with (l, w, h_1, h_2, w_m, s_m) = (30 mm, 20 mm, 0.168 mm, 0.168 mm, 0.3 mm, 0.32 mm) and the same dumbbell DGS filter with (a, b, c, g, w_c, l_c) = (3 mm, 3 mm, 3 mm, 0.2 mm, 11 mm, 5 mm) as in Section 11.3 are considered. Since the thickness of the cavity wall has little effect, it is fixed at 0.1 mm. w_m and s_m are chosen to achieve 100 Ω differential impedance. The material used for substrates 1 and 2 is Rogers 4350 (ε_r = 3.48, tanδ = 0.0037 @ 10 GHz).

11.4.2.1 Common-mode filtering characteristics

Figure 11.23 shows that the metallic cavity is able to wipe out the ground bounce noises. A stopband of 2.383 GHz with a center frequency at 16.46 GHz is obtained when the attenuation level is 10 dB while preserving good transmission for differential signals. The center frequency shifts to higher frequency due to addition of cavity. It can shift back by modifying the geometric parameters of DGS. However,

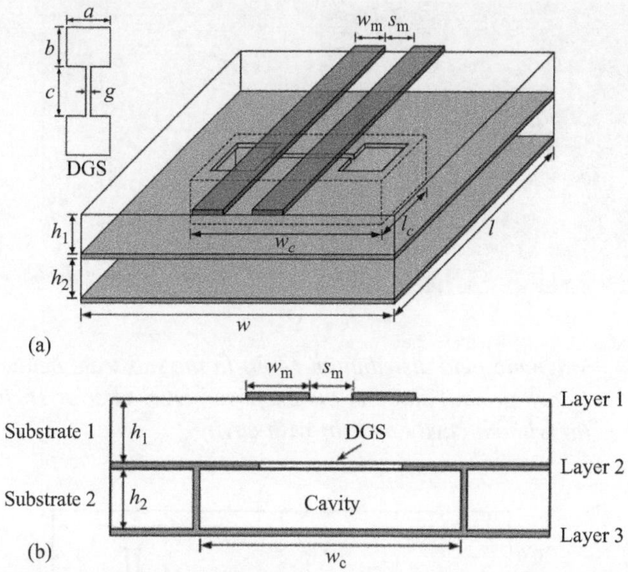

Figure 11.22 (a) The three-dimensional view and (b) the cross section of the cavity-backed dumbbell DGS filter in a three-layer PCB

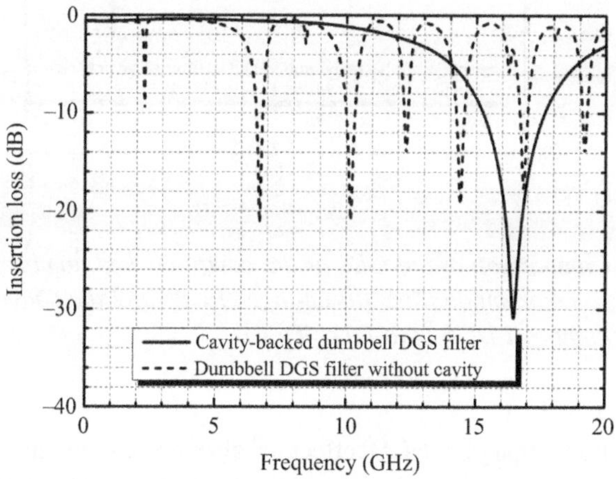

Figure 11.23 The comparison of insertion losses between cavity-backed DGS filter and noncavity-backed DGS filter

it is not the focus and has no impact on topics of the chapter. Thus it will not be discussed in depth.

It is illustrated in Figure 11.24 that the magnetic field distribution under the DGS is less affected by the surrounding fields caused by the ground bounce noises in the presence of the cavity, similar to electric field.

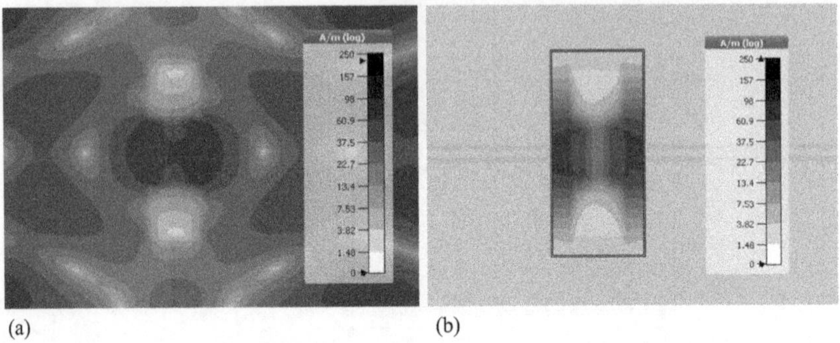

Figure 11.24 Magnetic field distribution (A/m) in the substrate below the DGS-etched ground plane in the proposed DGS filter at 16.46 GHz: (a) without cavity and (b) with cavity

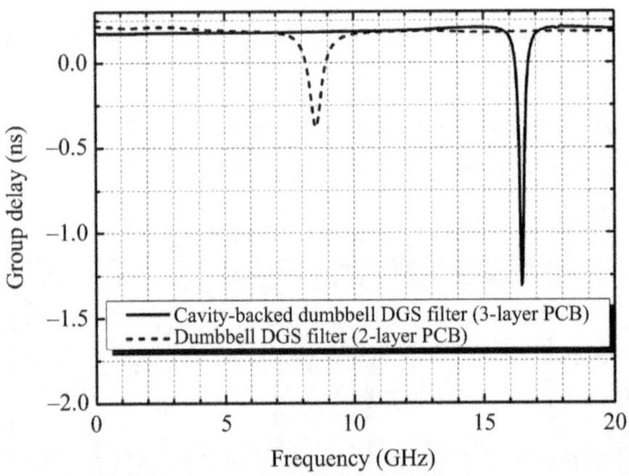

Figure 11.25 Comparison of the GDs of the cavity-backed dumbbell DGS common-mode filter (three-layer PCB) and the conventional dumbbell DGS (two-layer PCB)

11.4.2.2 NGD effect

Figure 11.25 shows that the NGD effect is also present in this cavity-backed dumbbell DGS filter (three-layer PCB) and the NGD time is longer than that of the conventional dumbbell DGS filter (two-layer PCB) in Section 11.3.1. An NGD time of 1.31 ns is obtained at the resonant frequency of 16.46 GHz with an NGD-BW of 504 MHz. Hence, this cavity-backed DGS filter design concept achieves common-mode noise suppression as well as NGD effect in multilayer PCBs.

11.4.3 Parametric study of cavity-backed dumbbell DGS filter design on common-mode filtering and NGD effect

As the cavity plays an instrumental role in the cavity-backed DGS filter for multilayer PCBs, the impact of varying each cavity design parameter on the common-mode insertion loss and NGD is studied. The cavity design parameters include physical parameters (refer Figure 11.22), namely, width w_c, length l_c, and depth h_2 as well as material property of the cavity. Since the resonant behavior is similar to that of a conventional DGS filter, the same parallel RLC circuit model (see Figure 11.4 in Section 11.2) can be used to model the cavity-backed DGS filter. Hence, the equivalent circuit model can be used to understand intuitively the correlation of cavity-backed DGS filter's behavior with its cavity design parameters.

11.4.3.1 Impact of cavity material on DGS filter

Figure 11.26 shows the common-mode insertion loss and GD time for different relative permittivity used in the cavity embedded in Substrate 2. The extracted resonant frequency, both BW and FBW at -10 dB, extracted equivalent resistance R, inductance L, and capacitance C are listed in Table 11.6. It is observed that the resonant frequency decreases as the relative permittivity increases. This is due to the increase in C and L as shown in the table according to relation, $f_0 = 1/\left(2\pi\sqrt{LC}\right)$. It is also worth to note that a dielectric material with higher relative permittivity can be used inside the cavity in order to reduce the overall size

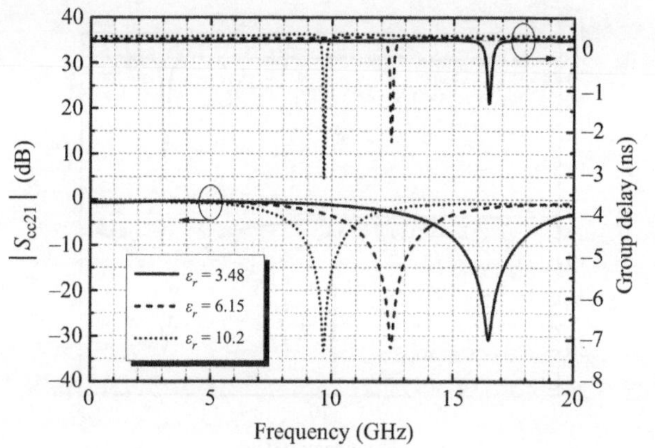

Figure 11.26 Common-mode insertion loss $|S_{cc21}|$ and GD for different relative permittivity of cavity material

Table 11.6 The resonant frequencies, bandwidths at −10 dB, extracted equivalent resistances, inductances, capacitances, NGD times, and NGD bandwidths obtained for different relative permittivity of cavity material. The other dimensions are fixed

ε_r	f_0/GHz	BW/GHz	FBW/%	R/kΩ	L/nH	C/pF	NGD/ns	NGD-BW/MHz
3.48	16.46	2.383	14.478	4.018	0.558	0.168	1.31	504
6.15	12.44	1.79	14.389	5.128	0.763	0.215	2.26	349
10.2	9.68	1.368	14.132	5.607	0.973	0.278	3.125	260

of the DGS filter so as to deal with the space constraint in practical PCBs. Figure 11.26 also shows that the *BW* and *FBW* decrease as the relative permittivity increases which is mainly due to the increase in *C* (refer to (11.19) and (11.20) in Section 11.3) as shown in Table 11.6.

As for NGD effect, the NGD time increases but NGD-BW decreases as the relative permittivity in the cavity increases as depicted in Figure 11.26. This is due to the increase in *R* and *C* since NGD time is directly proportional R^2C as observed in Table 11.6.

Hence, a larger relative permittivity of cavity material is recommended to increase NGD time but this reduces stopband and also NGD-BW. The impact of cavity material on NGD effect is significant.

11.4.3.2 Impact of cavity depth on DGS filter

Figure 11.27 illustrates the common-mode insertion loss and GD for cavity depth $h_2 = 0.168$ mm, 0.5 mm and 1.0 mm, respectively. Table 11.7 presents the values of the resonant frequencies f_0, both BW and FBW at −10 dB, extracted equivalent resistance *R*, inductance *L*, and capacitance *C* associated with different cavity

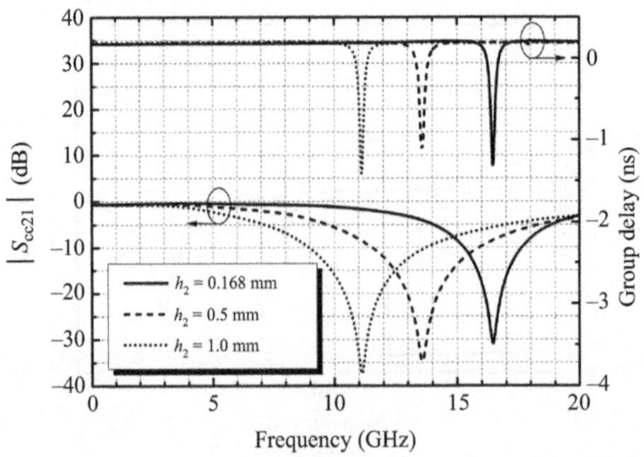

Figure 11.27 Common-mode insertion loss $|S_{cc21}|$ and GD for different cavity depths h_2

Table 11.7 The resonant frequencies, bandwidths at −10 dB, extracted equivalent resistances, inductances, capacitances, NGD times, and NGD bandwidths obtained for different cavity depths h_2. The other dimensions are fixed

h_2/mm	f_0/GHz	BW/GHz	FBW/%	R/kΩ	L/nH	C/pF	NGD/ns	NGD-BW/ MHz
0.168	16.46	2.383	14.478	4.018	0.558	0.168	1.31	504
0.5	13.6	4.378	32.191	6.406	1.383	0.099	1.115	559
1.0	11.12	4.794	43.112	8.684	2.417	0.085	1.406	536

depths. It is shown that as the depth increases, the resonant frequency decreases which is mainly due to the increase in L. Furthermore, the FBW generally increases as cavity depth increases which is due to the increase in L and decrease in C. Note that the cavity-backed DGS concept can be extended to more than three-layer PCB. If there are more layers, deeper cavity can be formed with the DGS size if needed.

For NGD effect, the NGD time generally slightly increases as cavity depth increases as shown in Figure 11.27. The trend is not so clear in this case as R increases but C decreases as presented in Table 11.7. In general, the impact of cavity depth is not significant on NGD effect.

11.4.3.3 Impact of cavity length and width on DGS filter

Figure 11.28 shows the common-mode insertion loss and GD for different cavity length l_c of 5, 7, and 10 mm, respectively. The values of the resonant frequencies f_0, both BW and FBW at −10 dB, extracted equivalent resistance R, inductance L, and capacitance C are listed in Table 11.8. It is observed that the resonant frequency

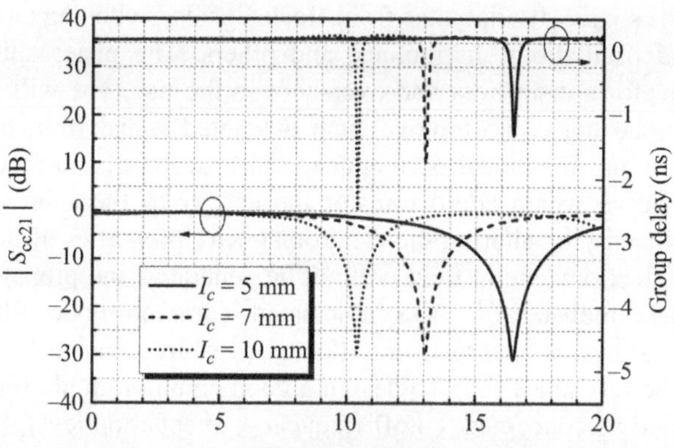

Figure 11.28 Common-mode insertion loss $|S_{cc21}|$ and GD for different cavity lengths l_c

Table 11.8 The resonant frequencies, bandwidths at −10 dB, extracted equivalent resistances, inductances, capacitances, NGD times, and NGD bandwidths obtained for different cavity lengths l_c. The other dimensions are fixed

l_2/mm	f_0/GHz	BW/GHz	FBW/%	R/kΩ	L/nH	C/pF	NGD/ns	NGD-BW/MHz
5	16.46	2.383	14.478	4.018	0.558	0.168	1.31	504
7	13.04	1.664	12.761	3.686	0.639	0.233	1.768	428
10	10.4	1.107	10.644	3.427	0.653	0.359	2.466	350

decreases as the cavity length increases, which corresponds to the increase of *L* and *C*. As for the *BW* and *FBW*, they decrease as cavity length increases which is mainly due to the increase of *C*.

Figure 11.28 also shows that the NGD time increases but NGD-BW decreases as the cavity length increases. This is mainly due to the increase of *C* as observed in Table 11.8.

Hence, a larger cavity length is recommended to achieve a longer NGD time but at the expense of having smaller stopband and NGD-BW. The impact of cavity length is significant for NGD effect. On the other hand, the variation of cavity width w_c does not affect the frequency response of the filter in terms of the resonant frequency and stopband because the current distribution is unaffected. This is the same case for NGD effect.

11.5 Conclusion

Common-mode noise is unwanted signal in high-speed electronic circuits and is one kind of EMI source. It originates from clock signals, switching circuit, parasitic resonances and oscillations, harmonics, and others. The phase shift, amplitude difference, waveform distortion, and overshoot in the rise time will also generate the common-mode noise. To remove such unwanted signal from the circuit for proper operation and functionality, metamaterial-based common-mode filter is an effective alternative which could transfer the energy to the sources or trap the energy at local cavity by altering the electromagnetic properties of the circuit thus the associated electromagnetic field will not be generated and propagated.

Within the stopband of DGS associated common-mode filter, most of common-mode signal is absorbed or reflected to source due to high impedance generated by the resonant DGS. To design a good common-mode filter, some key characters should be concerned: cutoff frequency, attenuation level, band gap, and band-pass ripple. Another feature being paid attention increasingly is NGD. It can shorten signal transmission delay time to ameliorate SI and EMI.

The introduction of metamaterial enable common-mode filter to be realized for high-speed electronic circuits in a compact and novel way. It is proved that integrating DGS in the common-mode filter design is very effective to tackle the common-mode noise associated issues in electronic circuits that operates at tens gigahertz or even higher frequency. The design not only provides wide noise attenuation band but also has an expected NGD. For instance, the filter with a simple dumbbell DGS has a common-mode stopband of 7.3 GHz and an NGD time of 0.397 ns. The NGD time can be extended with sacrifice of common mode stopband. The compromise is achieved based on the design requirement.

Furthermore, optimization technology could be beneficial for enhancing a DGS filter with wider common-mode stopband and larger NGD time. If the stop bandwidth and the negative delay time are expected not to be affected by the process variation, manufacturing tolerance and other unavoidable external influences, quality function deployment with integration of engineering design methodology are effective for the purpose. The variation of optimized performance would be to a minimum and the functionalities are preserved when the common-mode filter is confronted with difficult-to-be-control tolerances/variations of parameters in operation and unpredictable influence from external challenging environment. It is characterized as "robustness." To avoid the deviation of book content, it has not been discussed in this chapter.

References

[1] C. R. Paul, "A comparison of the contributions of common-mode and differential-mode currents in radiated emissions," IEEE Transactions on Electromagnetic Compatibility, Vol. 31, No. 2, pp. 189–193, 1989.

[2] V. G. Veselago, "The electrodynamics of substances with simultaneously negative values of ε and μ," Soviet Physics Uspekhi, Vol. 10, No. 4 pp. 509–514, 1968.

[3] J. B. Pendry, A. J. Holden, W. J. Sewart, and I. Youngs, "Extremely low frequency plasmons in metallic mesosstructure," Physical Review Letters, Vol. 76, No. 15, pp. 4773–4776, 1996.

[4] J. B. Pendry, A. J. Holden, D. J. Robbins, and W. J. Stewart, "Magnetic from conductors and enhanced nonlinear phenomena," IEEE Transactions on Microwave Theory and Techniques, Vol. 47, No. 11, pp. 2075–2084, 1999.

[5] D. R. Smith, W. J. Padilla, D. C. Vier, S. C. Nemat-Nasser, and S. Schultz, "Composite medium with simultaneously negative permeability and permittivity." Physical Review Letters, Vol. 84, No. 18, p. 4184, 2000.

[6] R. A. Shelby, D. R. Smith, and S. Schultz, "Experimental verification of a negative index of refraction," Science, Vol. 292, No. 5514, pp. 77–79, 2001.

[7] J. A. Kong, "Electromagnetic wave interaction with stratified negative isotropic media," Progress in Electromagnetics Research, Vol. 35, No. 10, pp. 1–52, 2002.

[8] M. F. Ubeid, M. M. Shabat, and M. O. Sid-Ahmed, "Numerical study of a structure containing left-handed material waveguide," Indian Journal of Physics (Springer), Vol. 86, No. 2, pp. 125–128, 2012.

[9] R. W. Ziolkowski and E. Heyman, "Wave propagation in media with negative permittivity and permeability," Physical Review E, Vol. 64, No. 5, p. 056625, 2001.

[10] H. Oraizi and A. Abdolali, "Mathematical formulation for zero reflection from multilayered metamaterial structures," IET Microwave Antennas Propagation, Vol. 3, No. 6, pp. 987–996, 2009.

[11] J. Yang, X. Hu, X. Li, Z. Liu, X. Jiang, and J. Zi, "Cancellation of reflection and transmission at metamaterial surfaces," Optics Letters, Vol. 35, No. 1, pp. 16–18, 2010.

[12] C. Sabah and S. Uckun, "High reflection coating with negative and positive refractive indices," Progress in Electromagnetics Research Online, Vol. 5, No. 7, pp. 601–604, 2009.

[13] H. Cory and C. Zach, "Wave propagation in metamaterial multi-layered structures," Microwave and Optical Technology Letter, Vol. 40, No. 6, pp. 460–465, 2004.
[14] S. Shahpania and O. M. Ramahi, "Electromagnetic interference (EMI) reduction from printed circuit boards (PCB) using electromagnetic bandgap structures," IEEE Transactions on Electromagnetic Compatibility, Vol. 46, No. 4, pp. 580–587, 2004.
[15] C. L. Wang, G. W. Shiue, W. D. Guo, and R. B. Wu, "A systematic design to suppress wideband ground bounce noise in high-speed circuits by electromagnetic-bandgap-enhanced split powers," IEEE Transactions on Microwave Theory and Techniques, Vol. 54, No. 12, pp. 4209–4217, 2006.
[16] J. Naqui, F. P. Armando, D. S. Miguel, *et al.*, "Common-mode suppression in microstrip differential lines by means of complementary split ring resonators: Theory and applications," IEEE Transactions on Microwave Theory and Techniques, Vol. 60, No. 10, pp. 3023–3034, 2012.
[17] S. J. Wu, C. H. Tsai, T. L. Wu, and T. Itoh, "A novel wideband common-mode suppression filter for gigahertz differential signals using coupled patterned ground structure," IEEE Transactions on Microwave Theory and Techniques, Vol. 57, No. 4, pp. 848–855, 2009.
[18] F. C. Huang, C. N. Chiu, T. L. Wu, and Y. P. Chiou, "A circular-ring miniaturized-element metasurface with many good features for frequency selective shielding applications," IEEE Transactions on Electromagnetic Compatibility, Vol. 57, No. 3, pp. 365–374, 2015.
[19] X. K. Gao, H. M. Lee, S. P. Gao, E. X. Liu, and C. E. Png, "A compact common-mode noise suppression filter for high speed differential signals using defected ground structure," 2015 Asia-Pacific International Symposium on EMC, pp. 685–688, Taipei, May 26–29, 2015.
[20] X. K. Gao, H. M. Lee, and S. P. Gao, "A robust parameter design of wide band DGS filter for common-mode noise mitigation in high speed electronics," IEEE Transactions on Electromagnetic Compatibility, Vol. 59, No. 6, pp. 1735–1740, 2017.
[21] G. Chaudhary, Y. Jeong, and J. Lim, "Miniaturized dual-band negative group delay circuit using dual-plane defected structures," IEEE Microwave and Wireless Components Letters, Vol. 24, No. 8, pp. 521–523, 2014.
[22] H. A. Lorentz, A. Einstein, H. Minkowski, and H. Weyl, The Principle of Relativity: A Collection of Original Memoirs on the Special and General Theory of Relativity. New York: Dover, 1952.
[23] L. Brillouin and A. Sommerfeld, Wave Propagation and Group Velocity. New York: Academic Press, 1960.
[24] M. Kandic and G. E. Bridges, "Limits of negative group delay phenomenon in linear causal media," Progress in Electromagnetics Research, Vol. 134, pp. 227–246, 2013.
[25] M. Mojahedi, E. Schamiloglu, F. Hegeler, and K. J. Malloy, "Time-domain detection of superluminal group velocity for single microwave pulses," Physical Review E, Vol. 62, No. 4, pp. 5758–5766, 2000.

[26] Y. Wang, Y. Zhang, L. He, F. Liu, H. Li, and H. Chen, "Direct observation of negative phase velocity and positive group velocity in time domain for composite right/left-handed transmission lines," Journal of Applied Physics, Vol. 100, No. 11, p. 113503, 2006.

[27] A. Ibraheem, J. Schoebel, and M. Koch, "Group delay characteristics in coplanar waveguide left-handed media," Journal of Applied Physics, Vol. 103, No. 2, p. 024903, 2008.

[28] M. Mojahedi, K. J. Malloy, G. V. Eleftheriades, J. Woodley, and R. Y. Chiao, "Abnormal wave propagation in passive media," IEEE Journal of Selected Topics in Quantum Electronics, Vol. 9, No. 1, pp. 30–39, 2003.

[29] L. Bolda, R. Y. Chiao, and J. C. Garrison, "Two theorems for the group velocity in dispersive media," Physical Review A, General Physics, Vol. 48, No. 5, pp. 3890–3894, 1993.

[30] D. Stenner, D. J. Gauthier, and M. A. Neifeld, "Fast causal information transmission in a medium with a slow group velocity," Physical Review Letters, Vol. 94, No. 5, 2005.

[31] S. Lucyszyn, I. D. Robertson, and A. H. Aghvami, "Negative group delay synthesizer," Electronics Letters, Vol. 29, No. 9, pp. 798–800, 1993.

[32] S. Lucyszyn and I. D. Robertson, "Analog reflection topology building blocks for adaptive microwave signal processing applications," IEEE Transactions on Microwave Theory and Techniques, Vol. 43, No. 3, pp. 601–611, 1995.

[33] G. Chaudhary, Y. Jeong, and J. Lim, "Microstrip line negative group delay filters for microwave circuits," IEEE Transactions on Microwave Theory and Techniques, Vol. 62, No. 2, pp. 234–243, 2014.

[34] O. F. Sidhiqui, M. Mojahedi, and G. V. Eleftheriades, "Periodically loaded transmission line with effective negative group delay index and negative group velocity," IEEE Transactions on Antenna Propagation, Vol. 51, No. 10, pp. 2619–2625, 2010.

[35] R. Garg, I. Bahl, and M. Bozzi, Microstrip Lines and Slotlines. Artech House, Boston, MA, 2013.

[36] W.-T. Liu, C.-H. Tsai, T.-W. Han, and T.-L. Wu, "An embedded common-mode suppression filter for GHz differential signals using periodic defected ground plane," IEEE Microwave and Wireless Components Letters, Vol. 18, No. 4, pp. 248–250, 2008.

[37] S.-P. Gao, H. M. Lee, X.-K. Gao, *et al.*, "Common-mode filter using cavity-backed defected ground structure for multilayer PCB," in 2016 Asia-Pacific Symposium on Electromagnetic Compatibility (APEMC 2016), 2016.

[38] S.-P. Gao, H. M. Lee, X.-K. Gao, E.-X. Liu, and C. E. Png, "Exploration of design guidelines for cavity-backed dumbbell-shaped DGS common-mode filters," in 2016 IEEE MTT-S International Conference on Numerical Electromagnetic and Multiphysics Modeling and Optimization (NEMO 2016), 2016.

[39] C.-C. Chang, C. Caloz, and T. Itoh, "Analysis of a compact slot resonator in the ground plane for microstrip structures," in Asia-Pacific Microwave Conference 2001 (APMC 2001), Vol. 3, pp. 1100–1103, 2001.

[40] D. Ahn, J.-S. Park, C.-S. Kim, J. Kim, Y. Qian, and T. Itoh, "A design of the low-pass filter using the novel microstrip defected ground structure," IEEE Transactions on Microwave Theory and Techniques, Vol. 49, No. 1, pp. 86–93, 2001.

[41] CST Microwave Studio, Computer Simulation Technology, www.cst.com, 2017.

[42] M. Kandic and G. E. Bridges, "Asymptotic limits of negative group delay in active resonator-based distributed circuits," IEEE Transactions on Circuits and Systems I: Regular Papers, Vol. 58, No. 8, pp. 1727–1735, 2011.

[43] B. Ravelo, A. Pérennec, M. Le Roy, and Y. G. Boucher, "Active microwave circuit with negative group delay," IEEE Microwave and Wireless Components Letters, Vol. 17, No. 12, pp. 861–863, 2007.

[44] Y. Jeong, H. Choi, and C. Kim, "Experimental verification for time advancement of negative group delay in RF electronic circuits," Electronics Letters, Vol. 46, No. 4, pp. 306–307, 2010.

[45] H. Choi, Y. Jeong, C. D. Kim, and J. S. Kenney, "Efficiency enhancement of feedforward amplifiers by employing a negative group-delay circuit," IEEE Transactions on Microwave Theory and Techniques, Vol. 58, No. 5, pp. 1116–1125, 2010.

[46] M. Kitano, T. Nakanishi, and K. Sugiyama, "Negative group delay and superluminal propagation: An electronic circuit approach," IEEE Journal of selected Topics in Quantum electronics, Vol. 9, No. 1, pp. 43–51, 2003.

[47] G. Chaudhary, Y. Jeong, and J. Lim, "Realization of negative group delay network using defected microstrip structure," International Journal of Antennas and Propagation, Vol. 2014, 2014.

Conclusion

The NGD concept remains one of the most counterintuitive effects found in input/output systems at all frequencies.

The present book presents different investigations performed by research groups worldwide. Various NGD concepts are developed in the 11 chapters. After recalling the analytical definition of the group delay from the transfer function approach, different possibilities to generate the NGD phenomenon are developed.

Deep investigations about analytical, numerical, and experimental aspects of NGD concept are introduced. Some fields of applications not only in electronics, RF, and microwave engineering but also in neural engineering for the prediction and design of human machine interactions are explored. The main knowledge and skills offered in the book could be helpful for any readers to practice and apply the NGD concept to engineering problems small and large. It is the firm belief of the editor and the contributing authors that applications of NGD in science and engineering are mostly limited by our hesitation to accept NGD as a real and useful concept. If this book can contribute its small part to change this situation, we would consider it a success.

Index